Earth and Environmental Sciences Library

Earth and Environmental Sciences Library (EESL) is a multidisciplinary book series focusing on innovative approaches and solid reviews to strengthen the role of the Earth and Environmental Sciences communities, while also providing sound guidance for stakeholders, decision-makers, policymakers, international organizations, and NGOs.

Topics of interest include oceanography, the marine environment, atmospheric sciences, hydrology and soil sciences, geophysics and geology, agriculture, environmental pollution, remote sensing, climate change, water resources, and natural resources management. In pursuit of these topics, the Earth Sciences and Environmental Sciences communities are invited to share their knowledge and expertise in the form of edited books, monographs, and conference proceedings.

More information about this series at http://www.springer.com/series/16637

Jose A. Raynal Villaseñor

Frequency Analyses of Natural Extreme Events

A Spreadsheets Approach

Springer

Jose A. Raynal Villaseñor
Department of Civil Engineering
Universidad de las Américas Puebla
Cholula, Puebla, Mexico

ISSN 2730-6674 ISSN 2730-6682 (electronic)
Earth and Environmental Sciences Library
ISBN 978-3-030-86389-0 ISBN 978-3-030-86390-6 (eBook)
https://doi.org/10.1007/978-3-030-86390-6

This Springer imprint is published by the registered company Springer Nature Switzerland AG
The registered company address is: Gewerbestrasse 11, 6330 Cham, Switzerland

This book is dedicated to my wife Maria Elena, my children Maria Elena and Jose Angel, my grandchildren Sara Elena, André, Matheus, Alexa and Iker; their support to produce this book was unvaluable and without their love this task would not be possible to be accomplished.

Preface

The very nature of natural extreme events, such as floods, droughts, maximum wave heights, annual maximum rainfalls, annual maximum winds, and earthquakes, among others, is composed by the occurrence of very complex natural phenomena. Some of these phenomena are related with meteorological variables, hydrological variables, soil and moisture variables and many other variables related to many fields of study.

One way to deal with these natural extreme events is to model them through the statistical tool known as frequency analysis; this tool is based on the use of mathematical models known as probability distribution functions, given that it is accepted that those phenomena are of a random nature, and they have a characteristic of independence.

Frequency analysis of natural extreme events is of paramount importance in the fields of engineering and applied sciences, given that through the values obtained by these procedures, many structures, like spillways of dams, are designed and constructed to provide protection to a fairly large number of human beings.

The motivation to produce this book was to provide procedures for the application of many probability distribution functions, all of them based in a common computational tool known as Excel®, which is available to any personal computer user. The computer procedures are given in enough detail so the user can develop its own Excel® spreadsheets.

All the probability distribution functions contained in the book have procedures to estimate its parameters, quantiles, and confidence limits through the methods of moments and maximum likelihood. For the case of the extreme value type I and the general extreme value distributions for the maxima, it is provided the method of probability weighted moments, in addition to the methods mentioned above.

Jose A. Raynal Villaseñor
Department of Civil Engineering
Universidad de las Américas Puebla
Cholula, Puebla, Mexico

Contents

Introduction

1

1.1 Introduction

Floods, droughts, maximum wave heights, annual maximum rainfall, annual maximum winds, storms, and earthquakes are natural phenomena that occur every year, in one place or another in the Planet Earth. While floods, maximum wave heights, annual maximum rainfall, annual maximum winds, storms, and earthquakes last up to a few days in a limited geographical area, droughts can last for decades in zones that comprise several countries. Even though floods, maximum wave heights, annual maximum rainfall, annual maximum winds, storms, and earthquakes are short-lived natural phenomena, they are part of the most catastrophic natural phenomena regarding human deaths that they cause every year. Figure 1.1 shows the relationship between the number of events, deaths, and people affected, expressed in percent, caused by several natural phenomena in year 2018, EM-DAT, CRED / UCLouvain, Brussels, Belgium—www.emdat.be (D. Guha-Sapir) (2019). The natural events considered were floods, storms, earthquakes, extreme temperatures, droughts, landslides, wildfires, volcanic activity, and mass movements. In year 2018 there were 315 natural events worldwide, 11,804 people were killed by these natural events and 68 million people were affected by these natural phenomena, EM-DAT, CRED/UCLouvain, Brussels, Belgium—www.emdat.be (D. Guha-Sapir) (2019).

In year 2018, floods ranked number one regarding total people affected and in number of events, ranked in second place in the number of people killed. Earthquakes ranked number one in people killed. Storms ranked in second place in number of events and people affected, in third place of people killed. Droughts were in third place in the number of people affected. Extreme temperatures were in third place in the number of events.

Again, in year 2019 floods ranked number one regarding total people affected and in number of events, ranked in second place in the number of people killed.

J. A. Raynal Villaseñor, *Frequency Analyses of Natural Extreme Events*,
Earth and Environmental Sciences Library,
https://doi.org/10.1007/978-3-030-86390-6_1

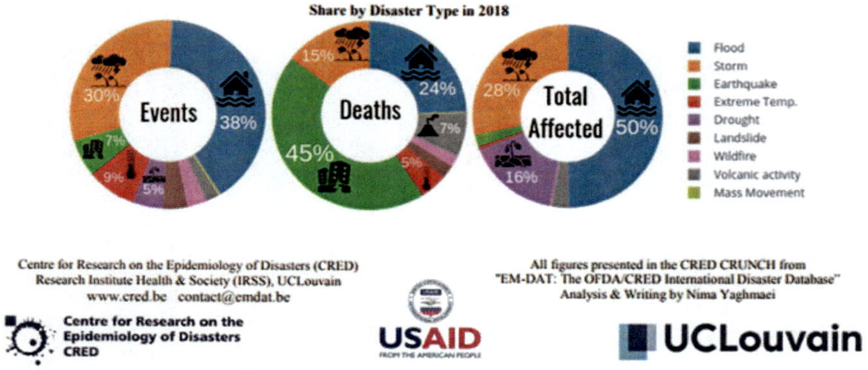

Fig. 1.1 Events, Number of Deaths, and Total of People Affected by Several Natural Phenomena in Year 2018, EM-DAT, CRED/UCLouvain, Brussels, Belgium—www.emdat.be (D. Guha-Sapir) (2019)

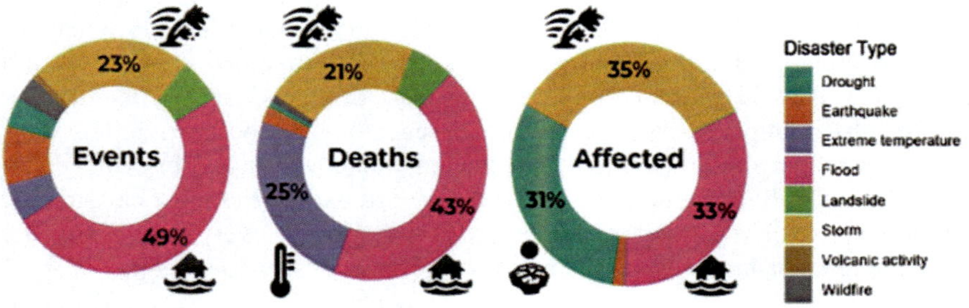

Fig. 1.2 Events, Number of Deaths, and Total of People Affected by Several Natural Phenomena in Year 2019, EM-DAT, CRED/UCLouvain, Brussels, Belgium—www.emdat.be (D. Guha-Sapir) (2020)

Storms ranked first place regarding number of people affected, in second place in number of events, and in third place in number of people killed. Extreme temperatures ranked in second place in number of people killed. Droughts ranked in third place regarding number of people affected.

Figure 1.2 shows the relationship between the number of events, deaths, and people affected, expressed in percent, caused by several natural phenomena in year 2019, EM-DAT, CRED/UCLouvain, Brussels, Belgium—www.emdat.be (D. Guha-Sapir) (2020).

A characteristic of economic losses due to floods, is that they are increasing with time, mainly due to the facts of population and economic growths, as well due to an increased occupation of the floodplains of rivers, mainly produced by the population growth in developing countries.

1.2 Brief History of Natural Extreme Events

The history of natural phenomena related to the human beings' dates to the very first moment of existence of the human beings in Planet Earth, the dependence of such species over water is a matter of physiological nature. The human beings are made of water in average 60% of their bodies, 55% in women and 63% in men. Also, to look for shelter when the extreme rainfall and winds occurred was a matter of live saving urgency. High tides were another natural phenomenon to keep an eye on them, for two main reasons, the first one was to protect the lives and goods of human beings and the other was to look for the best moment to sail when there were not engines to produce power to move the ships. The protection against earthquakes was another need of human beings to keep safe and sound.

This physiological dependence on water makes human beings to drink two liters of fresh water per day, if someone has no access to fresh water for three days, it is almost certain that he/she will die soon after.

In a greater scale, the most remarkable and long-lasting civilizations had been founded near a source of fresh water. The so-called "hydraulic civilizations", Biswas (1970), as those of the Mesopotamia (Tigris and Euphrates rivers), Egyptian (Nile river), Harappan (Indus river), Chinese (Yellow river), the Roman (Po river) and Aztec (lake of Mexico) civilizations were founded close to fresh water sources.

But during the thousands of years that passed between the first appearances of the human beings on Earth until they were able to secure their water needs, during the Nature's two extreme events related with water: floods and droughts, they had to survive between the scarcity and abundance of water that characterizes such extreme events.

The first record of the construction of a dam dates to 2800 BC, located in small creek near to El Cairo, Egypt in a place called Sadd El-Kafara ("Dam of the Pagans"), Biswas (1970). Also, the first record of water resources work comes from an imperial mace head of King Scorpion, 3200 BC. After him King Menes dammed the Nile and diverted its course in 3000 BC. In year 2750 BC, the water supply and drainage systems of the Indus Valley started. The first law code was that of Hammurabi's in 1760 BC. Sennacherib flooded Babylon as a part of his conquest in 689 BC, the flood was artificially produced by destroying a previously built dam on the Euphrates river.

Earthquakes and volcanic eruptions had been present in human beings' existence, some of them had the power to destroy kingdoms and human constructions like the Minoan in the Mediterranean Sea in the island of Crete in 1500 BC, that was destroyed by the eruption of the Santorini volcano and its associated earthquakes. Another remarkable volcanic eruption event was that destroyed Pompei, Herculaneum, and many villas in year 79 AD, and was due to an eruption of the Mount Vesuvius volcano. The Colossus of Rhodes, one of the Seven Wonders of the Ancient World, was destroyed by an earthquake in year 225/226 BC.

1.3 Motivation and Goals

There are many very good books on frequency analysis based in univariate probability distribution functions. A reduced list of this type of books is formed by the ones of Johnson and Kotz (1972 and 1994), Castillo (1988), Reiss (1992) and many others. With respect to flood frequency analysis, the books of Kite (1988) and Rao and Hammed (2000) are worth to be reviewed and applied. No formal books have been produced related with low flow frequency analyses.

The motivation to produce this book mainly comes from the need of having a book on maxima and minima natural extreme events frequency analyses with a direct aim on the use of personal computers, via the application of Excel© spreadsheets, to process every procedure contained in the most common natural extreme events frequency analyses.

Another motivation behind the writing of this book was the intention of producing a book on natural extreme events frequency analyses, written by an engineer and to be easily understood by engineers, too.

The main goal of this book is to provide a framework that can be used to develop a natural extreme events frequency analysis under the computational platform of a personal computer, and specifically related with applications based in the use of Excel© spreadsheets.

1.4 Chapter Outline

In Chap. 1, a brief description of the importance of modeling natural extreme events is given. The motivation in the writing of this book is explained and chapter outline is provided.

In Chap. 2, the main results produced by probability and statistics are brought to the attention to reader because they will be needed throughout the rest of the book. The chapter consists of several sections devoted to an introduction, basic notions of theory of probability, basic notions of statistics, methods for the estimation of parameters of probability distribution functions, quantile estimation and frequency factor, plotting position formulas, confidence limits, standard errors of estimates, goodness of fit test, tests for outliers, test for independence and stationarity, test for homogeneity and stationarity for samples of data. Several worked examples are included.

In Chap. 3, the Normal distribution is described with respect to the following topics: introductory remarks, probability distribution and density functions, estimation of parameters, estimation of quantiles, goodness of fit tests, estimation of standard errors, and estimation of confidence limits. Several worked examples are included, and several examples are developed by the application of Excel© spreadsheets.

In Chap. 4, the Two- parameters Log-Normal distribution is described with respect to the following topics: introductory remarks, probability distribution and density functions, estimation of parameters, estimation of quantiles, goodness of fit tests, estimation of standard errors, and estimation of confidence limits. Several worked examples are included, and several examples are developed by the application of Excel© spreadsheets.

In Chap. 5, the Three- parameters Log-Normal distribution is described with respect to the following topics: introductory remarks, probability distribution and density functions, estimation of parameters, estimation of quantiles, goodness of fit tests, estimation of standard errors, and estimation of confidence limits. Several worked examples are included, and several examples are developed by the application of Excel© spreadsheets.

In Chap. 6, the Gamma distribution is described with respect to the following topics: introductory remarks, probability distribution and density functions, estimation of parameters, estimation of quantiles, goodness of fit tests, estimation of standard errors, and estimation of confidence limits. Several worked examples are included, and several examples are developed by the application of Excel© spreadsheets.

In Chap. 7, the Pearson type III distribution is described with respect to the following topics: introductory remarks, probability distribution and density functions, estimation of parameters, estimation of quantiles, goodness of fit tests, estimation of standard errors, and estimation of confidence limits. Several worked examples are included, and several examples are developed by the application of Excel© spreadsheets.

In Chap. 8, the Log-Pearson with 3 parameters distribution is described with respect to the following topics: introductory remarks, probability distribution and density functions, estimation of parameters, estimation of quantiles, goodness of fit tests, estimation of standard errors, and estimation of confidence limits. Several worked examples are included, and several examples are developed by the application of Excel© spreadsheets.

In Chap. 9, the Extreme Value type I distribution is described with respect to the following topics: introductory remarks, probability distribution and density functions, estimation of parameters, estimation of quantiles, goodness of fit tests, estimation of standard errors, and estimation of confidence limits. Several worked examples are included, and several examples are developed by the application of Excel© spreadsheets.

In Chap. 10, the General Extreme Value distribution is described with respect to the following topics: introductory remarks, probability distribution and density functions, estimation of parameters, estimation of quantiles, goodness of fit tests, estimation of standard errors, and estimation of confidence limits. Several worked examples are included, and several examples are developed by the application of Excel© spreadsheets.

In Chap. 11, the Log-Normal with 3 parameters distribution, for the minima, is described with respect to the following topics: introductory remarks, probability distribution and density functions, estimation of parameters, estimation of quantiles, goodness of fit tests, estimation of standard errors, and estimation of confidence limits. Several worked examples are included, and several examples are developed by the application of Excel© spreadsheets.

In Chap. 12, the Pearson type III distribution, for the minima, is described with respect to the following topics: introductory remarks, probability distribution and density functions, estimation of parameters, estimation of quantiles, goodness of fit tests, estimation of standard errors, and estimation of confidence limits. Several worked examples are included, and several examples are developed by the application of Excel© spreadsheets.

In Chap. 13, the Extreme Value type III for the minima (Weibull) distribution, for the minima, is described with respect to the following topics: introductory remarks, probability distribution and density functions, estimation of parameters, estimation of quantiles, goodness of fit tests, estimation of standard errors, and estimation of confidence limits. Several worked examples are included, and several examples are developed by the application of Excel© spreadsheets.

In Chap. 14, the General Extreme Value distribution, for the minima, is described with respect to the following topics: introductory remarks, probability distribution and density functions, estimation of parameters, estimation of quantiles, goodness of fit tests, estimation of standard errors, estimation of confidence limits. Several worked examples are included, and several examples are developed by the application of Excel© spreadsheets.

Basic Notions of Probability and Statistics for Natural Extreme Events Frequency Analyses

2

"Sint ut sunt aut non sint" (Accept them as they are or deny their existence). Pope Clement XIII, as cited in Gumbel (1958).

2.1 Introduction

To understand better the procedures that will follow this chapter and that are the foundations of natural extreme events frequency analyses, a quick review of some basic notions of probability theory and statistics is needed. This chapter does not contain any theorems, lemmas, or axioms, it is focused on the use of probability models and their adjustment to real world data using statistical procedures.

First, a brief review of the basic concepts of theory of probability, related to frequency analysis, are given, such as the definition of probability, the notion of random variables, the definition of probability distribution and density functions, the definition of non-exceedance and exceedance probabilities, and the definition of the return period of random events.

For the case of the review of statistics, several topics are contained, such as measures of central tendency, measures of dispersion, measures of symmetry, measures of peakedness, methods of estimation of parameters, quantile and frequency factor estimation, confidence limits, goodness of fit tests, and tests for the data to show that they have independence and homogeneity and the possible existence of outliers.

J. A. Raynal Villaseñor, *Frequency Analyses of Natural Extreme Events*,
Earth and Environmental Sciences Library,
https://doi.org/10.1007/978-3-030-86390-6_2

2.2 Chapter Objectives

The objectives of the chapter are:

(1) To provide a quick review of probability theory dealing with the definitions of probability, random variables, probability distribution and density functions, exceedance, and non- exceedance probabilities, and return period.
(2) To define the population moments of a distribution and its sample moments counterpart
(3) To display useful statistical measures of central tendency, dispersion, symmetry, and peakedness and how to evaluate them using Excel® spreadsheets
(4) To define three methods of estimation of the parameters of a probability distribution function: moments, maximum likelihood, and probability weighted moments
(5) To establish how to calculate the quantiles of a probability distribution function and how to evaluate the frequency factors
(6) To provide a quick review of some of the plotting position formulas available nowadays
(7) To define how to calculate the confidence limits of a probability distribution function using the following three methods: moments, maximum likelihood, and probability weighted moments
(8) To provide a way to plot the hydrologic extreme value data and models used to fit such samples of natural extreme value data
(9) To establish two goodness of fit tests to compare and choose among the competing probability models
(10) To provide several tests for the data to guarantee that it is homogenous, stationary, and independent, and that it has not any low or high outliers
(11) To display in a graphical way several sets of actual natural extreme value data

2.3 Basic Notions of Theory of Probability

The theory of probability is that part of the mathematical statistics science that provides the mathematical infrastructure that supports the development and use of probability models like probability distribution functions and many others. It comprises many theorems, axioms, lemmas, etc.

2.3.1 Definition of Probability

The definition of probability used herein will be that given by the frequentist approach; so, the estimated probability or empirical probability of an event is the relative frequency of the occurrence of the event when the number of observations is very large. The limit of the relative frequency, as the number of observations increases indefinitely, is the probability itself.

The probability defined as above has the following properties, Woodroofe (1975):

(1) The probability P has the following limits: $0 \leq P \leq 1$. This inequality means that there is no negative probabilities and no probabilities bigger than one. $P = 0$ corresponds to the impossible event; on the contrary, $P = 1$ corresponds to the sure event.
(2) If A and B are mutually exclusive events, then:

$$P(A \cup B) = P(A) + P(B) \tag{2.1}$$

(3) If A_1, A_2... is an infinite sequence of mutually exclusive events, then:

$$P\left(\bigcup_{i=1}^{\infty} A_i\right) = \sum_{i=1}^{\infty} P(A_i) \tag{2.2}$$

in Eqs. (2.1) and (2.2) the symbol $\cup$ means the union of the events considered.

These are the three fundamental axioms of probability.

2.3.2 Random Variables

A random variable is a variable that will only be described in terms of its probabilistic occurrence not in absolute terms. For example, the area of a square is not a random variable but a deterministic variable, given that it is only needed to know the value of the length of a side of the square to obtain the area as: $A = b^2$. Here the deterministic variable is governed by a precise and well-defined law or formula; on the contrary, the amount of rain that will fall in a given area at a precise time of the year, is a random variable given that only can be described in probabilistic terms and such variable is not governed by a specific and well-defined formula. Up to now, it is not possible to know in advance the exact amount of rain that will fall in each area at a particular time, not even with meteorological radars and/or meteorological satellites and/or meteorological computer models.

Random variables may be continuous or discrete. A continuous random variable is that can theoretically assume any value within its domain; otherwise, is called a discrete random variable.

2.3.3 Probability Distribution Functions

By definition, the probability distribution function of a random variable X is defined as:

$$F(x) = \Pr(X \leq x) = \int_{-\infty}^{x} f(x)dx \tag{2.3}$$

where *F (x)* is the probability distribution function of the random variable *X*, *Pr (X ≤ x)* is the non-exceedance probability and *f (x)* is the probability density function of random variable *X*.

It is important to note that the probability distribution function is also a probability by itself, while the probability density function it is not a probability by itself. Another useful relationship when using probability distribution functions is:

$$0 \leq F(x) \leq 1 \tag{2.4}$$

This is a consequence of the first axiom of probability mentioned above.

2.3.4 Probability Density Functions

By definition, the probability density function of a random variable *X* is defined as:

$$f(x) = \frac{dF(x)}{dx} \tag{2.5}$$

The relationship between the probability distribution function and the probability density function can be established either by Eq. (2.3) or by Eq. (2.5).

2.3.5 Non-exceedance and Exceedance Probabilities

As it was defined before, the non-exceedance probability is:

$$\Pr(X \leq x) = F(x) = \int_{-\infty}^{x} f(x)dx \tag{2.6}$$

and its relationship with the exceedance probability is:

$$\Pr(X > x) = 1 - \Pr(X \leq x) = 1 - F(x) \tag{2.7}$$

given that:

$$\Pr(X \leq x) + \Pr(X > x) = 1 \tag{2.8}$$

Non-exceedance probabilities are often used in maxima natural extreme value frequency analysis, whereas exceedance probabilities are often used in minima natural extreme value frequency analysis.

2.3.6 Return Period

The return period, T_r, is defined as the mean value of the period of time that has been elapsed between consecutive occurrences of a certain event (maxima or minima). Its relationships with the non-exceedance and exceedance probabilities can be stated as:

$$\Pr(X \leq x) = F(x) = 1 - \frac{1}{Tr} \tag{2.9}$$

$$\Pr(X > x) = 1 - F(x) = \frac{1}{Tr} \tag{2.10}$$

For example, the 100-year flood will have a non-exceedance probability of 0.99. Given that the probabilities also can be expressed in percentage, such probability is 99%; so, there is only 1% chance, in the average, that the 100-year flood will be exceeded in a period of time of 100 years.

On the contrary, the 100-year drought will have an exceedance probability of 0.01 or 1%; so, there is a 99% chance, in the average, that the 100-year drought will be exceeded in a period of time of 100 years.

The return period is an important feature in flood and drought frequency analyses, it defines the level of protection that a given structure or non-structural measure will provide for the future occurrence of a certain event (maxima or minima). In engineering, many design criteria are based in the pre-assignment of the return period.

2.4 Basic Notions of Statistics

Statistics is the part of the mathematical statistics science that develops and applies scientific methods for the collection, organization, summarizing, presentation, and analysis of data as well producing valid conclusions and drawing reasonable decisions based in such scientific framework.

Statistics provides the means to use actual data to fit probabilistic models for a specific purpose, in this book that purpose is the probabilistic modeling of natural extreme events.

2.4.1 Moments of a Distribution

The definition of the r-th population moment about the origin of a distribution is:

$$\mu_r' = \int_{-\infty}^{\infty} x^r f(x)dx \tag{2.11}$$

where $\mu_r{}'$ is the r-th population moment about the origin of a distribution, x is the random variable, $f(x)$ is the probability density function. The corresponding r-th population moment about the mean, also called r-th population central moments, of a distribution is:

$$\mu_r = \int_{-\infty}^{\infty} (x-\mu)^r f(x)dx \tag{2.12}$$

where μ_r is the r-th population moment about the mean of a distribution.

The r-th sample moment about the origin is defined as:

$$m_r\prime = \frac{1}{N}\sum_{i=1}^{N} x_i^r \tag{2.13}$$

where $m_r{}'$ is the r-th sample moment about the origin, x is the random variable and N is the sample size. The corresponding r-th sample moment about the mean of a distribution is:

$$m_r = \frac{1}{N}\sum_{i=1}^{N}(x_i-\hat{\mu})^r \tag{2.14}$$

where m_r is the r-th sample moment about the mean.

2.4.2 Measures of Central Tendency

2.4.2.1 The Mean

One of the most useful measures of central tendency is the arithmetic mean, or briefly the mean, defined as:

$$\mu = \int_{-\infty}^{\infty} xf(x)dx \tag{2.15}$$

where μ is the population mean and *f(x)* is the probability density function of the random variable X. An estimator of the population mean, based in a sample size N, can obtained as:

$$\hat{\mu} = \overline{x} = \frac{1}{N}\sum_{i=1}^{N} x_i \tag{2.16}$$

where $\hat{\mu}$ or $\overline{x}$ is the sample mean, also called the arithmetic mean, N is the sample size and the x's represent the point values of the sample.

Excel® has a library function to compute the mean: **AVERAGE**(number1,number2, …). Number1, number2, … are 1 to 30 numeric arguments for which you want the average.

Example of Calculation of the Mean

Find the mean of the set of flood data of gauging station Huites, Mexico contained in Appendix A.

The mean can be calculated as:

$$\hat{\mu} = \overline{x} = \frac{1}{N}\sum_{i=1}^{N} x_i = \frac{1}{51}(127447.1) = 2498.9627m^3/s$$

In the case of this example, the AVERAGE function of Excel® will look like this:

= AVERAGE(b8:b58) equal to 2498.9627 m^3/s.

2.4.2.2 The Median

The median of a distribution is defined as the value that is located at the middle value of a probability distribution function:

$$F(x) = \int_{-\infty}^{M} f(x)dx = 0.5 \tag{2.17}$$

where $F(x)$ is the distribution function of the random variable X, $f(x)$ is its probability density function and M is the value of the population median. A sample estimator of the median will be the value that will leave about the same number of data points below and above the value of the median, when the data points have been ordered according to their magnitude, in either descending or ascending ways. When the number of data points is odd, the value of the sample median is unique, when the number of data points is even, the mean of the two middle values will be the sample median.

Excel® has a library function to compute the median: **MEDIAN**(number1, number2, …). Number1, number2, … are 1 to 30 numeric arguments for which you want the median.

Example of Calculation of the Median

Find the median of the set of flood data of gauging station Huites, Mexico contained in Appendix A.

The mean can be calculated as 1775.7 m^3/s, when ordering the sample in a descending order of magnitude, given that the sample size is an odd number, 51.

In the case of this example, the MEDIAN function of Excel® will look like this:

= MEDIAN(b8:b58) equal to 1775.7 m^3/s.

2.4.2.3 The Mode

The mode is defined as the most frequent value and a property associated with it is that the probability density function will be at its maximum at the value of the mode. This property is stated as:

$$\frac{d(f(x))}{dx} = 0 \tag{2.18}$$

for $x =$ mode.

The mode may not always exist and even if it exists it may not be unique. A unimodal distribution is that one with only one mode. If it has two modes is called bimodal.

Excel® has a library function to compute the mode: **MODE**(number1,number2, ...). Number1, number2, ... are 1 to 30 numeric arguments for which you want the mode.

Example of Calculation of the Mode

Find the mode of the set of flood data of gauging station St. Mary's River at Stillwater, Nova Scotia, Canada, contained in Appendix A.

The mode can be calculated as 232 m^3/s.

In the case of this example, the MODE function of Excel® will look like this:

= MODE (b8:b79) equal to 232 m^3/s.

2.4.2.4 The Geometric Mean

The geometric mean is defined as:

$$G = (x_1 x_2 x_3 \ldots x_N)^{\frac{1}{N}} \tag{2.19}$$

where G is the geometric mean, the x's are the data points and N is the sample size.

Excel® has a library function to compute the geometric mean: **GEOMEAN**(number1,number2, ...). Number1, number2, ... are 1 to 30 numeric arguments for which you want the geometric mean.

Example of Calculation of the Geometric Mean

Find the geometric mean of the set of flood data of gauging station Huites, Mexico contained in Appendix A.

The geometric mean can be calculated as:

$$G = (x_1 x_2 x_3 \ldots x_N)^{\frac{1}{N}} = (9.243\text{E} + 166)^{\frac{1}{51}} = 1878.621 m^3/s$$

In the case of this example, the GEOMEAN function of Excel® will look like this:

= GEOMEAN (b8:b58) equal to 1878.621 m^3/s.

2.4.2.5 The Root Mean Square

The root mean square or quadratic mean is defined as:

$$RMS = \sqrt{\overline{X}^2} = \sqrt{\frac{\sum_{i=1}^{N} x_i^2}{N}} \tag{2.20}$$

where *RMS* is the root mean square, the *x*'s are the data points and *N* is the sample size.

Excel® has no library function to compute the root mean square or quadratic mean directly.

2.4.2.6 Example of Calculation of the Root Mean Square

Find the root mean square of the set of flood data of gauging station Huites, Mexico contained in Appendix A.

The root mean square can be calculated as:

$$RMS = \sqrt{\overline{X}^2} = \sqrt{\frac{\sum_{i=1}^{N} x_i^2}{N}} = \sqrt{\frac{565208142}{51}} = 3329.0408 \text{ m}^3/\text{s}$$

2.4.2.7 Quartiles, Deciles, Percentiles and Quantiles

As in the definition of the median, it is the value that divides by half the distribution function, the quartiles are those values that divide the distribution function into four equal parts. The first, second and third quartiles are denoted by Q_1, Q_2, and Q_3, respectively, being Q_2 equal to the median.

In a similar fashion the values that divide the distribution function into ten equal parts are called deciles and they are denoted by D_1, D_2, …, D_9.

The percentiles are those values that divide the distribution function into one hundred equal parts, and they are denoted as P_1, P_2…, P_{99}.

The fifth decile and the fiftieth percentile correspond to the median. The twenty-fifth and seventy-fifth percentiles correspond to the first and third quartiles, respectively.

Quartiles, deciles, percentiles and other values obtained by equal subdivisions of the distribution function are called quantiles, collectively.

Excel® has a library function to compute the percentiles: **PERCENTILE(array,k)**. Array is the array or range of data that defines relative standing. K is the percentile value in the range (0,1), inclusive.

Excel® has a library function to compute the quartiles: **QUARTILE(array,quart)**. Array is the array or cell range of numeric values for which you want the quartile value. Quart indicates which value to return.

Using the above two Excel® statistical functions, one may obtain the deciles and any other quantiles.

Example of Calculation of the Quartiles, Deciles, Percentiles, and Quantiles
Using the flood data contained in Appendix A of gauging station Huites, Mexico (1942–1992), find the three quartiles corresponding to a probability of 75%, 50% and 25%. Find the deciles corresponding to a probability of 10%, 20%, 30% and 50%. Finally find the percentiles corresponding to a probability of 7%, 45% and 95%.

Ordering the sample in descending order of magnitude and assigning the number one to the largest value of the flood data and 51 to the least value of the flood data, then the cumulative probability can be found as 1–1/(N/m), where N is the sample size and m is the order number, so the approximate values are obtained as:

Q_1 = 1176.1 (25.49%); Q_2 = 1820.2 (50.98%); Q_3 = 2722.4 (74.51%).

D_1 = 853.6 (9.80%); D_2 = 1071.3 (19.81%); D_3 = 1312.7 (29.47%); D_5 = 1820.2 (50.98%).

P_7 = 728.6 (7.84%); P_{45} = 1627.5 (44.23%); P_{95} = 8600.4 (94.11%).

By using the statistical functions contained in Excel®, the results are as follows:

Q_1 = 1155 (25%); Q_2 = 1775.7 (50%); Q_3 = 2691.45 (75%).

D_1 = 853.6 (10%); D_2 = 1071.3 (20%); D_3 = 1312.7 (30%); D_5 = 1775.7 (50%).

P_7 = 785.55(7%); P_{45} = 1627.6 (45%); P_{95} = 7679.4 (95%).

2.4.3 Measures of Dispersion

2.4.3.1 The Range

The range of a sample of data points is the difference between the largest and the smallest values in the sample. It is stated as:

$$R = x_N - x_1 \tag{2.21}$$

where R is the range, and x_N and x_I are the largest and the smallest values, respectively, in the population or in a sample, when the x_i´s, i = 1, 2, …, N, have been rank ordered from x_I to x_N: $x_I < x_2 < \ldots < x_N$.

Excel® has no library function to compute the range directly.

Example of Calculation of the Range

Find the range of the set of flood data of gauging station Huites, Mexico contained in Appendix A.

The range can be computed as:

R = 10,129.3 – 421.8 = 9707.5 m^3/s.

2.4.3.2 The Mean Deviation

The mean deviation or average deviation is defined as:

$$MD = \frac{\sum_{i=1}^{N} |x_i - \overline{X}|}{N} \tag{2.22}$$

where *MD* is the mean deviation, $\overline{X}$ is the sample mean, *N* is the sample size and the *x's* represent the point values of the sample.

Excel® has a library function to compute the mean deviation: **AVEDEV(number1**,number2, ...). Number1, number2, ... are 1 to 30 arguments for which you want the average of the absolute deviations.

Example of the Computation of the Mean Deviation

Find the mean deviation of the set of flood data of gauging station Huites, Mexico contained in Appendix A.

Using Eq. (2.22) the mean deviation can be found:

$$MD = \frac{\sum_{i=1}^{N} |x_i - \overline{X}|}{N} = \frac{79429.19}{51} = 1557.435$$

In the case of this example, the AVEDEV function of Excel® will look like this:

= AVEDEV(B8:B58) equal to 1557.435.

2.4.3.3 The Semi-Interquartile Range

The semi-interquartile range, or quartile deviation, is defined by:

$$Q = \frac{Q_3 - Q_1}{2} \tag{2.23}$$

where Q is the semi-interquartile range, Q_1 and Q_3 are the first and third quartiles, respectively. No function is available in Excel® to compute directly the semi-interquartile range.

2.4.3.4 The 10–90 Percentile Range

The 10–90 Percentile Range is defined as:

$$10 - 90\ Percentile\ Range = P_{90} - P_{10} \tag{2.24}$$

where P_{90} and P_{10} are the tenth and ninetieth percentiles, respectively.

2.4.3.5 The Variance and Standard Deviation

A very useful measure of dispersion is the variance; it measures the dispersion of the values of the x's around its central value represented by the mean. The variance is defined as:

$$\sigma^2 = \int_{-\infty}^{\infty} (x - \mu)^2 f(x) dx \tag{2.25}$$

where σ^2 is the population variance, μ is the population mean, and $f(x)$ is the probability density function of the random variable x.

The squared root of the variance, namely the standard deviation, is also a very useful measure of dispersion. It is defined as:

$$\sigma = \sqrt{\sigma^2} \tag{2.26}$$

A sample estimator for the variance is given by:

$$\hat{\sigma}^2 = s^2 = \frac{1}{N} \sum_{i=1}^{N} (x_i - \hat{\mu})^2 \tag{2.27}$$

where $\hat{\sigma}^2$ or s^2 is the sample variance, $\hat{\mu}$ is the sample mean, N is the sample size and the x's represent the point values of the sample.

Excel® has library functions to compute the population and the sample variance: **VARP(number1**,number2,...). Number1,number2,... are 1 to 30 number arguments corresponding to a population. **VAR(number1**,number2,...). Number1,number2,... are 1 to 30 number arguments corresponding to a sample of a population.

A sample estimator for the standard deviation is:

$$\hat{\sigma} = s = \left[\frac{1}{N} \sum_{i=1}^{N} (x_i - \hat{\mu})^2 \right]^{1/2} \tag{2.28}$$

In the two preceding equations, when the number of data points is less than 30, a bias correction must be applied by changing the factor N by its corrected value of $(N - 1)$.

Excel® has a library function to compute the population and the sample standard deviation: **STDEVP(number1**,number2,...). Number1,number2,... are 1 to 30 number arguments corresponding to a population. You may use a single array or a reference to an array instead of arguments separated by commas, too.

STDEV(number1,number2,...). Number1, number2,... are 1 to 30 number arguments corresponding to a sample of a population. As before, you may use a single array or a reference to an array instead of arguments separated by commas, too.

2.4.3.6 The Absolute and Relative Dispersions and the Coefficient of Variation

The actual variation or dispersion, determined by the standard deviation or other measure of dispersion, is called absolute dispersion. The relative dispersion is defined by:

$$Relative\ dispersion = \frac{Absolute\ dispersion}{average} \tag{2.29}$$

When the absolute dispersion is the standard deviation and the average is the mean, then the relative dispersion is called the coefficient of variation or coefficient of dispersion and the coefficient of variation (CV) is defined as the proportion of the standard deviation over the mean:

$$CV = \frac{\hat{\sigma}}{\hat{\mu}} \tag{2.30}$$

The CV is usually expressed as a percentage.

2.4.3.7 Standardized Variables

A standardized variable is defined as:

$$z = \frac{x - \hat{\mu}}{\hat{\sigma}} = \frac{x - \overline{x}}{s} \tag{2.31}$$

where z is the standardized variable, $\hat{\mu}$ or $\overline{x}$ is the sample mean, $\hat{\sigma}$ or s is the sample standard deviation and x is the original value from the data sample. The standardized variable z is dimensionless, that is, independent of the units used.

2.4.4 Measures of Symmetry

2.4.4.1 The Skewness Coefficient

The skewness coefficient measures the lack of symmetry of a probability density function. If the skewness coefficient is positive, then the probability density function will be skewed to the left, that is has a longer tail to the right of the central maximum than to the left, see Fig. 2.1. If the skewness coefficient is negative, then the probability density function will be skewed to the right, that is it has a longer tail to the left of the central maximum than to the right, see Fig. 2.2. In the case of the normal distribution, given that this is a symmetrical probability density function, the value of the skewness coefficient is zero, see Fig. 2.3.

The skewness coefficient is defined as:

$$\gamma = \frac{1}{\sigma^3} \int_{-\infty}^{\infty} (x - \mu)^3 f(x) dx \tag{2.32}$$

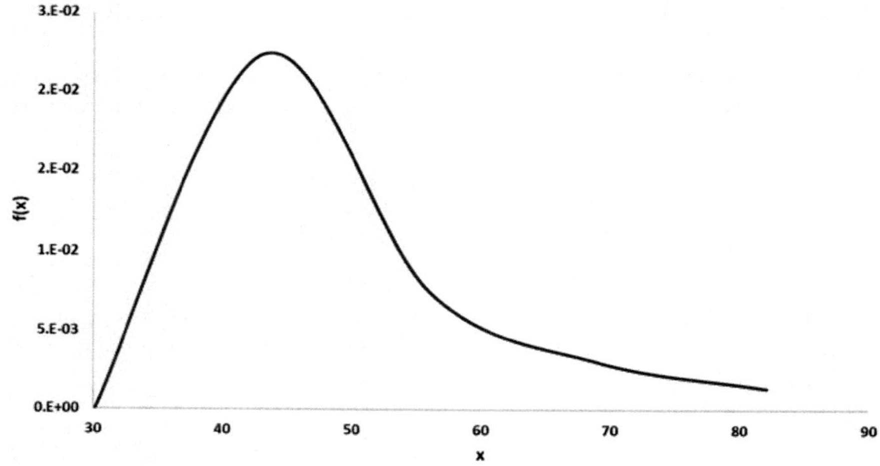

Fig. 2.1 Probability density function with positive skewness

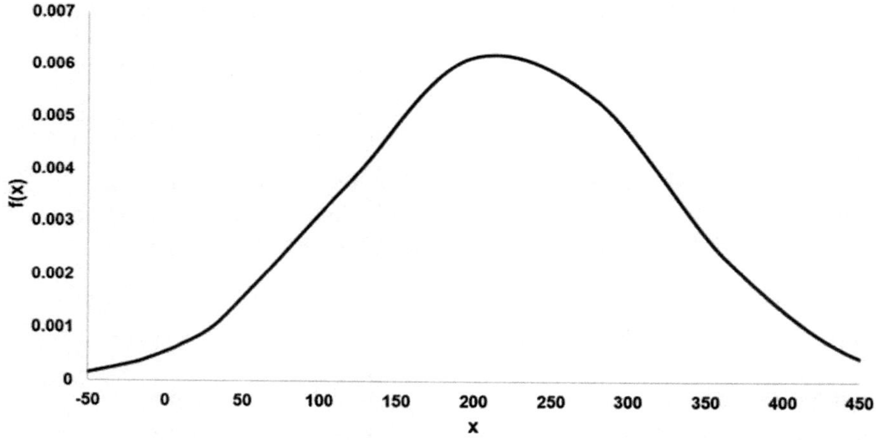

Fig. 2.2 Probability density function with negative skewness

where γ is the population skewness coefficient.

A sample unbiased estimator of the population skewness coefficient is:

$$\hat{\gamma} = g = \frac{N}{(N-1)(N-2)\hat{\sigma}^3} \sum_{i=1}^{N} (x_i - \hat{\mu})^3 \tag{2.33}$$

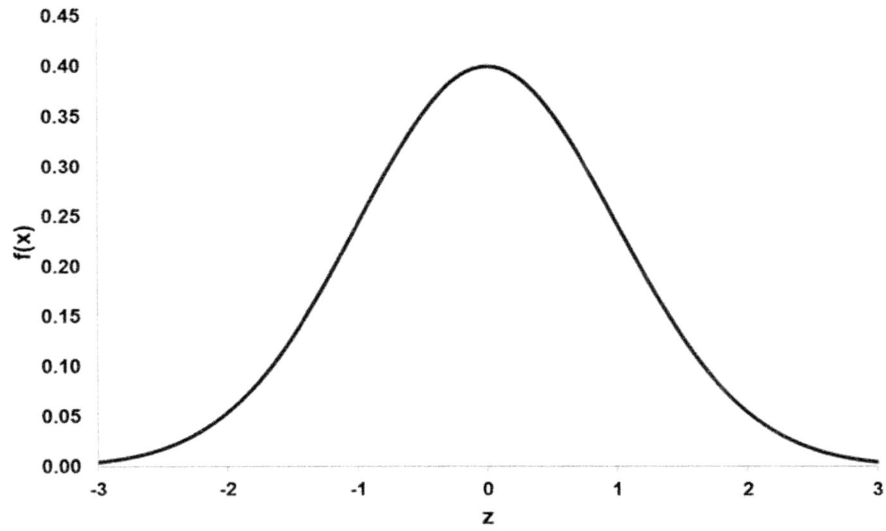

Fig. 2.3 Probability density function with zero skewness

2.4.5 Measures of Peakedness

2.4.5.1 The Kurtosis Coefficient

The kurtosis coefficient measures the peakedness or flatness of a probability density function. A positive value on the kurtosis coefficient indicates that the probability density function will be a peaked one. On the contrary, a negative value on the kurtosis coefficient indicates that the probability density function will be a flat one.

The kurtosis coefficient is defined as:

$$\kappa = \frac{1}{(\sigma^2)^2} \int_{-\infty}^{\infty} (x-\mu)^4 f(x)dx \tag{2.34}$$

where κ is the population coefficient of kurtosis.

A sample estimator of the population skewness coefficient is:

$$\hat{\kappa} = k = \frac{N(N+1)}{(N-1)(N-2)(N-3)(\hat{\sigma}^2)^2} \sum_{i=1}^{N} \left(x_i - \hat{\mu}\right)^4 - \frac{3(N-1)^2}{(N-2)(N-3)}. \tag{2.35}$$

A probability density function showing a relatively high peak is called leptokurtic, on the contrary a probability density function showing a relatively flat-topped peak is called platykurtic. The normal distribution being not very peaked or very

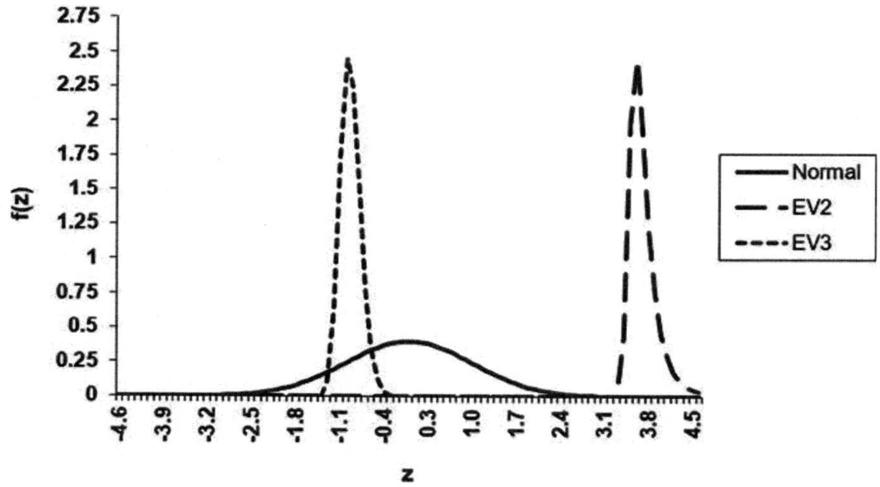

Fig. 2.4 The normal distribution compared with the EV2 and EV3 probability density functions regarding skewness and peakedness

flat-topped is called mesokurtic. The normal distribution has a population kurtosis coefficient with value of 3.

A graphic comparison between the standard normal distribution and the reduced variate extreme value types II and III distributions is made in Fig. 2.4, in which the extreme value distributions are more skewed and peaked than the normal distribution.

2.4.6 Descriptive Statistics

Several of the previous statistical measures, are contained within a library function in the data analysis section contained in any Excel® spreadsheet, it designated as Descriptive Statistics and it is shown in Fig. 2.5.

2.5 Methods for the Estimation of Parameters of Probability Distribution Functions

2.5.1 The Method of Moments (MOM)

The method of moments is based on the procedure devised to obtain the moments of inertia in statics. Fisher-Tippet (1928) adapted the method of moments to be used in statistics by considering the probability density function as the body to which the moments of inertia must be computed. In this method, either the r-th moments about the origin or the r-th central moments of a probability distribution

Fig. 2.5 Descriptive statistics

	A	B
1	*Column1*	
2		
3	Mean	42.04543478
4	Standard Error	1.97843545
5	Median	40.24
6	Mode	#N/A
7	Standard Deviation	13.41840208
8	Sample Variance	180.0535143
9	Kurtosis	-0.725243585
10	Skewness	0.359116378
11	Range	54.3
12	Minimum	17.12
13	Maximum	71.42
14	Sum	1934.09
15	Count	46
16	Largest(1)	71.42
17	Smallest(1)	17.12
18	Confidence Level(95.0%)	3.984773545

function are used. The population r-th moment about the origin of a probability distribution function is defined as:

$$\mu_r' = \int_{-\infty}^{\infty} x^r f(x)dx \tag{2.36}$$

where μ_r' is the population r-th moment about the origin and f (x) is the probability density function of the x's. The population central r-th moment is defined as:

$$\mu_r = \int_{-\infty}^{\infty} (x - \mu_1')^r f(x)dx \tag{2.37}$$

where μ_r is the population r-th central moment and μ_1' is the first moment about the origin, namely the population mean.

The sample r-th moment about the origin of a probability distribution function is defined as:

$$\hat{\mu}_r' = \frac{1}{N}\sum_{i=1}^{N} x_i^r \tag{2.38}$$

where $\hat{\mu}_r'$ is the sample r-th moment about the origin.

The definition of the sample central r-th moment of a probability distribution function is:

$$\hat{\mu}_r = m_r = \frac{1}{N}\sum_{i=1}^{N}(x_i - \overline{x})^r \tag{2.39}$$

where $\hat{\mu}_r$ or m_r is the sample central r-th moment.

First, equating the population moments with the sample moments, and then simultaneously solving the resulting system of equations, one may obtain the estimators of the parameters. It is very common that the sample mean, and sample variance are also represented by $\hat{\mu}$ and $\hat{\sigma}^2$, respectively.

As an example, for a two-parameter probability distribution function the following equations must be set and solved:

$$\mu = \hat{\mu} = \overline{x}$$

$$\sigma^2 = \hat{\sigma}^2 = s^2$$

where μ and σ^2 are the population mean and variance, respectively, and $\overline{x}$ or $\hat{\mu}$ and $\hat{\sigma}^2$ or s^2 are the sample mean and variance, respectively. The sample mean and variance were defined before as:

$$\overline{x} = \hat{\mu} = \frac{1}{N}\sum_{i=1}^{N} x_i$$

$$s^2 = \hat{\sigma}^2 = \frac{1}{N}\sum_{i=1}^{N}(x_i - \hat{\mu})^2$$

In the last equation, when the number of data points is less than 30, a bias correction must be applied by changing the factor N by its corrected value of (N − 1).

2.5.2 The Method of Maximum Likelihood (ML)

The method of maximum likelihood is based on the maximization of either the likelihood or the log-likelihood functions with respect to the parameters to be estimated, given the mathematical property that the maximum occur at the same point for both functions.

Such likelihood function of the random variable X is found as the joint probability density function of N independent identically distributed random variables, x_i's:

$$L(x, \theta) = \prod_{i=1}^{N} f(x_i) \tag{2.40}$$

where $L(x, \theta)$ is the likelihood function and $f(x_i)$ are the probability densities functions of the x's.

If the natural logarithms are taken in both sides of the previous equation, the result is the log-likelihood function, which is commonly used to estimate the parameters of a probability distribution function:

$$LL(x,\theta) = \sum_{i=1}^{N} Ln[f(x_i)] \tag{2.41}$$

where $LL(x, \theta)$ is the log-likelihood function and $f(x_i)$ are the probability densities functions of the x's.

Now, either the likelihood or the log-likelihood functions must be maximized to obtain the maximum likelihood estimators for the parameters of the probability distribution function. There are many ways to accomplish this; one of them is to directly maximize either the likelihood or the log-likelihood functions with a direct approach that obtains the maximum likelihood estimators of the parameters, by using any of the optimization procedures available now. Care must be taken that the optimization procedure is applicable to either the likelihood or the log-likelihood functions and its restrictions on the parameters. Usually, a multivariable constrained non-linear optimization procedure is required to obtain the maximum likelihood estimators of the parameters of many of the probability distribution functions used in natural extreme events frequency analyses. This approach will not be used in this book.

A classical approach is the indirect method to maximize either the likelihood or the log-likelihood functions. This approach proceeds to obtain the first-order partial derivatives of either the likelihood or the log-likelihood functions, with respect to every parameter that the distribution function has, and then equated them to zero. The system of equations must be solved in closed form or in approximated form to obtain the maximum likelihood estimators of the parameters of the probability distribution function. This approach will be the one used herein.

The system of equations has the form:

$$\frac{\partial LL(x,\theta)}{\partial \theta_1} = 0 \tag{2.42}$$

$$\frac{\partial LL(x,\theta)}{\partial \theta_2} = 0 \tag{2.43}$$

$$\frac{\partial LL(x,\theta)}{\partial \theta_n} = 0 \tag{2.44}$$

2.5.3 The Method of Probability Weighted Moments (PWM)

A distribution function may be characterized by its probability weighted moments (PWM) defined as, Greenwood et al. (1979):

$$M_{l,j,k} = E\left[X^l F^j (1-F)^k\right] = \int_0^1 [x(f)]^l F^j (1-F)^k dF \tag{2.45}$$

There are two different expressions to evaluate the population PWM defined in the previous equation, they are, Greenwood et al. (1979):

$$M_{l,0,k} = \sum_{j=0}^{k} \binom{k}{j} (-1)^j M_{l,j,0} \tag{2.46}$$

and:

$$M_{l,j,0} = \sum_{k=0}^{j} \binom{j}{k} (-1)^k M_{l,0,k} \tag{2.47}$$

$M_{1,0,0}$, represents the conventional moments about the origin. The following convention is taken to simplify the notation, Greenwood et al. (1979):

$$M_{(k)} = M_{1,0,k} \tag{2.48}$$

An unbiased estimator of $M_{(k)}$ is, Maciunas Landwher et al. (1979):

$$M_{(k)} = \frac{1}{N} \sum_{i=1}^{N-k} x_i \frac{\binom{N-i}{k}}{\binom{N-1}{k}} \tag{2.49}$$

and k is a non-negative integer and the x_i's, i = 1, 2, ..., N, have been rank ordered from x_1 to x_N: $x_1 < x_2 < \ldots < x_N$.

2.6 Quantile Estimation and Frequency Factor

Once the parameters of the probability distribution function have been computed, it is possible to estimate of the quantiles (x_T) that correspond to a return period of interest. As it was shown before, the return period is related with the non-exceedance probability by the following relationship:

$$\Pr(X \le x) = F(x) = 1 - \frac{1}{Tr} \tag{2.50}$$

and the return period is related with the exceedance probability by the following expression:

$$\Pr(X > x) = 1 - F(x) = \frac{1}{Tr} \tag{2.51}$$

So, in any case, the quantiles can be obtained by inverting the probability distribution function in a way that the quantiles are a function of either the probability distribution function or the return period, that is:

$$x_T = \varphi_1(F(x)) = \varphi_2(T_r) \tag{2.52}$$

where x_T is the desired quantile and $\phi_1(F(x))$ or $\phi_2(T_r)$ will depend on each of the probability distribution function that will be used. There are some cases, for example the normal, Two and Three parameters Log-Normal, Gamma, Pearson III, and Log-Pearson type III probability distribution functions which cannot be expressed in inverse form; so, in these cases a numerical procedure will be required to invert the probability distribution function and to obtain the desired quantiles.

Chow (1964) proposed a general form to calculate the quantiles x_T as:

$$x_T = \mu_1' + K_T\sqrt{\mu_2} = m_1' + K_T\sqrt{m_2} \tag{2.53}$$

where μ_1' and μ_2 are the population moments and m_1' and m_2 are the sample moments of the distribution and K_T is the so-called frequency factor.

2.7 Plotting Position Formulas

To be able to plot the points of natural extreme data into a probability paper, Gumbel's probability paper is the most widely used probability paper in engineering natural extreme event studies, there is a need for a formula that relates the data points with their return period or with their non-exceedance or exceedance probability, this is provided by the plotting position formulas. There are many plotting positions formulas that have been proposed in the literature, Cunnane (1989); some of them are shown in Table 2.1.

To apply these formulas, in the maxima case, the data must be ranked in descending order, that is $m = 1$ for the largest value in the sample, $m = N$ for the smallest value in the sample. For the case of the minima, the data must be ranked in ascending order, that is $m = 1$ for the smallest value in the sample, $m = N$ for the largest value in the sample.

To choose a plotting position formula depends heavily on which probability distribution function paper is being used to fit the natural extreme event data. For example, the Blom's formula is recommended when using the normal probability paper for several purposes and the Gringorten's formula is appropriate for the extreme value type I (Gumbel) distribution, Castillo (1988). The use of the probability weighted moments (PWM) plotting position formula is recommended for

Table 2.1 Plotting positions formulas

Plotting Position Formula	Return Period Tr	Non-exceedance Probability P (X ≤ x) = 1 – 1/Tr	Exceedance Probability P (X > x) = 1/Tr
Weibull	$\frac{N+1}{m}$	$1-\left(\frac{m}{N+1}\right)$	$\frac{m}{N+1}$
Gringorten	$\frac{N+0.12}{m-0.44}$	$1-\left(\frac{m-0.44}{N+0.12}\right)$	$\frac{m-0.44}{N+0.12}$
Foster or Hazen	$\frac{N}{m-0.5}$	$1-\left(\frac{m-0.5}{N}\right)$	$\frac{m-0.5}{N}$
Blom	$\frac{N+0.25}{m-0.375}$	$1-\left(\frac{m-0.375}{N+0.25}\right)$	$\frac{m-0.375}{N+0.25}$
Cunnane	$\frac{N+0.2}{m-0.4}$	$1-\left(\frac{m-0.4}{N+0.2}\right)$	$\frac{m-0.4}{N+0.2}$
California	$\frac{N}{m}$	$1-\left(\frac{m}{N}\right)$	$\frac{m}{N}$
Chegodayeb	$\frac{N+0.4}{m-0.3}$	$1-\left(\frac{m-0.3}{N+0.4}\right)$	$\frac{m-0.3}{N+0.4}$
Adamowski	$\frac{N+0.5}{m-0.24}$	$1-\left(\frac{m-0.24}{N+0.5}\right)$	$\frac{m-0.24}{N+0.5}$
PWM	$\frac{N}{m-0.35}$	$1-\left(\frac{m-0.35}{N}\right)$	$\frac{m-0.35}{N}$
EWSD	$\frac{N+1-\alpha}{m-\alpha}$	$1-\left(\frac{m-\alpha}{N+1-\alpha}\right)$	$\frac{m-\alpha}{N+1-\alpha}$

the general extreme value (Hosking et al., 1985), generalized Pareto (Hosking and Wallis, 1987) and Wakeby (Landwher et al., 1979b).

2.8 Confidence Limits

It can be shown, by using the asymptotic theory, that the distribution of the variates is asymptotically normal with mean x_T and variance ${S_T}^2$ when $N \rightarrow \infty$, Rao and Hamed (2000). By using distribution-sampling experiments, Kite (1975) derived distributions of extreme events generated from probability distributions functions commonly used in hydrology. These derived distributions proved to be statistically indistinguishable from a normal distribution. So, by assuming that the quantiles are normally distributed, the following expression can be used to set an approximate two-sided (1 – α) confidence limits on such quantiles:

$$x_l = x_T \pm z_{\alpha/2} S_T \tag{2.54}$$

where x_l is the upper or lower confidence limits, the plus sign is used for the upper confidence limit; whereas the negative sign is used for the lower confidence limit, z_α is the standard normal variate corresponding to a significance level α, and S_T is the squared root of the standard error of the estimate.

The values of z_α are given in Table 2.2 for the most common values of (1-α) confidence level.

Table 2.2 Values of the standard normal variate for different confidence levels for two-sided confidence limits

α	Confidence Level (1-α) (%)	z_α
0.10	90	1.65
0.05	95	1.96
0.02	98	2.33
0.01	99	2.58

The evaluation procedures of the standard errors of the estimates will be described in the following subsections.

2.9 Standard Errors of Estimates

The standard error of estimates depends upon the method of parameter estimation.

2.9.1 MOM Method

The MOM estimator of the standard error of estimates for a three-parameters probability distribution function, with parameters α, β and γ, and the quantile x_T depends upon the first three moments of a distribution is, Kite (1988):

$$\begin{aligned} S_T^2 = & \left(\frac{\partial x_T}{\partial m_1'}\right)^2 \mathrm{var}(m_1') + \left(\frac{\partial x_T}{\partial m_2}\right)^2 \mathrm{var}(m_2) \\ & + \left(\frac{\partial x_T}{\partial m_3}\right)^2 \mathrm{var}(m_3) + 2\left(\frac{\partial x_T}{\partial m_1'}\right)\left(\frac{\partial x_T}{\partial m_2}\right)\mathrm{cov}(m_1', m_2) \\ & + 2\left(\frac{\partial x_T}{\partial m_1'}\right)\left(\frac{\partial x_T}{\partial m_3}\right)\mathrm{cov}(m_1', m_3) \\ & + 2\left(\frac{\partial x_T}{\partial m_2}\right)\left(\frac{\partial x_T}{\partial m_3}\right)\mathrm{cov}(m_2, m_3). \end{aligned} \tag{2.55}$$

The variances and covariances of sample moments are obtained as, Kendall and Stuart (1994):

$$\mathrm{var}(m_1') = \frac{\hat{\mu}_2}{N} \tag{2.56}$$

$$\mathrm{var}(m_2) = \frac{1}{N}\left(\hat{\mu}_4 - \hat{\mu}_2^2\right) \tag{2.57}$$

$$\mathrm{var}(m_3) = \frac{1}{N}\left(\hat{\mu}_6 - \hat{\mu}_3^2 - 6\hat{\mu}_4\hat{\mu}_2 + 9\hat{\mu}_2^3\right) \tag{2.58}$$

$$\mathrm{covar}\left(m_1', m_2\right) = \frac{\hat{\mu}_3}{N} \tag{2.59}$$

$$\mathrm{covar}\left(m_1', m_3\right) = \frac{1}{N}\left(\hat{\mu}_4 - 3\hat{\mu}_2^2\right) \tag{2.60}$$

$$\mathrm{covar}(m_2, m_3) = \frac{1}{N}\left(\hat{\mu}_5 - 4\hat{\mu}_3\hat{\mu}_2\right) \tag{2.61}$$

And the relationship between x_T and the moments have been stated as, Kite (1988):

$$x_T = m_1\prime + K_T\sqrt{m_2} \tag{2.62}$$

where K_T is a function of the skewness coefficient, $\hat{\gamma}$ or g, and it is defined as:

$$\hat{\gamma} = \frac{m_3}{m_2^{\frac{3}{2}}} \tag{2.63}$$

Now the first-order partial derivatives of x_T with respect to the three first sample moments, can be obtained as, Kite (1988):

$$\frac{\partial x_T}{\partial m_1\prime} = 1 \tag{2.64}$$

$$\frac{\partial x_T}{\partial m_2} = \frac{K_T}{2\sqrt{m_2}} + \sqrt{m_2}\frac{\partial K_T}{\partial m_2} = \frac{K_T}{2\sqrt{m_2}} - \frac{3\hat{\gamma}}{2\sqrt{m_2}}\frac{\partial K_T}{\partial \hat{\gamma}} \tag{2.65}$$

$$\frac{\partial x_T}{\partial m_3} = \left(\frac{\partial x_T}{\partial K_T}\right)\left(\frac{\partial K_T}{\partial \hat{\gamma}}\right)\left(\frac{\partial \hat{\gamma}}{\partial m_3}\right) = \frac{1}{m_2}\left(\frac{\partial K_T}{\partial \hat{\gamma}}\right) \tag{2.66}$$

And inserting the preceding results into Eq. (2.55) transforms it to:

$$S_T^2 = \frac{\hat{\mu}_2}{N}\left\{1 + K_T\hat{\gamma} + \frac{\partial K_T}{\partial \gamma}\left[2\hat{\kappa} - 3\hat{\gamma}^2 - 6 + K_T\left(\hat{\lambda}_1 - \frac{6\hat{\gamma}\hat{\kappa}}{4} - \frac{10\hat{\gamma}}{4}\right)\right]\right.$$
$$\left. + \frac{K_T^2}{4}(\hat{\kappa} - 1) + \left(\frac{\partial K_T}{\partial \hat{\gamma}}\right)^2\left(\hat{\lambda}_2 - 3\hat{\lambda}_1\hat{\gamma} - 6\hat{\kappa} - \frac{9\hat{\gamma}^2\hat{\kappa}}{4} + \frac{35\hat{\gamma}^2}{4} + 9\right)\right\} \tag{2.67}$$

where K_T is the frequency factor, and $\hat{\gamma}$ and $\hat{\kappa}$ are the sample skewness and kurtosis coefficients, respectively, and $\hat{\lambda}_1$ and $\hat{\lambda}_2$ are defined as:

$$\hat{\lambda}_1 = \frac{\hat{\mu}_5}{\hat{\mu}_2^{5/2}} \tag{2.68}$$

$$\hat{\lambda}_2 = \frac{\hat{\mu}_6}{\hat{\mu}_2^3} \tag{2.69}$$

When the probability distribution function only has two parameters, Eq. (2.55) simplifies to:

$$S_T^2 = \left(\frac{\partial x_T}{\partial \alpha}\right)^2 \text{var}(\alpha) + \left(\frac{\partial x_T}{\partial \beta}\right)^2 \text{var}(\beta) + 2\left(\frac{\partial x_T}{\partial \alpha}\right)\left(\frac{\partial x_T}{\partial \beta}\right)\text{cov}(\alpha, \beta) \quad (2.70)$$

And a further simplification provides the following expression:

$$S_T^2 = \left(\frac{\partial x_T}{\partial m_1'}\right)^2 \text{var}(m_1') + \left(\frac{\partial x_T}{\partial m_2}\right)^2 \text{var}(m_2) + 2\left(\frac{\partial x_T}{\partial m_1'}\right)\left(\frac{\partial x_T}{\partial m_2}\right)\text{cov}(m_1', m_2) \quad (2.71)$$

And the previous expression can also be set in terms of the frequency factor as:

$$S_T^2 = \frac{\mu_2}{N}\left[1 + K_T\hat{\gamma} + \frac{K_T^2}{4}(\hat{\kappa} - 1)\right] \quad (2.72)$$

2.9.2 ML Method

The ML estimator of the standard error of estimates for a three-parameters probability distribution function with parameters α, β and γ is, Kite (1988):

$$\begin{aligned} S_T^2 = {} & \left(\frac{\partial x_T}{\partial \alpha}\right)^2 \text{var}(\alpha) + \left(\frac{\partial x_T}{\partial \beta}\right)^2 \text{var}(\beta) \\ & + \left(\frac{\partial x_T}{\partial \gamma}\right)^2 \text{var}(\gamma) + 2\left(\frac{\partial x_T}{\partial \alpha}\right)\left(\frac{\partial x_T}{\partial \beta}\right)\text{cov}(\alpha, \beta) \\ & + 2\left(\frac{\partial x_T}{\partial \alpha}\right)\left(\frac{\partial x_T}{\partial \gamma}\right)\text{cov}(\alpha, \gamma) + 2\left(\frac{\partial x_T}{\partial \beta}\right)\left(\frac{\partial x_T}{\partial \gamma}\right)\text{cov}(\beta, \gamma) \end{aligned} \quad (2.73)$$

and its variance–covariance matrix for the parameters for a three-parameter distribution is:

$$\begin{aligned} [V] &= \begin{bmatrix} \text{var}(\alpha) & \text{cov}(\alpha, \beta) & \text{cov}(\alpha, \gamma) \\ \text{cov}(\alpha, \beta) & \text{var}(\beta) & \text{cov}(\beta, \gamma) \\ \text{cov}(\alpha, \gamma) & \text{cov}(\beta, \gamma) & \text{var}(\gamma) \end{bmatrix} \\ &= \begin{bmatrix} E\left(-\frac{\partial^2 LL}{\partial \alpha^2}\right) & E\left(-\frac{\partial^2 LL}{\partial \alpha \partial \beta}\right) & E\left(-\frac{\partial^2 LL}{\partial \alpha \partial \gamma}\right) \\ E\left(-\frac{\partial^2 LL}{\partial \alpha \partial \beta}\right) & E\left(-\frac{\partial^2 LL}{\partial \beta^2}\right) & E\left(-\frac{\partial^2 LL}{\partial \beta \partial \lambda}\right) \\ E\left(-\frac{\partial^2 LL}{\partial \alpha \partial \gamma}\right) & E\left(-\frac{\partial^2 LL}{\partial \beta \partial \lambda}\right) & E\left(-\frac{\partial^2 LL}{\partial \gamma^2}\right) \end{bmatrix}^{-1} \end{aligned} \quad (2.74)$$

$$[I] = \begin{bmatrix} \left(-\frac{\partial^2 LL}{\partial \alpha^2}\right) & \left(-\frac{\partial^2 LL}{\partial \alpha \partial \beta}\right) & \left(-\frac{\partial^2 LL}{\partial \alpha \partial \gamma}\right) \\ \left(-\frac{\partial^2 LL}{\partial \alpha \partial \beta}\right) & \left(-\frac{\partial^2 LL}{\partial \beta^2}\right) & \left(-\frac{\partial^2 LL}{\partial \beta \partial \lambda}\right) \\ \left(-\frac{\partial^2 LL}{\partial \alpha \partial \gamma}\right) & \left(-\frac{\partial^2 LL}{\partial \beta \partial \lambda}\right) & \left(-\frac{\partial^2 LL}{\partial \gamma^2}\right) \end{bmatrix} \tag{2.75}$$

where [I] is the so-called Fisher's information matrix. The variance–covariance matrix is obtained as the inverse of the expected values of the Fisher's information matrix.

When the probability distribution function only has two parameters, the preceding equation simplifies to:

$$S_T^2 = \left(\frac{\partial x_T}{\partial \alpha}\right)^2 \text{var}(\alpha) + \left(\frac{\partial x_T}{\partial \beta}\right)^2 \text{var}(\beta) + 2\left(\frac{\partial x_T}{\partial \alpha}\right)\left(\frac{\partial x_T}{\partial \beta}\right)\text{cov}(\alpha, \beta) \tag{2.76}$$

and its variance–covariance matrix of the parameters is:

$$[V] = \begin{bmatrix} \text{var}(\alpha) & \text{cov}(\alpha, \beta) \\ \text{cov}(\alpha, \beta) & \text{var}(\beta) \end{bmatrix} = \begin{bmatrix} E\left(-\frac{\partial^2 LL}{\partial \alpha^2}\right) & E\left(-\frac{\partial^2 LL}{\partial \alpha \partial \gamma}\right) \\ E\left(-\frac{\partial^2 LL}{\partial \alpha \partial \gamma}\right) & E\left(-\frac{\partial^2 LL}{\partial \beta^2}\right) \end{bmatrix}^{-1} \tag{2.77}$$

and the Fisher's information matrix is:

$$[I] = \begin{bmatrix} \left(-\frac{\partial^2 LL}{\partial \alpha^2}\right) & \left(-\frac{\partial^2 LL}{\partial \alpha \partial \beta}\right) \\ \left(-\frac{\partial^2 LL}{\partial \alpha \partial \beta}\right) & \left(-\frac{\partial^2 LL}{\partial \beta^2}\right) \end{bmatrix} \tag{2.78}$$

2.9.3 PWM Method

The PWM estimator of the standard error of estimates for a three parameters probability distribution function with parameters α, β and γ is, Kite (1988):

$$\begin{aligned} S_T^2 = &\left(\frac{\partial x_T}{\partial \alpha}\right)^2 \text{var}(\alpha) + \left(\frac{\partial x_T}{\partial \beta}\right)^2 \text{var}(\beta) \\ &+ \left(\frac{\partial x_T}{\partial \gamma}\right)^2 \text{var}(\gamma) + 2\left(\frac{\partial x_T}{\partial \alpha}\right)\left(\frac{\partial x_T}{\partial \beta}\right)\text{cov}(\alpha, \beta) \\ &+ 2\left(\frac{\partial x_T}{\partial \alpha}\right)\left(\frac{\partial x_T}{\partial \gamma}\right)\text{cov}(\alpha, \gamma) + 2\left(\frac{\partial x_T}{\partial \beta}\right)\left(\frac{\partial x_T}{\partial \gamma}\right)\text{cov}(\beta, \gamma) \end{aligned} \tag{2.79}$$

When the probability distribution function only has two parameters, the preceding equation simplifies to:

$$S_T^2 = \left(\frac{\partial x_T}{\partial \alpha}\right)^2 \text{var}(\alpha) + \left(\frac{\partial x_T}{\partial \beta}\right)^2 \text{var}(\beta) + 2\left(\frac{\partial x_T}{\partial \alpha}\right)\left(\frac{\partial x_T}{\partial \beta}\right)\text{cov}(\alpha, \beta) \tag{2.80}$$

2.10 Plotting the Extreme Value Data and Models

There are several ways to plot the extreme value data and the models used to fit such sets of natural extreme value data.

2.10.1 Normal Probability Paper

The Normal probability paper was constructed to identify, graphically, if the data was coming from a Normal distribution, that is, if the data points were aligned along a straight line represented by such Normal distribution. This paper was widely used to avoid the computation of the mean and the standard deviation, which are the parameters that define the Normal distribution. This probability paper can be found in almost any office supplies store. This paper will not be used within this book given that it is not used for frequency analysis of natural extreme values.

2.10.2 Gumbel's Probability Paper

The Gumbel's probability paper was created by Emil. J. Gumbel and it is, by far, the most popular probability paper in the field of frequency analysis of natural extreme values. The Gumbel's probability paper consists of two axes, one of them is a natural scale axis where the scale of the natural extreme values is contained, and it is the axis of the ordinates. The axis of the abscissas is the probability scale, given as probability scale at the top of the paper and a return period scale at the bottom part of the paper. This paper is depicted in Fig. 2.6. It is possible to construct a Gumbel's probability paper using the graphical resources contained in Excel®, the only thing is that Excel® cannot handle a probability axis neither as a probability scale or as return period scale.

This difficulty can be avoided by using the Gumbel's reduced variate:

$$y = -Ln[-Ln(F(x)] \tag{2.81}$$

where y is the Gumbel's reduced variate and $F(x)$ is the probability distribution function of x. By using such reduced variate, the scale of the abscissas is now linear and can be handled by Excel®. An example of the construction of Gumbel's probability paper using Excel® spreadsheet, is shown in Fig. 2.7.

2.11 Goodness of Fit Tests

There are several measures of goodness of fit test that have been proposed in the literature. Only the standard error of fit (SEF), Kite (1988), and the mean absolute relative deviation (MARD), Jain and Singh (1987), will be used from now on in

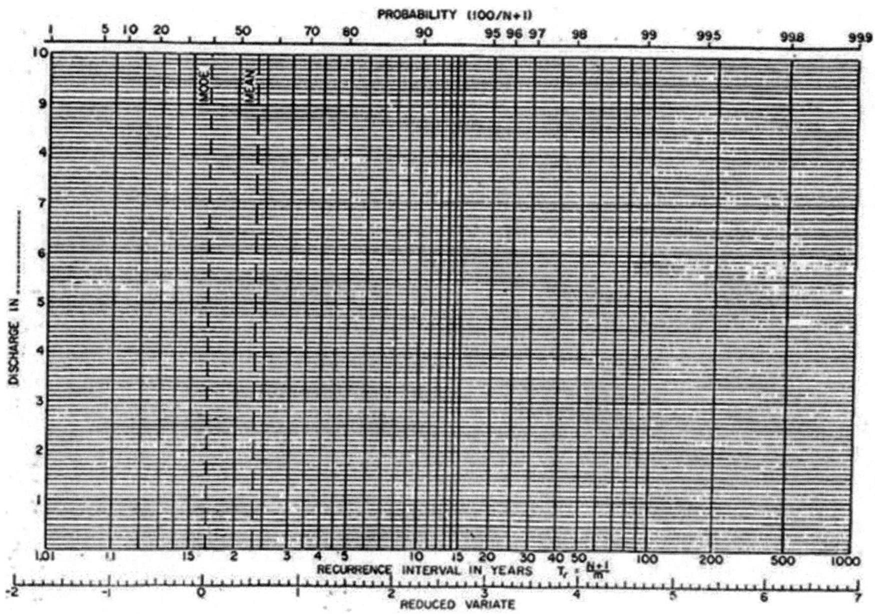

Fig. 2.6 Original form of Gumbel's probability paper

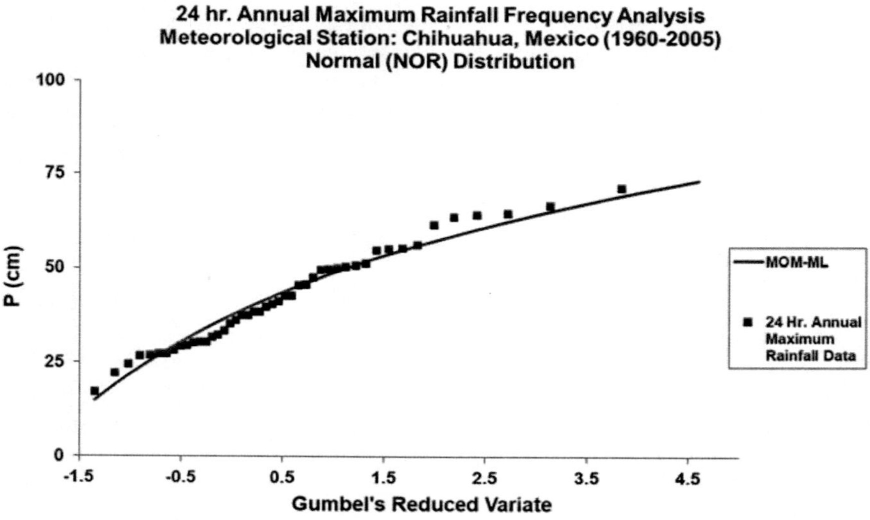

Fig. 2.7 Excel® Spreadsheet Gumbel's probability paper using Gumbel's reduced variate as the Abscissa axis

all the probability distribution functions and methods of estimation of parameters. For the specific case of the maximum likelihood method, the Akaike Criterion, Akaike (1974), will be used, too.

2.11.1 The Standard Error of Fit

The standard error of fit (SEF) is, Kite (1988):

$$SEF = \left[\frac{\sum_{i=1}^{N} (xi - yi)^2}{\left(N - n_p\right)} \right]^{1/2} \tag{2.82}$$

where x_i are the sample historical values, y_i are the distribution function values, corresponding to the same return periods of the historical values, N is the sample size, and n_p is the number of parameters of the distribution function.

Then $n_p = 2$ for the Normal, Two- parameters Log-Normal, Gamma and Extreme Value type I distributions. For the Three- parameters Log-Normal, Pearson type III, Log-Pearson type III and General Extreme Value the value of n_p will be 3.

2.11.2 The Mean Absolute Relative Deviation

The mean absolute relative deviation (MARD) is, Jain and Singh (1987):

$$MARD = \frac{100}{N} \sum_{i=1}^{N} \left| \frac{(x_i - y_i)}{x_i} \right| \tag{2.83}$$

where x_i are the sample historical values, y_i are the distribution function values, corresponding to the same return periods of the historical values, N is the sample size.

2.11.3 The Akaike's Information Criterion

The Akaike's information criterion (IAC) is, Akaike (1974):

$$AIC = NLn(LL) + 2n_p \tag{2.84}$$

where N is the sample size, LL is the log-likelihood function and n_p is the number of parameters of the distribution function. This measure is only useful when using the maximum likelihood method for the estimation of parameters of a distribution function.

2.12 Outliers Tests

There are several tests to detect high or low outliers, from these only the Grubbs and Beck test will be presented here as a measure to detect high or low outliers in natural extreme samples of data.

2.12.1 The Grubbs and Beck Test

The Grubbs and Beck Test (GBT) is an easy way to detect high and/or low outliers on natural extreme events samples of data. The high and low outliers are detected by using the following equations:

$$x_H = \exp(\overline{x} + k_N s) \tag{2.85}$$

$$x_L = \exp(\overline{x} - k_N s) \tag{2.86}$$

where $\overline{x}$ and s are the sample mean and standard deviation of the natural logarithms of the data, respectively, and k_N is the GBT statistic which is tabulated for various sample sizes and significance levels in Grubbs and Beck (1972). With a 10% significance level, the following approximation was proposed for k_N, Pilon and Adamowski (1993):

$$\begin{aligned} k_N = & -3.62201 + 6.28446N^{0.25} - 2.49835N^{0.5} \\ & + 0.491436N^{0.75} - 0.037911N \end{aligned} \tag{2.87}$$

where N is the sample size. So, if there are values greater that x_H then they are considered as high outliers, if there is values smaller that x_L then they are considered low outliers.

2.13 Test for Independence and Stationarity

The Wald-Wolfowitz test (WWT) can be used to prove the independence of a sample of data and to test the existence of a trend on it. For a sample of data x_1, x_2..., x_N the statistic R may be computed as, Wald-Wolfowitz (1943):

$$R = \sum_{i=1}^{N-1} x_i x_{i+1} + x_i x_N \tag{2.88}$$

When the members of a sample data are independent, R is normally distributed with the mean and variance defined as, Wald-Wolfowitz (1943):

$$\overline{R} = \frac{\left(s_1^2 - s_2\right)}{(N-1)} \tag{2.89}$$

$$\mathrm{var}(R) = \frac{\left(s_2^2 - s_4\right)}{(N-1)} - \overline{R}^2 + \frac{\left(s_1^4 - 4s_1^2 s_2 + 4s_1 s_3 + s_2^2 - 2s_4\right)}{(N-1)(N-2)} \tag{2.90}$$

where $s_r = N m_r{}'$ and $m_r{}'$ is the sample r-th moment about the origin. The following statistic is approximately standard normally distributed (N(0,1)) and is used to test the hypothesis of independence at a significance level α by comparing such statistic with the standard normal variate $z_{\alpha/2}$ corresponding to a probability of exceedance $\alpha/2$:

$$u = \frac{\left(R - \overline{R}\right)}{[\mathrm{var}(R)]^{\frac{1}{2}}} \tag{2.91}$$

2.14 Test for Homogeneity and Stationarity

In the Mann–Whitney (MWT) test two samples of size p and q with $p \leq q$ are compared. Then the combined data of size $N = p + q$ is ranked in increasing order of magnitude. The statistic U used for this test is, Mann–Whitney (1947):

$$U = pq - \left\{R_1 - \frac{[p(p+1)]}{2}\right\} \tag{2.92}$$

where R_2 is the sum of the ranks of the elements of the first sample, with sample size p, in the combined series, with sample size N, and U can be computed form R, p and q.

The term $\{R - [p(p + 1)]/2\}$ represents the number of times an item in sample 1 follows an item in sample 2 in the ranking. Similarly, U can be calculated for sample 2 following sample 1:

$$U = pq - \left\{R_2 - \frac{[q(q+1)]}{2}\right\} \tag{2.93}$$

When $N > 20$ and $p > 3$ and $q > 3$, under the null hypothesis that the two samples came from the same population, the M-W test statistic U is approximately normally distributed with mean equal to:

$$\overline{U} = \frac{pq}{2} \tag{2.94}$$

and variance equal to:

$$\mathrm{var}(U) = \left[\frac{pq}{N(N-1)}\right]\left[\frac{N^3 - N}{12} - \sum T\right] \tag{2.95}$$

where:

$$T = \frac{(J^3 - J)}{12} \tag{2.96}$$

and J is the number of observations tied at a given rank. T is summed over all groups of tied observations in both samples of sizes p and q. The standardized statistic u, defined as:

$$u = \frac{(U - \overline{U})}{[\mathrm{var}(U)]^{\frac{1}{2}}} \tag{2.97}$$

is used to test the homogeneity, at significance level α, by making a comparison with the standard normal variate for such significance level.

3 Normal Distribution

Life is beautiful, magnificent thing, even to a jellyfish. C. Chaplin

3.1 Introduction

The Normal distribution, (NOR), was used since the Eighteen Century by the French mathematicians De Moivre and Laplace and later by the German mathematician Friederick Gauss. The normal distribution is also well-known as the Gaussian Bell. It is, by far, the most widely used probability distribution function in the world and where most of the research is done for what probability distribution functions are concerned, too.

Hazen (1914) introduced the NOR distribution for hydrologic data analysis, Markovic (1965) applied to NOR distribution to fit annual data of rainfall and runoff and Slack et al. (1975) considered the NOR distribution as the best option when there is a lack of information about the distribution of floods and economic losses associated with the design of flood reduction measures.

3.2 Chapter Objectives

After reading this chapter, you will know how to:

1. Recognize the distribution and density functions of the NOR distribution
2. Estimate the parameters of the NOR distribution
3. Compute the quantiles and confidence limits of the NOR distribution
4. Make a graphic display of your data and the NOR distribution
5. Develop an application of all the above using Excel® spreadsheets

J. A. Raynal Villaseñor, *Frequency Analyses of Natural Extreme Events*,
Earth and Environmental Sciences Library,
https://doi.org/10.1007/978-3-030-86390-6_3

3.3 Probability Distribution and Density Functions

The probability distribution function of the NOR distribution is:

$$F(x) = \frac{1}{\sigma\sqrt{2\pi}} \int_{-\infty}^{x} \exp\left[-\frac{1}{2\sigma^2}(x-\mu)^2\right] dx \tag{3.1}$$

where μ and σ, are the location and scale parameters. *F(x)* is the probability distribution function of the random variable x and in the case of maxima natural extreme event frequency analysis is equal to the non-exceedance probability, $Pr(X \leq x)$. The domain of variable x in this distribution is $-\infty < x < \infty$.

The probability density function for the NOR distribution is:

$$f(x) = \frac{1}{\sigma\sqrt{2\pi}} \exp\left[-\frac{1}{2\sigma^2}(x-\mu)^2\right] \tag{3.2}$$

where $f(x)$ is the probability density function of the random variable x.

The standard normal variate z is a normal variable with zero mean and variance equal to one, that is z is distributed as N(0,1). Also, the variate z is defined as:

$$z = \frac{x-\mu}{\sigma} \tag{3.3}$$

The standard NOR distribution function is:

$$F(z) = \frac{1}{\sqrt{2\pi}} \int_{-\infty}^{z} \exp\left(-\frac{z^2}{2}\right) dz \tag{3.4}$$

The domain of variable z in this distribution is $-\infty < x < \infty$. The corresponding probability density function is:

$$f(z) = \frac{1}{\sqrt{2\pi}} \exp\left(-\frac{z^2}{2}\right) \tag{3.5}$$

A graphical representation of the N(0,1) probability density is given in Fig. 3.1. The probability distribution and density functions tabulated in statistical and engineering books all are referred to the standard NOR distribution; otherwise, a specific set of tables will be needed for every combination of values of the location and scale parameters in Eqs. (3.1) and (3.2).

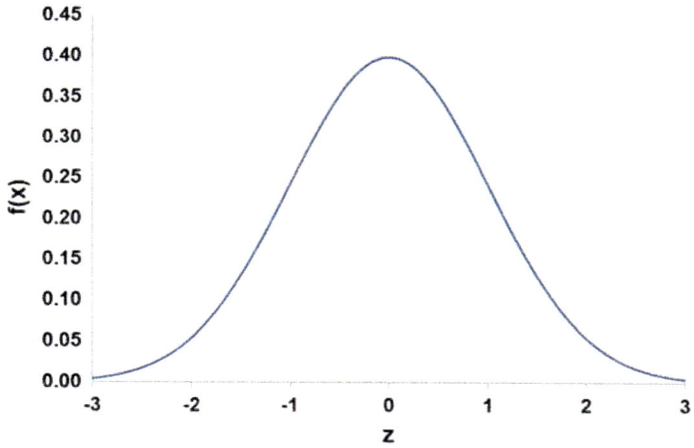

Fig. 3.1 The standard normal density function

3.4 Estimation of Parameters

3.4.1 MOM Method

The MOM estimators of the parameters for the NOR distribution are:

$$\hat{\mu} = \frac{1}{N}\sum_{i=1}^{N} x_i \tag{3.6}$$

$$\hat{\sigma} = \left[\frac{1}{N}\sum_{i=1}^{N}(x_i - \hat{\mu})^2\right]^{1/2} \tag{3.7}$$

3.4.1.1 Example of Application of Estimation of the Parameters of the NOR Distribution Using the MOM Method

Find the MOM estimators for the parameters of the NOR distribution for the flood sample data of gauging station Huites, Mexico (1942–1992), contained in Appendix A.

The following statistics have already been obtained:

$$\hat{\mu} = \overline{x} = \frac{1}{N}\sum_{i=1}^{N} x_i = 2498.9627 \text{ m}^3/\text{s}$$

$$\hat{\sigma} = s = \left[\frac{1}{N}\sum_{i=1}^{N}(x_i - \hat{\mu})^2\right]^{1/2} = 2199.4767 \text{ m}^3/\text{s}$$

Finally, the MOM estimators of the parameters of the NOR distribution for the flood sample data of gauging station Huites, Mexico are:
location parameter:

$$\hat{\mu} = 2498.9627 \text{ m}^3/\text{s}$$

and:
scale parameter:

$$\hat{\sigma} = 2199.4767 \text{ m}^3/\text{s}$$

3.4.2 ML Method

The likelihood function for the NOR distribution is:

$$L(x, \mu, \sigma) = \left(\frac{1}{\sigma\sqrt{2\pi}}\right)^N \prod_{i=1}^{N} \exp\left[-\frac{1}{2\sigma^2}(x_i - \mu)^2\right] \tag{3.8}$$

By taking the natural logarithm of the previous equation, the log-likelihood function is obtained as:

$$LL(x, \mu, \sigma) = -NLn(\sigma) - NLn\left(\sqrt{2\pi}\right) - \frac{1}{2\sigma^2}\sum_{i=1}^{N}(x_i - \mu)^2 \tag{3.9}$$

Now, the ML method requires the computation of the first-order partial derivatives of the log-likelihood function with respect to each of its parameters, equating them equal to zero and then solving the resulting system of equations. So, the first-order partial derivatives are obtained as follows:

$$\frac{\partial LL(x, \mu, \sigma)}{\partial \mu} = \frac{1}{2\sigma^2}\sum_{i=1}^{N} 2(x_i - \mu) = 0 \tag{3.10}$$

$$\frac{\partial (x, \mu, \sigma)}{\partial \sigma} = -\frac{N}{\sigma} + \frac{2}{2\sigma^3}\sum_{i=1}^{N}(x_i - \mu)^2 = 0 \tag{3.11}$$

By solving the system of Eqs. (3.10) and (3.11), the ML estimators of the parameters for the NOR distribution are obtained as follows:

$$\hat{\mu} = \frac{1}{N}\sum_{i=1}^{N} x_i \tag{3.12}$$

$$\hat{\sigma} = \left[\frac{1}{N}\sum_{i=1}^{N}(x_i - \hat{\mu})^2\right]^{1/2} \tag{3.13}$$

which happen to be the same as the estimators produced by the MOM method.

3.5 Estimation of Quantiles for the NOR Distribution

The quantiles for the NOR distribution can be obtained by inverting the NOR distribution to obtain the quantiles given by the following expression, Abramowitz and Stegun (1965):

$$z_T = w - \left(\frac{c_0 + c_1 w + c_2 w^2}{1 + d_1 w + d_2 w^2 + d_3 w^3} \right) \tag{3.14}$$

where z_T is the standard normal deviate that corresponds to a certain return period, T_r, which is linked to a certain value of the standard normal distribution function, Abramowitz and Stegun (1965):

$$F(z_T) = 1 - \frac{1}{T_r} \tag{3.15}$$

The coefficients c and d are as follows, Abramowitz and Stegun (1965):

$$c_0 = 2.515517;\ c_1 = 0.802853;\ c_2 = 0.010328.$$
$$d_1 = 1.432788;\ d_2 = 0.189269;\ d_3 = 0.001308.$$

and, Abramowitz and Stegun (1965):

$$w = \sqrt{Ln\left(\frac{1}{1 - F(z_T)} \right)^2} \tag{3.16}$$

where $F(z_T)$ is the normal standard distribution function. If $F(z_T) < 0.5$, then put $1 - F(z_T)$ instead $F(z_T)$ in Eq. (3.16) and change the sign in the resulting values of z_T on Eq. (3.14).

Then, the quantile values in the standard NOR distribution are those produced by Eq. (3.14), so they need the following transformation to have the quantiles of the NOR distribution with location parameter $\hat{\mu}$ and scale parameter $\hat{\sigma}$:

$$x_T = \hat{\mu} + \hat{\sigma} z_T \tag{3.17}$$

By using engineering terms, the following expression is widely used:

$$Q_T = \hat{\mu} + \hat{\sigma} z_T \tag{3.18}$$

where Q_T is design value corresponding to a certain value of the return period.

3.5.1 Examples of Estimation of MOM and ML Quantiles for the NOR Distribution

Find the MOM and ML estimator of the quantiles 2, 5, 10, 20, 50, 100, 500, and 1000 years of return period, for the NOR distribution for the flood sample data of gauging station Huites, Mexico (1942–1992), contained in Appendix A.

The estimation of MOM and ML quantiles for the NOR distribution is made by using the parameters estimated in the preceding sections and then inserted the parameters into Eq. (3.18). Table 3.1 summarizes these results.

3.6 Goodness of Fit Test

The standard error of fit (SEF) for the NOR has the following form:

$$SEF = \left[\frac{\sum_{i=1}^{N} (xi - yi)^2}{(N-2)} \right]^{1/2} \tag{3.19}$$

while the mean absolute relative deviation (MARD) remains the same as it was defined in Chap. 2:

$$MARD = \frac{100}{N} \sum_{i=1}^{N} \left| \frac{(x_i - y_i)}{x_i} \right| \tag{3.20}$$

where x_i are the sample historical values, y_i are the distribution function values corresponding to the same return periods of the historical values, N is the sample size.

Table 3.1 Estimation of MOM and ML Quantiles for the NOR Distribution for Gauging Station Huites, Mexico

Tr (years)	MOM-ML
	Q_T (m^3/s)
2	2499
5	4369
10	5346
20	6153
50	7061
100	7667
500	8892
1000	9363

3.6.1 Examples of Application of the SEF and MARD to the MOM-ML Estimators of the Parameters of the NOR Distribution

Find the values of the SEF and MARD of the MOM-ML estimators of the parameters of the NOR distribution for the flood sample data of gauging station Huites, Mexico (1942–1992), contained in Appendix A.

The values of SEF and MARD for the MOM-ML estimators of the parameters of the NOR distribution for the flood sample data of gauging station Huites, Mexico, have been obtained through the application of Eqs. (3.10) and (3.20) using the parameters obtained in previous examples. The results are as follows:

$$SEF = \left[\frac{\sum_{i=1}^{N}(xi - yi)^2}{(N-2)}\right]^{1/2} = \frac{1369153.6858}{(51-2)} = 1170$$

$$MARD = \frac{100}{N}\sum_{i=1}^{N}\left|\frac{(x_i - y_i)}{x_i}\right| = \frac{(100)(32.7371)}{51} = 64$$

When the NOR distribution is used, the SEF and MARD tests are the same for the methods of moments and maximum likelihood given that the estimators of the parameters of the NOR distribution of such methods are the same for the two mentioned methods.

3.7 Estimation of the Confidence Limits for the NOR Distribution

By assuming that the quantiles are normally distributed, the following form can be used to set the confidence limits on such quantiles:

$$x_l = x_T \pm z_\alpha S_T \tag{3.21}$$

where x_l is the upper or lower confidence limit, depending on the sign in the preceding formula (+) for the upper confidence limit and (−) for the lower confidence limit, z_α is the standard normal variate corresponding to a confidence level α, and S_T is the squared root of the standard error of the estimate. The evaluation procedures of the standard errors of the estimates will be described in the following subsections.

3.8 Estimation of the Standard Errors for the NOR Distribution

3.8.1 MOM Method

The general form of the MOM estimator of the standard error of the estimate of a two-parameter distribution is, Kite (1988):

$$S_T^2 = \left(\frac{\partial x_T}{\partial m_1^{'}}\right)^2 \mathrm{var}(m_1^{'}) + \left(\frac{\partial x_T}{\partial m_2}\right)^2 \mathrm{var}(m_2) + 2\left(\frac{\partial x_T}{\partial m_1^{'}}\right)\left(\frac{\partial x_T}{\partial m_2}\right)\mathrm{cov}(m_1^{'}, m_2) \tag{3.22}$$

and Eq. (3.22) can be simplified in terms of the frequency factor, K_T, as, Kite (1988):

$$S_T^2 = \frac{\mu_2}{N}\left[1 + K_T\hat{\gamma} + \frac{K_T^2}{4}(\hat{\kappa} - 1)\right] \tag{3.23}$$

and given that the NOR distribution has a zero-skewness coefficient, $\hat{\gamma}$, a kurtosis coefficient, $\hat{\kappa}$, equal to 3, and the frequency factor equal to z_T, then the moment estimator of the standard error of the estimate for the NOR distribution is, Kite (1988):

$$S_T^2 = \frac{\sigma^2}{N}\left(1 + \frac{z_T^2}{2}\right) \tag{3.24}$$

where z_T is the standard normal variate corresponding to the return period T.

3.8.1.1 Example of Estimation of MOM Standard Errors and the Two-Sided 95% Confidence Limits for the NOR Distribution

Find the MOM estimators of the standard errors and the two-sided 95% confidence limits for the NOR distribution for the flood sample data of gauging station Huites, Mexico (1942–1992), contained in Appendix A.

The estimation of MOM estimators of the standard errors and the two-sided 95% confidence limits for the NOR distribution is made by using the moment estimators of the parameters and then inserted into Eqs. (3.24) and (3.18), using selected values of the return intervals. Table 3.2 summarizes these results.

3.8.2 ML Method

The general form of the ML estimator of the standard error of the estimate of a two-parameter distribution is:

$$S_T^2 = \left(\frac{\partial x_T}{\partial \alpha}\right)^2 \mathrm{var}(\alpha) + \left(\frac{\partial x_T}{\partial \beta}\right)^2 \mathrm{var}(\beta) + 2\left(\frac{\partial x_T}{\partial \alpha}\right)\left(\frac{\partial x_T}{\partial \beta}\right)\mathrm{cov}(\alpha, \beta) \tag{3.25}$$

Table 3.2 Estimation of MOM standard errors and the two-sided 95% confidence limits for the NOR distribution for gauging station Huites, Mexico

Tr (years)	S_T (m^3/s)	95% Lower limit (m^3/s)	Q_T (m^3/s)	95% Upper limit (m^3/s)
2	307.9883	1895	2499	3103
5	358.3831	3647	4350	5052
10	415.6604	4503	5318	6133
20	472.4746	5192	6118	7044
50	543.1299	5953	7017	8082
100	592.9847	6454	7617	8779
500	698.4547	7461	8830	10,199
1000	740.1764	7846	9296	10,747

and the variance–covariance matrix of the parameters for the NOR distribution is known to be:

$$[V] = \begin{bmatrix} \mathrm{var}(\mu) & \mathrm{cov}(\mu, \sigma^2) \\ \mathrm{cov}(\mu, \sigma^2) & \mathrm{var}(\sigma^2) \end{bmatrix} = \begin{bmatrix} \frac{\sigma^2}{N} & 0 \\ 0 & \frac{2(\sigma^2)^2}{N} \end{bmatrix} \tag{3.26}$$

and for the NOR distribution the partial derivatives in Eq. (3.25) are, Kite (1988):

$$\frac{\partial x_T}{\partial \mu} = 1 \tag{3.27}$$

and:

$$\frac{\partial x_T}{\partial \sigma^2} = \frac{z_T}{2\sigma} \tag{3.28}$$

The substitution of Eqs. (3.26) to (3.28) provides the maximum likelihood estimator of the standard error of the estimate for the NOR distribution is:

$$S_T^2 = \frac{\sigma^2}{N}\left(1 + \frac{z_T^2}{2}\right) \tag{3.29}$$

where z_T is the standard normal variate corresponding to the return period T.

The standard errors of the estimate are the same for the methods of MOM and ML in the case of the NOR distribution.

3.9 Examples of Application for the NOR Distribution Using Excel® Spreadsheets

3.9.1 Flood Frequency Analysis

By using the flood data from gauging station Huites, Mexico the following results were obtained. The descriptive statistics of the sample are shown in Fig. 3.2

3.9.1.1 MOM and ML Methods

By using an Excel® worksheet, the MOM and ML estimators of the parameters and of the standard errors, quantiles, and confidence limits for gauging station Huites, Mexico, are shown in Fig. 3.3.

A graphical comparison between the histogram and the theoretical NOR density is shown in Fig. 3.4.

A graph containing the empirical and theoretical NOR frequency curves is shown in Fig. 3.5.

	A	B
1	FLOOD FREQUENCY ANALYSIS	
2	Gauging Station: Huites, Sinaloa, Mexico (1942-1992)	
3		
4	*Descriptive Statistics*	
5	Mean	2498.9627
6	Sample Standard Error	311.0530
7	Median	1775.7000
8	Mode	#N/D
9	Standard Deviation	2221.3626
10	Sample Variance	4934451.7500
11	Kurtosis	3.6701
12	Skewness	2.0123
13	Range	9707.5000
14	Minimum	421.8000
15	Maximum	10129.3000
16	Sum	127447.1000
17	Count	51.0000
18	Largest(1)	10129.3000
19	Smallest(1)	421.8000
20	Mean Confidence Level(95.0%)	624.7683
21		

Fig. 3.2 Descriptive statistics for flood sample data at Huites, Mexico

	A	B	C	D	E	F	G
1				FLOOD FREQUENCY ANALYSIS			
2				Gauging Station: Huites, Sinaloa, Mexico (1942-1992)			
3				Normal Distribution			
4				Methods: MOM-ML			
5							
6				Estimation of Parameters and Goodness of Fit Tests			
7	Location Parameter =		2498.9627	m^3/s			
8	Scale Parameter =		2221.3626	m^3/s			
9	Log-Likelihood function =		-464.0835				
10	SEF =	1174		MARD =	65	AIC =	932
11							
12				Design Values and Confidence Limits			
13				Squared Root	Two-sided	Design	Two-sided
14				Standard Error	95%	Values	95%
15	Tr (years)	F(x)	z	S_T	Lower Limit	Q_T (m^3/s)	Upper Limit
16	2	0.5	0.0000	311.0530	1889	2499	3109
17	5	0.8	0.8416	361.9677	3659	4369	5078
18	10	0.9	1.2816	419.7703	4523	5346	6169
19	20	0.95	1.6449	477.1163	5218	6153	7088
20	50	0.98	2.0537	548.4546	5986	7061	8136
21	100	0.99	2.3263	598.8030	6493	7667	8840
22	500	0.998	2.8782	705.3367	7510	8892	10275
23	1000	0.999	3.0902	747.4835	7898	9363	10829
24	Two-sided Confidence Limits:		z_α =	1.96			

Fig. 3.3 MOM and ML Estimators for the Parameters, Standard Errors, Quantiles, and confidence limits of the NOR distribution for flood data at Huites, Mexico

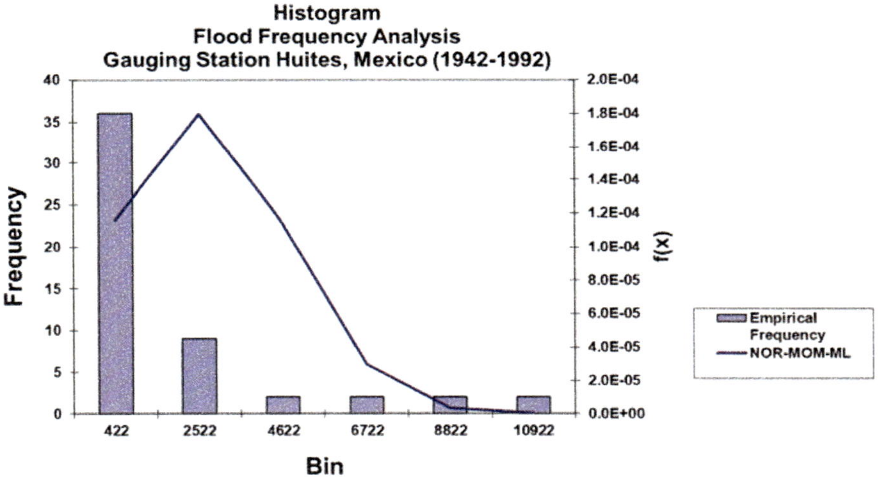

Fig. 3.4 Histogram and theoretical density for flood data at Huites, Mexico

A graphical depiction of the empirical and theoretical frequency curves and the confidence limits of the best fit provided, in this case that of MOM-ML, is shown in Fig. 3.6.

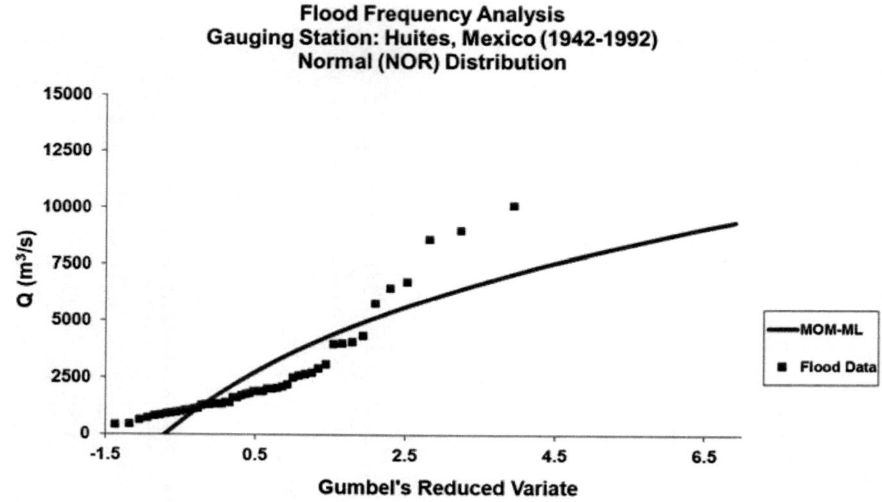

Fig. 3.5 Empirical and MOM and ML theoretical curves for flood data at Huites, Mexico

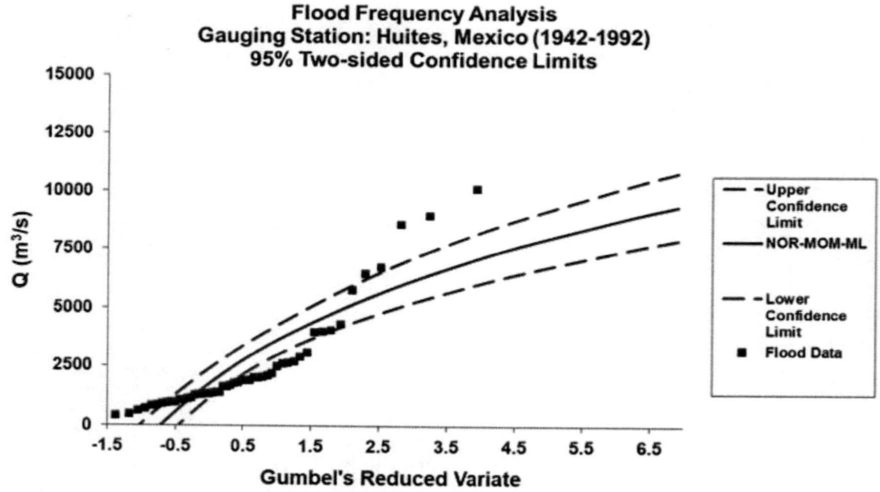

Fig. 3.6 Empirical and MOM and ML theoretical frequency curves and MOM and ML confidence limits for the flood data at Huites, Mexico

3.9.2 Rainfall Frequency Analysis

By using the 24-h. annual maximum rainfall data at meteorological station Chihuahua, Mexico the following results were obtained. The descriptive statistics of the sample are shown in Fig. 3.7.

	A	B	C
1	RAINFALL FREQUENCY ANALYSIS		
2	Meteorological Station: Chihuahua, Mexico (1960-2005)		
3			
4	*Descriptive Statistics*		
5	Mean	42.0454	
6	Sample Standard Error	1.9784	
7	Median	40.2400	
8	Mode	#N/D	
9	Standard Deviation	13.4184	
10	Sample Variance	180.0535	
11	Kurtosis	-0.7252	
12	Skewness	0.3591	
13	Range	54.3000	
14	Minimum	17.1200	
15	Maximum	71.4200	
16	Sum	1934.0900	
17	Count	46.0000	
18	Largest(1)	71.4200	
19	Smallest(1)	17.1200	
20	Mean Confidence Level(95.0%)	3.9848	

Fig. 3.7 Descriptive statistics for 24 h, annual maximum rainfall sample data at meteorological station Chihuahua, Mexico

3.9.2.1 Methods of MOM and ML

By using an Excel® worksheet, the results contained in Figs. 3.8, were obtained for the methods MOM and ML applied to the NOR distribution for meteorological station Chihuahua, Mexico.

In Fig. 3.9 a comparison is made between the histogram and NOR theoretical density for 24 h. Rainfall Data at meteorological station Chihuahua, Mexico.

A graph containing the empirical and theoretical NOR frequency curves is shown in Fig. 3.10.

A graphical depiction of the empirical and theoretical frequency curves and the confidence limits of the best fit provided, in this case that of MOM-ML, is shown in Fig. 3.11.

3.9.3 Wave Height Frequency Analysis

By using the data of maximum significant wave heights contained in Castillo (1988), the following results were obtained. The descriptive statistics of the sample are shown in Fig. 3.12.

	A	B	C	D	E	F	G
1			RAINFALL FREQUENCY ANALYSIS				
2			Meteorological Station: Chihuahua, Mexico (1960-2005)				
3			Normal Distribution				
4			Methods: MOM-ML				
5							
6			Estimation of Parameters and Goodness of Fit Tests				
7	Location Parameter =		42.0454	cm			
8	Scale Parameter =		13.4184	cm			
9	Log-Likelihood function =		-292.4109				
10	SEF =	2		MARD =	5	AIC =	589
11							
12			Design Values and Confidence Limits				
13–16	Tr (years)	F(x)	z	Squared Root Standard Error S_T	Two-sided 95% Lower Limit (cm)	Design Values P_T (cm)	Two-sided 95% Upper Limit (cm)
17	2	0.5	0.0000	1.9784	38	42	46
18	5	0.8	0.8416	2.3023	49	53	58
19	10	0.9	1.2816	2.6699	54	59	64
20	20	0.95	1.6449	3.0347	58	64	70
21	50	0.98	2.0537	3.4884	63	70	76
22	100	0.99	2.3263	3.8087	66	73	81
23	Two-sided Confidence Limits:		z_α =	1.96			

Fig. 3.8 MOM and ML estimators for the parameters, standard errors, quantiles, and confidence limits of the NOR Distribution for the 24 h, Annual maximum rainfall data at meteorological station Chihuahua, Mexico

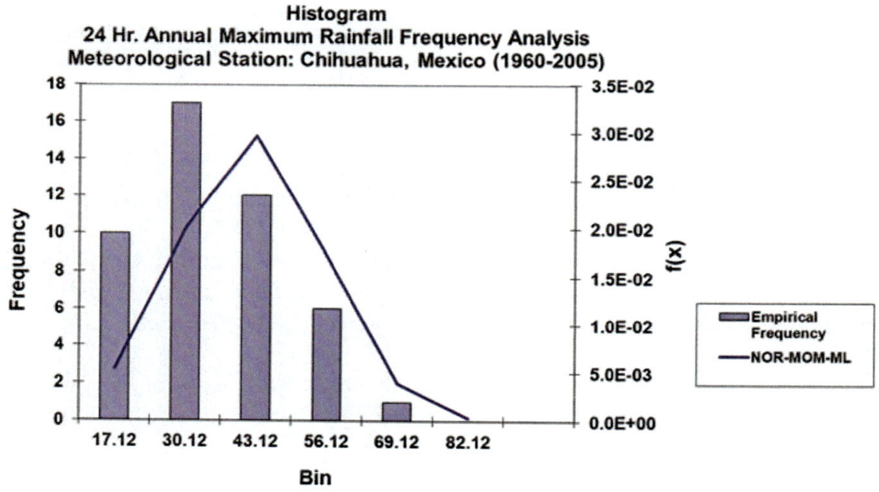

Fig. 3.9 Histogram and theoretical NOR Density for 24 h, Annual maximum meteorological station rainfall data at Chihuahua, Mexico

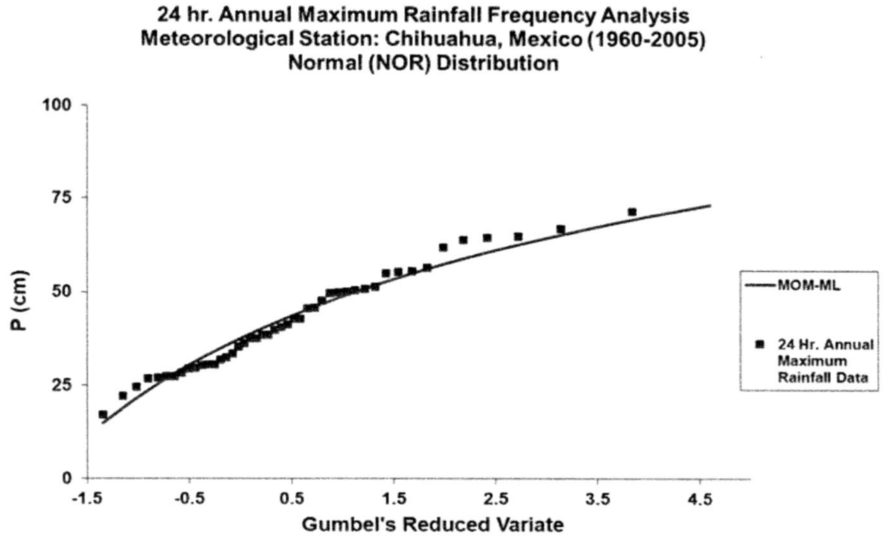

Fig. 3.10 Empirical and theoretical NOR distributions for 24 h, Annual maximum rainfall data at Chihuahua, Mexico

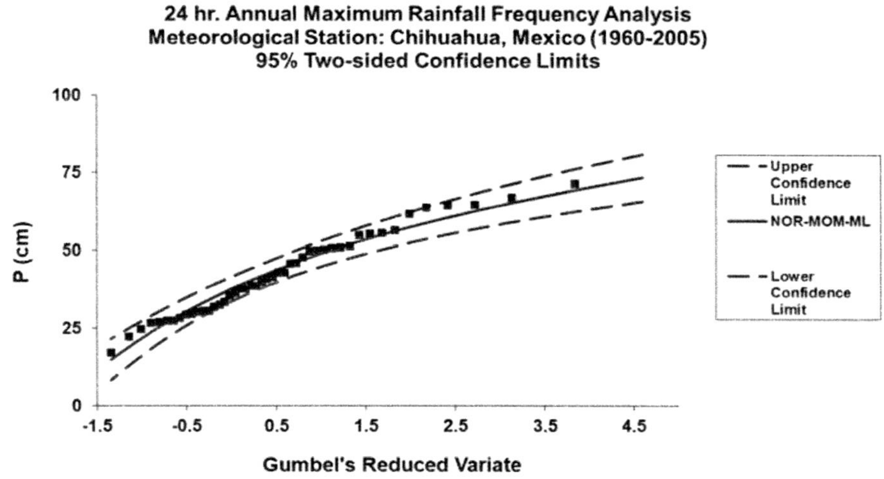

Fig. 3.11 Empirical and NOR MOM-ML theoretical frequency curves and MOM-ML confidence limits for 24 h, annual maximum rainfall data at meteorological station Chihuahua, Mexico

	A	B	C	D
1	**MAXIMUM WAVE HEIGHT FREQUENCY ANALYSIS**			
2				
3	***Descriptive Statistics***			
4				
5	**Mean**	**4.4247**		
6	**Standard Error**	**0.3053**		
7	**Median**	**4.1666**		
8	**Mode**	**3.5509**		
9	**Standard Deviation**	**2.1585**		
10	**Sample Variance**	**4.6591**		
11	**Kurtosis**	**1.0740**		
12	**Skewness**	**0.9052**		
13	**Range**	**10.4485**		
14	**Minimum**	**0.8870**		
15	**Maximum**	**11.3355**		
16	**Sum**	**221.2330**		
17	**Count**	**50.0000**		
18	**Largest(1)**	**11.3355**		
19	**Smallest(1)**	**0.8870**		
20	**Confidence Level(95.0%)**	**624.7686**		

Fig. 3.12 Descriptive statistics for maximum significant wave height sample data

3.9.3.1 MOM and ML Methods

By using an Excel® worksheet, the results contained in Fig. 3.13, were obtained for the methods MOM and ML applied to the NOR distribution for the maximum significant wave height data.

	A	B	C	D	E	F	G	H
1	MAXIMUM SIGNIFICANT WAVE HEIGHT FREQUENCY ANALYSIS							
2	Normal (NOR) Distribution							
3	Methods: MOM-ML							
4								
5	Estimation of Parameters and Goodness of Fit Tests							
6	Location Parameter =		4.4247	(m)				
7	Scale Parameter =		2.1585	(m)				
8	SEF =	0.48		MARD =	7.75			
9								
10	Design Values and Confidence Limits							
11–13	Tr (years)	F(x)	v	z	Squared Root Stand. Error S_T	Two-sided 95% Lower Limit	Design Values H_T (m)	Two-sided 95% Upper Limit
14	2	0.5	1.1774	0.0000	0.3053	4	4	5
15	5	0.8	1.7941	0.8415	0.3552	6	6	7
16	10	0.9	2.1460	1.2817	0.4120	6	7	8
17	20	0.95	2.4477	1.6452	0.4683	7	8	9
18	50	0.98	2.7971	2.0542	0.5383	8	9	10
19	100	0.99	3.0349	2.3268	0.5877	8	9	11
20	Two-sided Confidence Limits:		z_α =	1.96				

Fig. 3.13 MOM and ML estimators for the parameters, standard errors, quantiles, and confidence limits of the NOR distribution for the maximum significant wave height frequency analysis

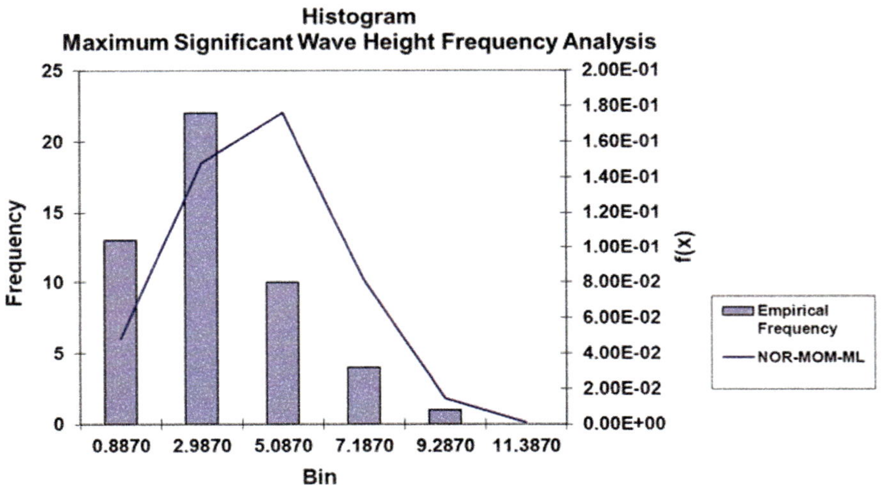

Fig. 3.14 Histogram and theoretical NOR density for the maximum significant wave height frequency analysis

In Fig. 3.14 a comparison is made between the histogram and NOR theoretical density for maximum significant wave height data.

A graph containing the empirical and theoretical NOR frequency curves is shown in Fig. 3.15.

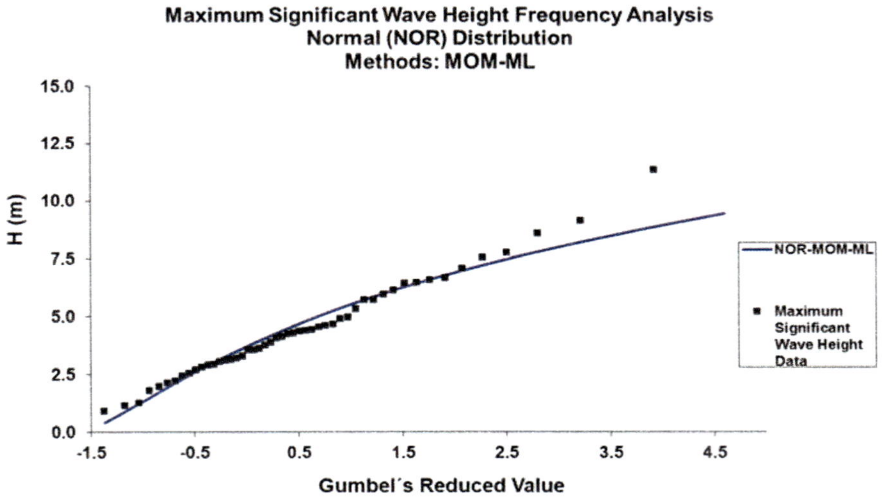

Fig. 3.15 Empirical and theoretical NOR distributions for maximum significant wave height data

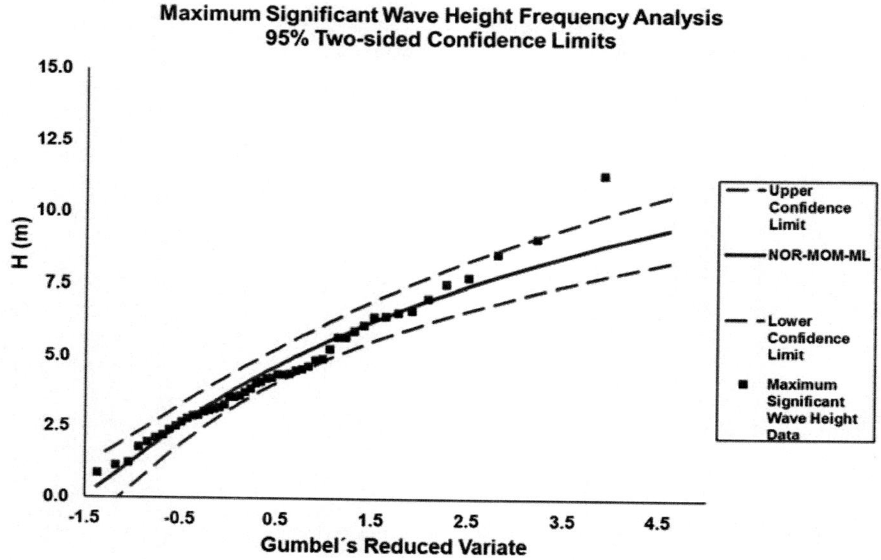

Fig. 3.16 Empirical and MOM and ML theoretical frequency curves and MOM and ML confidence limits for maximum significant wave height data

A graphical depiction of the empirical and theoretical frequency curves and the confidence limits of the best fit provided, in this case that of MOM-ML, is shown in Fig. 3.16.

3.9.4 Maximum Annual Wind Speed Frequency Analysis

By using the data of maximum wind speed contained in Castillo (1988), the following descriptive statistics were obtained and are shown in Fig. 3.17.

3.9.4.1 MOM and ML Methods

By using an Excel® worksheet, the results contained in Figs. 3.18, were obtained for the methods MOM and ML applied to the NOR distribution for the maximum wind speed.

In Fig. 3.19 a comparison is made between the histogram and NOR theoretical density for maximum wind speed.

A graph containing the empirical and theoretical NOR frequency curves for the maximum wind speed is shown in Fig. 3.20.

A graphical depiction of the empirical and theoretical frequency curves of the maximum wind speed and their confidence limits of the best fit provided, in this case that of MOM-ML, is shown in Fig. 3.21.

	A	B	C	D	E
1	MAXIMUM ANNUAL WIND SPEED FREQUENCY ANALYSIS				
2					
3	*Descriptive Statistics*				
4	Mean	54.7964			
5	Standard Error	2.9320			
6	Median	47.9965			
7	Mode	#N/D			
8	Standard Deviation	20.7327			
9	Sample Variance	429.8442			
10	Kurtosis	12.5699			
11	Skewness	3.0813			
12	Range	121.5117			
13	Minimum	36.4278			
14	Maximum	157.9394			
15	Sum	2739.8213			
16	Count	50.0000			
17	Largest(1)	157.9394			
18	Smallest(1)	36.4278			
19	Confidence Level(95.0%)	624.7686			

Fig. 3.17 Descriptive statistics for annual maximum wind speed sample data

	A	B	C	D	E	F	G	H
1	MAXIMUM ANNUAL WIND SPEED FREQUENCY ANALYSIS							
2	Normal (NOR) Distribution							
3	Methods: MOM-ML							
4								
5	Estimation of Parameters and Goodness of Fit Tests							
6	Location Parameter =		54.7964	(km/hr)				
7	Scale Parameter =		20.7327	(km/hr)				
8	SEF =	12		MARD =	14			
9								
10	Design Values and Confidence Limits							
11					Squared Root	Two-sided	Design	Two-sided
12	Tr (years)	F(x)	v	z	Stand. Error	95%	Values	95%
13					S_T	Lower Limit	V_T (km/hr)	Upper Limit
14	2	0.5	1.1774	0.0000	2.9320	49	55	61
15	5	0.8	1.7941	0.8415	3.4118	66	72	79
16	10	0.9	2.1460	1.2817	3.9571	74	81	89
17	20	0.95	2.4477	1.6452	4.4980	80	89	98
18	50	0.98	2.7971	2.0542	5.1706	87	97	108
19	100	0.99	3.0349	2.3268	5.6452	92	103	114
20	Two-sided Confidence Limits:		z_α =	1.96				

Fig. 3.18 MOM and ML estimators for the parameters, standard errors, quantiles, and confidence limits of the NOR Distribution for the annual maximum wind speed frequency analysis

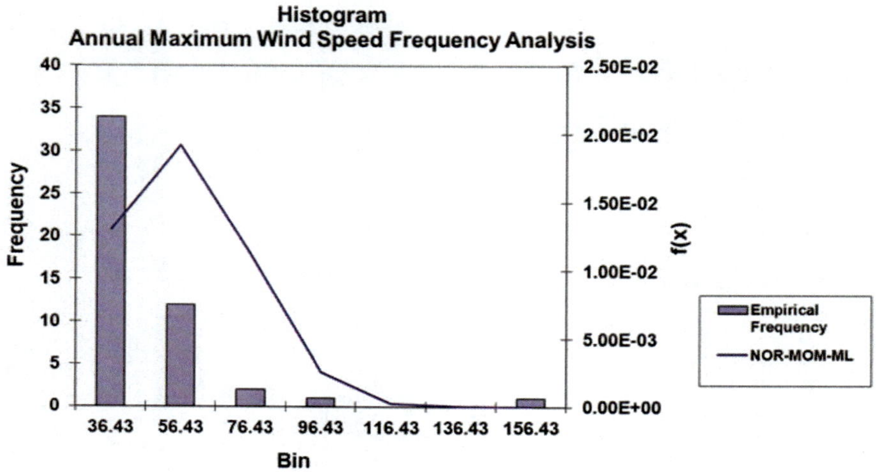

Fig. 3.19 Histogram and theoretical NOR density for the annual maximum wind speed frequency analysis

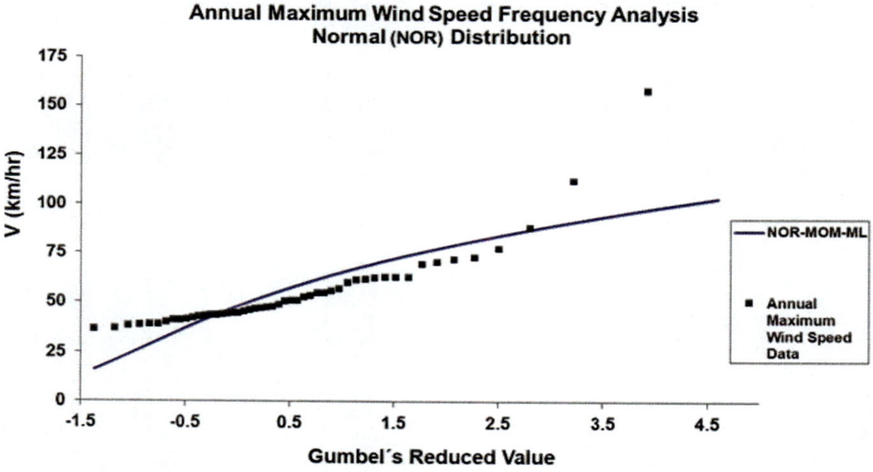

Fig. 3.20 Empirical and theoretical NOR distributions for annual wind speed data

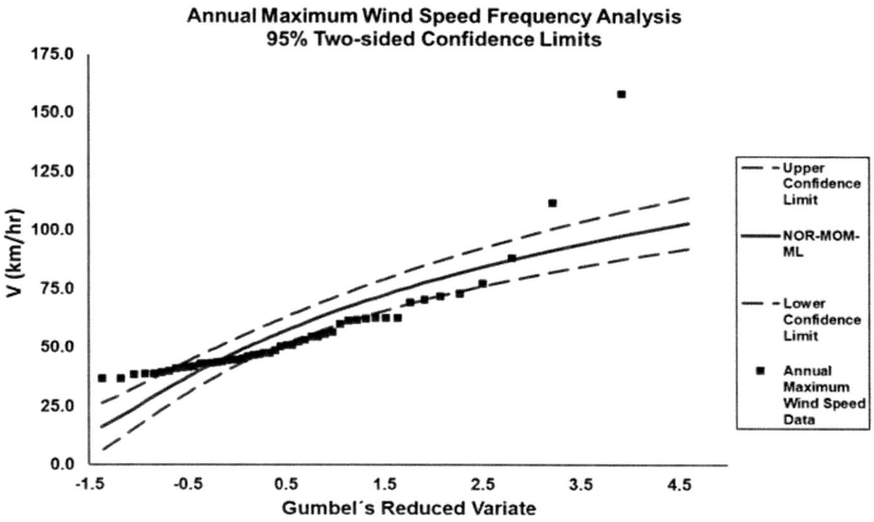

Fig. 3.21 Empirical and MOM and ML theoretical frequency curves and MOM and ML confidence limits for annual maximum wind speed data

4 Two-Parameters Log-Normal Distribution

If you don't remember history accurately, how can you learn? M. Lin

4.1 Introduction

If y = Ln(x) is normally distributed, then the random variable x has a Two-parameters Log-Normal (LN2) distribution.

Chow (1954, 1959, and 1964), made an extensive work with the Log-Normal distribution. He proposed a graphical method to compute the frequency factor for the LN2 distribution.

Singh and Sinclair (1972) derived a mixed probability distribution made of two LN2 distributions.

Yevjevich (1972), have demonstrated the applicability of Log-Normal distributions with 2 and 3 parameters to solve natural extreme events problems.

Kite (1988) provided the form of the first two moments of the LN2 distribution.

The LN2 distribution can be applied only to non-zero positive data, the involvement of natural logarithms in all the formulation of such probability distribution puts such a condition in its use. To avoid the zero values in a sample of data, Kilmartin and Peterson (1972), suggested to ignore zero values or to replace them with values equal to 1.

4.2 Chapter Objectives

After reading this chapter, you will know how to:

J. A. Raynal Villaseñor, *Frequency Analyses of Natural Extreme Events*,
Earth and Environmental Sciences Library,
https://doi.org/10.1007/978-3-030-86390-6_4

(a) Recognize the distribution and density functions of the LN2 distribution
(b) Estimate the parameters of the LN2 distribution
(c) Compute the quantiles and confidence limits of the LN2 distribution
(d) Make a graphic display of your data and the LN2 distribution
(e) Develop an application of all the above using Excel® spreadsheets.

4.3 Probability Distribution and Density Functions

The probability distribution function of the Two-parameters Log-Normal (LN2) distribution is:

$$F(x) = \int_0^x \frac{1}{x\sigma_y\sqrt{2\pi}} \exp\left[-\frac{\left(Ln(x) - \mu_y\right)^2}{2\sigma_y^2}\right] dx \tag{4.1}$$

where μ_y and σ_y are the location and scale parameters. *F(x)* is the probability distribution function of the random variable x and in the case of flood frequency analysis is equal to the non-exceedance probability, $Pr(X \leq x)$. The domain of variable x in this distribution is $0 < x < \infty$.

The probability density function for the LN2 distribution is:

$$f(x) = \frac{1}{x\sigma_y\sqrt{2\pi}} \exp\left[-\frac{\left(Ln(x) - \mu_y\right)^2}{2\sigma_y^2}\right] \tag{4.2}$$

where *f(x)* is the probability density function of the random variable x. A graphical representation of the LN2 distribution is shown in Fig. 4.1.

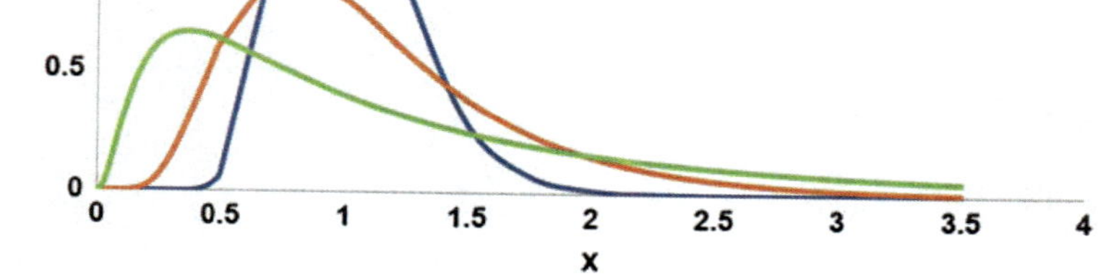

Fig. 4.1 LN2 distribution functions

4.4 Estimation of the Parameters

4.4.1 MOM Method

The general equation of the moments with respect to the origin for the LN2 distribution is, Kite (1988):

$$\mu_r' = \int_0^\infty x^r \left(\frac{1}{x\sigma_y\sqrt{2\pi}} \right) \exp\left[-\frac{\left(Ln(x) - \mu_y\right)^2}{2\sigma_y^2} \right] dx = \exp\left(r\mu_y + \frac{r^2\sigma_y^2}{2} \right) \tag{4.3}$$

So, the population mean and variance of the LN2 distribution are, Kite (1988):

$$\mu = \exp\left(\mu_y + \frac{\sigma_y^2}{2} \right) \tag{4.4}$$

$$\sigma^2 = \left[\exp\left(\sigma_y^2\right) - 1 \right] \left[\exp\left(\mu_y + \frac{\sigma_y^2}{2} \right) \right]^2 \tag{4.5}$$

and the following relationships exists between the LN2 and NOR distributions, Yevjevich (1972):

$$\mu_y = \frac{1}{2} \left[\frac{\mu^2}{1 + \frac{\sigma^2}{\mu^2}} \right] = \frac{1}{2} \left[\frac{\mu^2}{1 + CV^2} \right] \tag{4.6}$$

$$\sigma_y^2 = Ln\left(1 + \frac{\sigma^2}{\mu^2} \right) = Ln\left(1 + CV^2\right) \tag{4.7}$$

where CV is the coefficient of variation of the x's.

The MOM estimators for the parameters of the LN2 distribution are:

$$\hat{\mu}_y = \frac{1}{N} \sum_{i=1}^{N} Ln(x_i) \tag{4.8}$$

$$\hat{\sigma}_y = \left[\frac{\sum_{i=1}^{N} \left(Ln(x_i) - \hat{\mu}_y\right)^2}{N} \right]^{1/2} \tag{4.9}$$

4.4.1.1 Example of Application of Estimation of the Parameters of the LN2 Distribution Using the MOM Method

Find the MOM estimators for the parameters of the LN2 distribution for the flood sample data of gauging station Huites, Mexico (1942–1992), contained in Appendix A.

The following statistics can be obtained:

$$\hat{\mu}_y = \frac{1}{N}\sum_{i=1}^{N} Ln(x_i) = 7.5383$$

$$\hat{\sigma}_y = \left[\frac{1}{N}\sum_{i=1}^{N}\left(Ln(x_i) - \hat{\mu}_y\right)^2\right]^{1/2} = 0.7240$$

Finally, the moment estimators of the parameters of the LN2 distribution for the flood sample data of gauging station Huites, Mexico, are:

location parameter:

$$\hat{\mu}_y = 7.5383$$

and:

scale parameter:

$$\hat{\sigma}_y = 0.7240$$

4.4.2 ML Method

The likelihood function of the LN2 distribution is:

$$L\left(x, \mu_y, \sigma_y\right) = \prod_{i=1}^{N}\left\{\frac{1}{x\sigma_y\sqrt{2\pi}}\exp\left[-\frac{\left(Ln(x) - \mu_y\right)^2}{2\sigma_y^2}\right]\right\} \tag{4.10}$$

which simplifies to:

$$L\left(x, \mu_y, \sigma_y\right) = \left(\frac{1}{\sigma_y\sqrt{2\pi}}\right)^N \prod_{i=1}^{N}\left(\frac{1}{x_i}\right)\exp\left[-\sum_{i=1}^{N}\frac{\left(Ln(x_i) - \mu_y\right)^2}{2\sigma_y^2}\right] \tag{4.11}$$

The Log-Likelihood function of the LN2 distribution is:

$$LL\left(x, \mu_y, \sigma_y\right) = N\left(-Ln\left(\sigma_y\right) - \frac{1}{2}Ln(2\pi)\right) - \sum_{i=1}^{N} Ln(x_i)$$

$$-\frac{1}{\left(2\sigma_y^2\right)}\sum_{i=1}^{N}\left(Ln(x_i)-\mu_y\right)^2 \tag{4.12}$$

by taking the partial derivatives of Eq. (4.10) with respect to the parameters μ_y and σ_y, equated them equal to zero and then solving the system of equations, the maximum likelihood estimators of the parameters of the LN2 distribution can estimated as:

$$\frac{\partial LL}{\partial \mu_y}=\frac{1}{\sigma_y^2}\left[\sum_{i=1}^{N}Ln(x_i)-N\mu_y\right]=0 \tag{4.13}$$

$$\frac{\partial LL}{\partial \sigma_y}=-\frac{N}{\sigma_y}+\frac{1}{\sigma_y^3}\sum_{i=1}^{N}\left[Ln(x_i)-\mu_y\right]^2=0 \tag{4.14}$$

finally, the maximum likelihood estimators of the parameters of the LN2 distribution are, Yevjevich (1972):

$$\hat{\mu}_y=\frac{1}{N}\sum_{i=1}^{N}Ln(x_i) \tag{4.15}$$

$$\hat{\sigma}_y=\left[\frac{\sum_{i=1}^{N}\left(Ln(x_i)-\hat{\mu}_y\right)^2}{N}\right]^{1/2} \tag{4.16}$$

and happens to be the same as the estimators produced by the MOM method.

4.5 Estimation of Quantiles for the LN2 Distribution

The quantiles for the LN2 distribution can be obtained by inverting the Normal distribution to obtain the quantiles given by the following expression, Abramowitz and Stegun (1965):

$$z_T=w-\left(\frac{c_0+c_1w+c_2w^2}{1+d_1w+d_2w^2+d_3w^3}\right) \tag{4.17}$$

where z_T is the standard normal deviate that corresponds to a certain return period, T_r, which is linked to a certain value of the standard normal distribution function, Abramowitz and Stegun (1965):

$$F(z_T)=1-\frac{1}{T_r} \tag{4.18}$$

The coefficients c_i and d_i are as follows, Abramowitz and Stegun (1965):

$$c_0 = 2.515517;\ c_1 = 0.802853;\ c_2 = 0.010328$$
$$d_1 = 1.432788;\ d_2 = 0.189269;\ d_3 = 0.001308$$

and, Abramowitz and Stegun (1965):

$$w = \sqrt{Ln\left(\frac{1}{1 - F(z_T)}\right)^2} \tag{4.19}$$

where $F(z_T)$ is the normal standard distribution function. If $F(z_T) < 0.5$, then substitute $(1 - F(z_T))$ instead $F(z_T)$ in Eq. (4.19) and change the sign in the resulting values of z_T.

Now, the quantiles for the LN2 distribution are obtained as:

$$x_T = \exp\left(\mu_y + z_T\sigma_y\right) \tag{4.20}$$

By using engineering terms, the following expression is widely used:

$$Q_T = \exp\left(\mu_y + z_T\sigma_y\right) \tag{4.21}$$

where Q_T is design value corresponding to a certain value of the return period T_r.

4.5.1 Examples of Estimation of MOM and ML Quantiles for the LN2 Distribution

Find the MOM and ML estimators of the quantiles 2, 5, 10, 20, 50, 100, 500, and 1,000 years of return period, for the LN2 distribution for the flood sample data of gauging station Huites, Mexico (1942–1992), contained in Appendix A.

The estimation of MOM and ML quantiles for the LN2 distribution is made by using the parameters estimated in the preceding sections and then inserting the parameters into Eq. (4.21). Table 4.1 summarizes these results.

4.6 Goodness of Fit Test

The standard error of fit (SEF) for the LN2 has the following form:

$$SEF = \left[\frac{\sum_{i=1}^{N}(xi - yi)^2}{(N - 2)}\right]^{1/2} \tag{4.22}$$

Table 4.1 Estimation of MOM and ML quantiles for the LN2 distribution for gauging station Huites, Mexico

Tr (Years)	MOM-ML
	Q_T (m^3/s)
2	1879
5	3455
10	4752
20	6182
50	8313
100	10,127
500	15,099
1000	17,604

while the mean absolute relative deviation (MARD) remains the same as it was defined in Chap. 2:

$$MARD = \frac{100}{N}\sum_{i=1}^{N}\left|\frac{(x_i - y_i)}{x_i}\right| \tag{4.23}$$

where x_i are the sample historical values, y_i are the distribution function values, corresponding to the same return periods of the historical values, N is the sample size.

4.6.1 Examples of Application of the SEF and MARD to the MOM and ML Estimators of the Parameters of the LN2 Distribution

Find the values of the SEF and MARD of the MOM and ML estimators of the parameters of the LN2 distribution for the flood sample data of gauging station Huites, Mexico (1942–1992), contained in Appendix A.

The values of SEF and MARD for the MOM and ML estimators of the parameters of the LN2 distribution for the flood sample data of gauging station Huites, Mexico, have been obtained through the application of Eqs. (4.22) and (4.23) using the parameters obtained in previous examples. The results are as follows:

$$SEF = \left[\frac{\sum_{i=1}^{N}(xi - yi)^2}{(N-2)}\right]^{1/2} = \left[\frac{483301.4288}{(51-2)}\right]^{\frac{1}{2}} = 695$$

$$MARD = \frac{100}{N}\sum_{i=1}^{N}\left|\frac{(x_i - y_i)}{x_i}\right| = \frac{(100)(4.9334)}{51} = 10$$

When the LN2 distribution is used, the SEF and MARD tests are the same for the MOM and ML methods given that the estimators of the parameters of the LN2 distribution of such methods are the same for the two mentioned methods.

4.7 Estimation of the Confidence Limits for the LN2 Distribution

By assuming that the quantiles are normally distributed, the following form can be used to set the confidence limits on such quantiles:

$$x_l = x_T \pm z_\alpha S_T \tag{4.24}$$

where x_l is the upper or lower confidence limit, depending on the sign in the preceding formula (+) for the upper confidence limit and (-) for the lower confidence limit, z_α is the standard normal variate corresponding to a confidence level α, and S_T is the squared root of the standard error of the estimate. The evaluation procedures of the standard errors of the estimates will be described in the following subsections.

4.8 Estimation of the Standard Errors for the LN2 Distribution

4.8.1 MOM Method

The general form of the MOM estimator of the standard error of the estimate of a two-parameters distribution is, Kite (1988):

$$S_T^2 = \left(\frac{\partial x_T}{\partial m_1'}\right)^2 \text{var}(m_1') + \left(\frac{\partial x_T}{\partial m_2}\right)^2 \text{var}(m_2) + \left(\frac{\partial x_T}{\partial m_3}\right)^2 \text{var}(m_3)$$
$$+ 2\left(\frac{\partial x_T}{\partial m_1'}\right)\left(\frac{\partial x_T}{\partial m_2}\right)\text{cov}(m_1', m_2) \tag{4.25}$$

and Eq. (4.25) can be simplified in terms of the frequency factor, K_T, as, Kite (1988):

$$S_T^2 = \frac{\mu_2}{N}\left[1 + K_T\hat{\gamma} + \frac{K_T^2}{4}(\hat{\kappa} - 1)\right] \tag{4.26}$$

and given that computing the skewness and kurtosis coefficients for the logarithms will produce values of zero and 3, respectively, also the frequency factor will be equal to z_T in the LN2 distribution, and the moment estimator of the standard error of the estimate for the LN2 distribution is, Kite (1988):

$$S_T^2 = \frac{\sigma_y^2}{N}\left(1 + \frac{z_T^2}{2}\right) \tag{4.27}$$

where z_T is the standard normal variate corresponding to the return period Tr. To introduce the squared root standard error of the estimate into Eq. (4.24), the following expressions must be used for the positive value associated with the upper confidence limit and the negative value associated with the lower confidence limit, these expressions transform squared root standard error of the estimate into linear units, Kite (1988):

$$S_T(+) = x_T(\exp(S_T) - 1) \tag{4.28}$$

$$S_T(-) = -x_T(\exp(-S_T) - 1) \tag{4.29}$$

4.8.1.1 Example of Estimation of Moment Standard Errors and the Two-Sided 95% Confidence Limits for the LN2 Distribution

Find the MOM estimators of the standard errors and the two-sided 95% confidence limits for the LN2 distribution for the flood sample data of gauging station Huites, Mexico (1942–1992), contained in Appendix A.

The estimation of moment estimators of the standard errors and the two-sided 95% confidence limits for the LN2 distribution is made by using the moment estimators of the parameters and then inserted in the Eqs. (4.28), (4.29) and (4.24), using selected values of the return intervals. Table 4.2 summarizes these results.

4.8.2 ML Method

The general form of the ML estimator of the standard error of the estimate of a two-parameter distribution is:

$$S_T^2 = \left(\frac{\partial x_T}{\partial \alpha}\right)^2 \mathrm{var}(\alpha) + \left(\frac{\partial x_T}{\partial \beta}\right)^2 \mathrm{var}(\beta) + 2\left(\frac{\partial x_T}{\partial \alpha}\right)\left(\frac{\partial x_T}{\partial \beta}\right)\mathrm{cov}(\alpha, \beta) \tag{4.30}$$

and the variance–covariance matrix of the parameters for the LN2 distribution is known to be:

$$[V] = \begin{bmatrix} \mathrm{var}(\mu_y) & \mathrm{cov}(\mu_y, \sigma_y^2) \\ \mathrm{cov}(\mu_y, \sigma_y^2) & \mathrm{var}(\sigma_y^2) \end{bmatrix} = \begin{bmatrix} \frac{\sigma_y^2}{N} & 0 \\ 0 & \frac{2\left(\sigma_y^2\right)^2}{N} \end{bmatrix} \tag{4.31}$$

Table 4.2 Estimation of MOM standard errors and the two-sided 95% confidence limits for the LN2 distribution for gauging station Huites, Mexico

Tr (Years)	Logarithmic S_T (m^3/s)	Two-sided 95% lower limit (m^3/s)	Q_T (m^3/s)	Two-sided 95% upper limit (m^3/s)
2	0.1014	1524	1879	2272
5	0.1180	2701	3455	4303
10	0.1368	3561	4752	6117
20	0.1555	4437	6182	8221
50	0.1788	5645	8313	11,503
100	0.1952	6607	10,127	14,405
500	0.2299	9020	15,099	22,749
1000	0.2436	10,143	17,604	27,124

Two-sided Limits: $z_\alpha = 1.96$

and for the LN2 distribution the partial derivatives in Eq. (4.30) are, Kite (1988):

$$\frac{\partial x_T}{\partial \mu_y} = \exp(\mu_y + z_T\sigma_y) \tag{4.32}$$

and:

$$\frac{\partial x_T}{\partial \sigma_y^2} = \frac{z_T \exp(\mu_y + z_T\sigma_y)}{2\sigma_y} \tag{4.33}$$

The substitution of Eqs. (4.31)–(4.33) provides the ML estimator of the standard error of the estimate for the LN2 distribution is, Kite (1988):

$$S_T^2 = \frac{\sigma_y^2}{N}\left(1 + \frac{z_T^2}{2}\right)\exp[2(\mu_y + z_T\sigma_y)] \tag{4.34}$$

4.8.2.1 Example of Estimation of ML Standard Errors and the Two-Sided 95% Confidence Limits for the LN2 Distribution

Find the ML estimators of the standard errors and the two-sided 95% confidence limits for the LN2 distribution for the flood sample data of gauging station Huites, Mexico (1942–1992), contained in Appendix A.

The estimation of ML estimators of the standard errors and the two-sided 95% confidence limits for the LN2 distribution is made by using the ML estimators of the parameters and then inserted in the Eqs. (4.34) and (4.24), using selected values of the return intervals. Table 4.3 summarizes these results.

Table 4.3 Estimation of ML standard errors and the two-sided 95% confidence limits for the LN2 distribution for gauging station Huites, Mexico

Tr (Years)	Logarithmic S_T (m^3/s)	Two-sided 95% lower limit (m^3/s)	Q_T (m^3/s)	Two-sided 95% upper limit (m^3/s)
2	192.3549	1502	1879	2256
5	414.1987	2664	3476	4288
10	662.6090	3497	4795	6094
20	982.2955	4329	6254	8180
50	1522.6836	5450	8434	11,419
100	2029.1821	6317	10,295	14,272
500	3578.2498	8398	15,411	22,425
1000	4428.1479	9317	17,997	26,676

4.9 Examples of Application for the LN2 Distribution Using Excel® Spreadsheets

4.9.1 Flood Frequency Analysis

By using the flood data from gauging station Huites, Mexico the descriptive statistics were obtained and are displayed in Fig. 4.2.

FLOOD FREQUENCY ANALYSIS

Gauging Station: Huites, Sinaloa, Mexico (1942-1992)

Descriptive Statistics		*Descriptive Statistics of the Natural Logarithms*	
Mean	2498.9627	Mean	7.5383
Sample Standard Error	311.0530	Sample Standard Error	0.1024
Median	1775.7000	Median	7.4819
Mode	#N/D	Mode	#N/D
Standard Deviation	2221.3626	Standard Deviation	0.7312
Sample Variance	4934451.7500	Sample Variance	0.5347
Kurtosis	3.6701	Kurtosis	0.0081
Skewness	2.0123	Skewness	0.4627
Range	9707.5000	Range	3.1787
Minimum	421.8000	Minimum	6.0445
Maximum	10129.3000	Maximum	9.2232
Sum	127447.1000	Sum	384.4530
Count	51.0000	Count	51.0000
Largest(1)	10129.3000	Largest(1)	9.2232
Smallest(1)	421.8000	Smallest(1)	6.0445
Mean Confidence Level(95.0%)	624.7683	Mean Confidence Level(95.0%)	0.2057

Fig. 4.2 Descriptive statistics for the flood sample of Huites, Mexico

4.9.1.1 MOM-ML Methods

By using an Excel® worksheet, the results contained in Fig. 4.3, were obtained for the MOM-ML methods applied to the LN2 distribution for gauging station Huites, Mexico.

In Fig. 4.4, a comparison is made between the histogram and LN2 theoretical density for flood sample of Huites, Mexico.

FLOOD FREQUENCY ANALYSIS
Gauging Station: Huites, Mexico (1942-1992)
Two-Parameters Log-Normal (LN2) Distribution
Method: MOM-ML

Estimation of Parameters and Goodness of Fit Tests				
Location Parameter =		7.5383		
Scale Parameter =		0.7312		
SEF =	673		MARD =	10

Design Values and Confidence Limits							
Return Period			Standard Normal Value	Logarithms Squared root Stand. Error	Two-sided 95% Lower Limit	Design Values Q_T	Two-sided 95% Upper Limit
Tr (years)	F(x)	v	z_T	S_T	(m³/s)	(m³/s)	(m³/s)
2	0.5	1.1774	0.0000	0.1024	1520	1879	2276
5	0.8	1.7941	0.8415	0.1191	2711	3476	4338
10	0.9	2.1460	1.2817	0.1382	3583	4796	6189
20	0.95	2.4477	1.6452	0.1571	4474	6256	8342
50	0.98	2.7971	2.0542	0.1806	5705	8437	11709
100	0.99	3.0349	2.3268	0.1971	6686	10298	14696
500	0.998	3.5255	2.8785	0.2322	9155	15415	23313
1000	0.999	3.7169	3.0905	0.2461	10304	18000	27844
Two-sided Confidence Limits:		z_α =	1.96				

Fig. 4.3 MOM-ML estimators for the parameters, standard errors, quantiles, and confidence limits of the LN2 distribution for the flood sample of Huites, Mexico

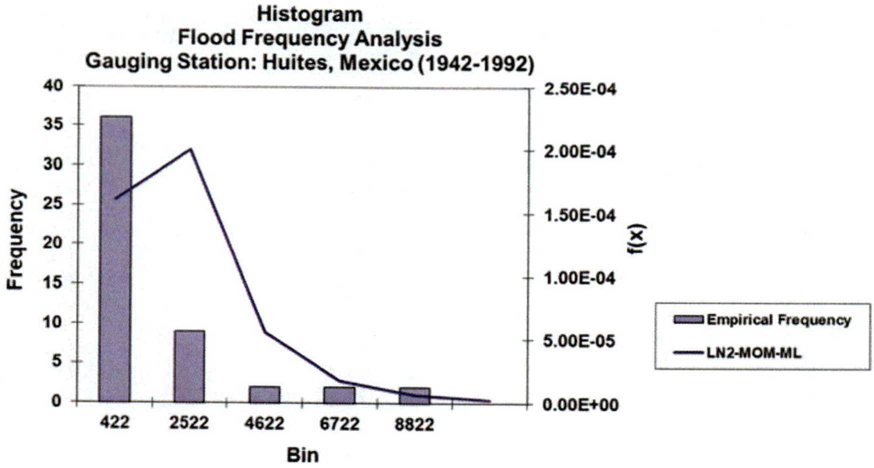

Fig. 4.4 Histogram and theoretical MOM-ML LN2 density for flood sample of Huites, Mexico

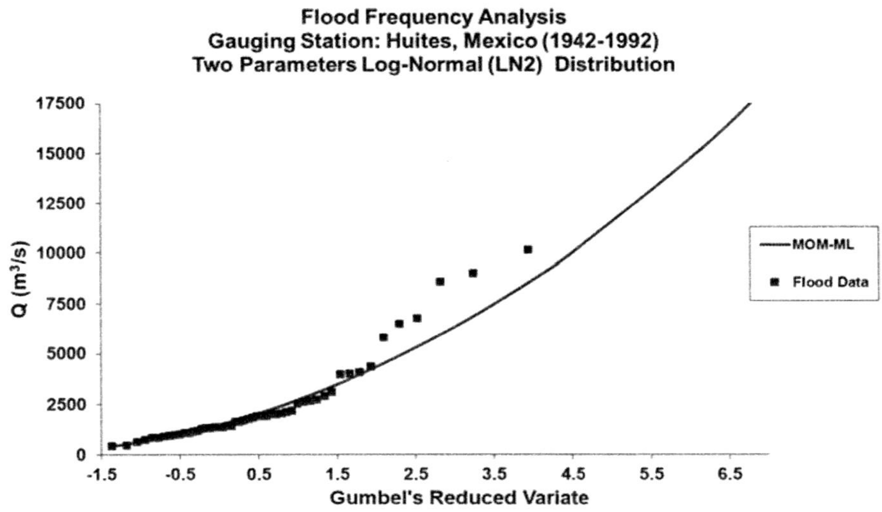

Fig. 4.5 Empirical and MOM-ML LN2 theoretical curves for flood data at Huites, Mexico

In Fig. 4.5, it is shown a graph containing the empirical and theoretical LN2 frequency curves.

A graphical depiction of the empirical and theoretical frequency curves and the confidence limits of the best fit provided, in this case that of MOM-ML, is shown in Fig. 4.6.

4.9.2 Rainfall Frequency Analysis

By using 24 h, annual maximum rainfall data from rainfall gauging station Chihuahua, Mexico the following descriptive statistics were obtained and are shown in Fig. 4.7.

4.9.2.1 MOM-ML Methods

By using an Excel® worksheet, the results contained in Fig. 4.8, were obtained for the MOM and ML methods applied to the LN2 distribution for 24 h, annual maximum rainfall gauging station Chihuahua, Mexico.

In Fig. 4.9 a comparison is made between the histogram and LN2 theoretical density for 24 h, annual maximum rainfall data at Chihuahua, Mexico.

A graph containing the empirical and theoretical LN2 frequency curves is shown in Fig. 4.10.

A graphical depiction of the empirical and theoretical frequency curves and the confidence limits of the best fit provided, in this case that of MOM-ML, is shown in Fig. 4.11.

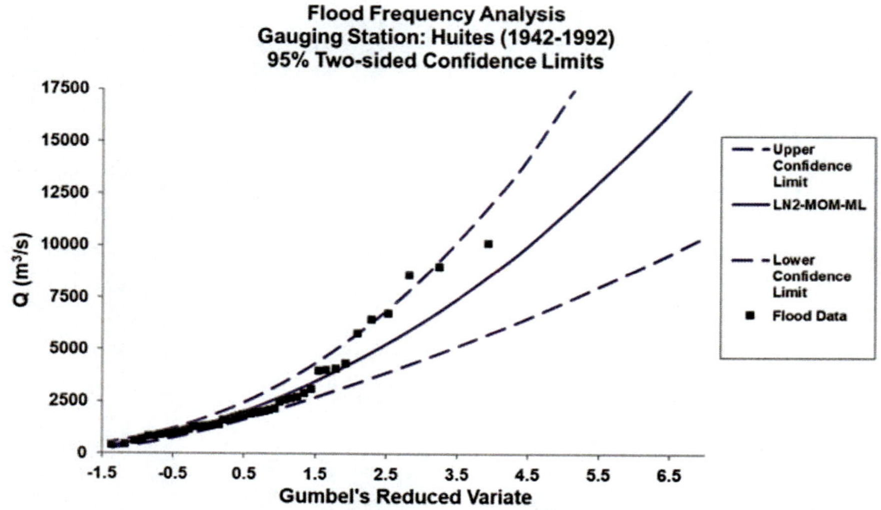

Fig. 4.6 Empirical and MOM and ML theoretical frequency curves and MOM and ML confidence limits of the LN2 distribution for the flood data at Huites, Mexico

	A	B	C	D
1	24 HR. ANNUAL MAXIMUM RAINFALL FREQUENCY ANALYSIS			
2	Meteorological Station: Chihuahua, Mexico (1960-2005)			
3				
4	*Descriptive Statistics*		*Descriptive Statistics of the Natural Logarithms*	
5	Mean	42.0454	Mean	3.6871
6	Sample Standard Error	1.9784	Sample Standard Error	0.0624
7	Median	40.2400	Median	3.6948
8	Mode	#N/A	Mode	#N/A
9	Standard Deviation	13.4184	Standard Deviation	0.3302
10	Sample Variance	180.0535	Sample Variance	67.6667
11	Kurtosis	-0.7252	Kurtosis	-0.4776
12	Skewness	0.3591	Skewness	-0.2507
13	Range	54.3000	Range	27.0000
14	Minimum	17.1200	Minimum	1978.0000
15	Maximum	71.4200	Maximum	2005.0000
16	Sum	1934.0900	Sum	55762.0000
17	Count	46.0000	Count	28.0000
18	Largest(1)	71.4200	Largest(1)	2005.0000
19	Smallest(1)	17.1200	Smallest(1)	1978.0000
20	Mean Confidence Level(95.0%)	3.9848	Mean Confidence Level(95.0%)	0.1280

Fig. 4.7 Descriptive statistics for the 24 h, annual maximum rainfall data at Chihuahua, Mexico

4.9.3 Maximum Significant Wave Height Frequency Analysis

By using the maximum significant wave height data, Castillo (1988), the following descriptive statistics for the data and its natural logarithms were obtained and are shown in Fig. 4.12.

	A	B	C	D	E	F	G	H
1	RAINFALL FREQUENCY ANALYSIS							
2	Meteorological Station: Chihuahua, Mexico (1960-2005)							
3	Two Parameters Log-Normal (LN2) Distribution							
4	Method: MOM-ML							
5								
6	Estimation of Parameters and Goodness of Fit Tests							
7	Location Parameter=		3.6871					
8	Scale Parameter =		0.3302					
9	SEF =	2.07		MARD =	2.33			
10								
11	Design Values and Confidence Limits							
12				Standard	Logarithmic	Two-sided	Design	Two-sided
13	Return	F(x)		Normal	Squared Root	95%	Values	95%
14	Period			Variate	Stand. Error	Lower Limit	P_T	Upper Limit
15	Tr (years)		v	z_T	S_T	(cm)	(cm)	(cm)
16	2	0.5	1.1774	0.0000	0.0129	39	40	41
17	5	0.8	1.7941	0.8415	0.0150	51	53	54
18	10	0.9	2.1460	1.2817	0.0174	59	61	63
19	20	0.95	2.4477	1.6452	0.0198	66	69	71
20	50	0.98	2.7971	2.0542	0.0228	75	79	82
21	100	0.99	3.0349	2.3268	0.0248	82	86	90
22	Two-sided Confidence Limits:			z_α =	1.96			

Fig. 4.8 MOM and ML estimation of the parameters, standard errors, quantiles, and confidence limits for the 24 h, annual maximum rainfall data at Chihuahua, Mexico

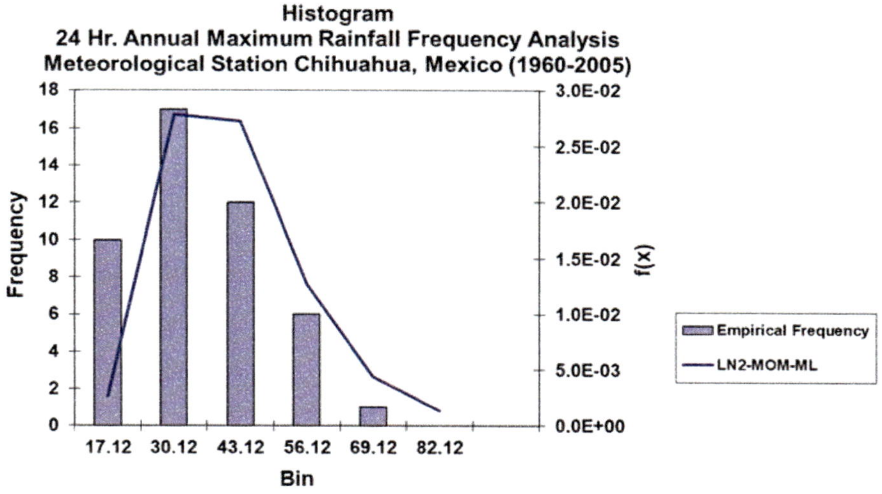

Fig. 4.9 Histogram and theoretical LN2 Density for 24 h, annual maximum rainfall data at Chihuahua, Mexico

4.9.3.1 MOM-ML Methods

By using an Excel® worksheet, the results contained in Fig. 4.13, were obtained for the MOM-ML methods applied to the LN2 distribution for maximum significant wave height data.

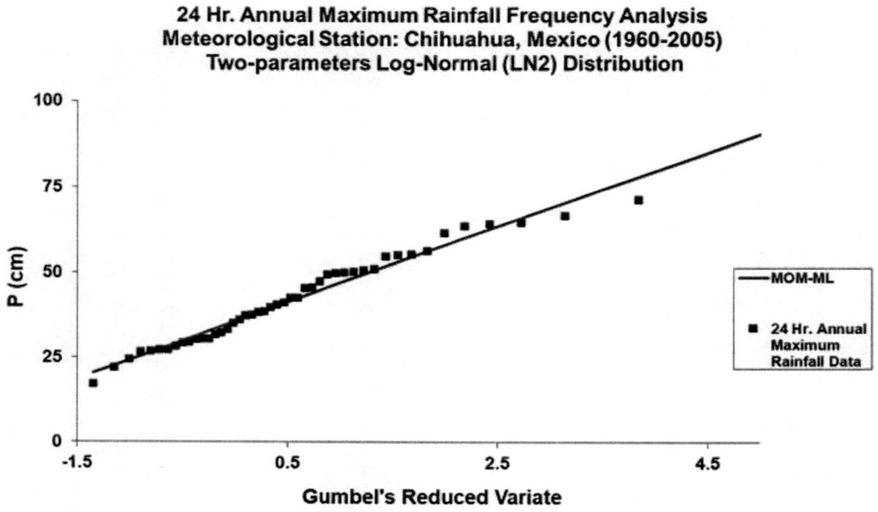

Fig. 4.10 Empirical and theoretical LN2 distributions for 24 h, annual maximum rainfall data at Chihuahua, Mexico

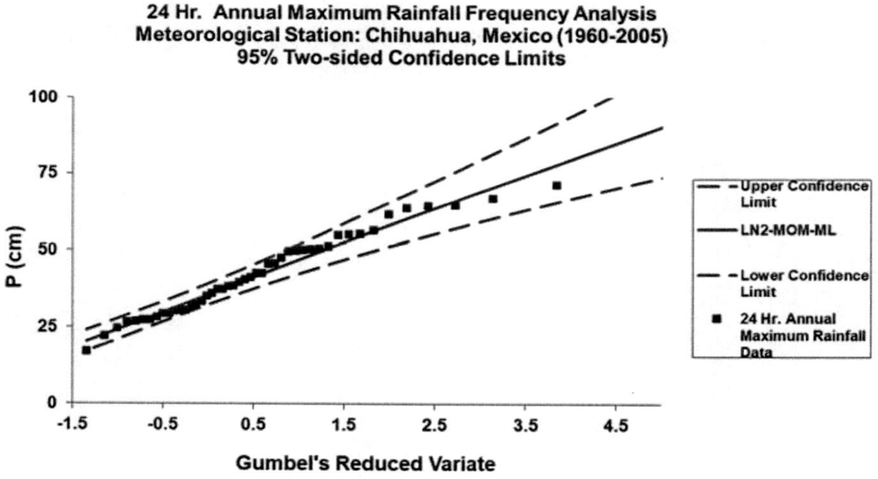

Fig. 4.11 Empirical and MOM and ML theoretical frequency curves and MOM and ML confidence limits for 24 h, annual maximum rainfall data at Chihuahua, Mexico

In Fig. 4.14 a comparison is made between the histogram and LN2 theoretical density for maximum significant wave height data sample.

A graph containing the empirical and theoretical LN2 frequency curves is shown in Fig. 4.15.

	A	B	C	D
1	**Maximum Significant Wave Height Frequency Analysis**			
2				
3	*Descriptive Statistics*		*Descriptive Statistics of the Natural Logarithms*	
4	Mean	4.4247	Mean	1.3628
5	Sample Standard Error	0.3053	Sample Standard Error	0.0748
6	Median	4.1666	Median	1.4270
7	Mode	3.5509	Mode	1.2672
8	Standard Deviation	2.1585	Standard Deviation	0.5287
9	Sample Variance	4.6591	Sample Variance	0.2796
10	Kurtosis	1.0740	Kurtosis	0.6324
11	Skewness	0.9053	Skewness	-0.6102
12	Range	10.4485	Range	2.5478
13	Minimum	0.8870	Minimum	-0.1199
14	Maximum	11.3355	Maximum	2.4279
15	Sum	221.2326	Sum	68.1403
16	Count	50.0000	Count	50.0000
17	Largest(1)	11.3355	Largest(1)	2.4279
18	Smallest(1)	0.8870	Smallest(1)	-0.1199
19	Mean Confidence Level(95.0%)	0.6134	Mean Confidence Level(95.0%)	0.1503
20				

Fig. 4.12 Descriptive statistics for the maximum significant wave height data sample

	A	B	C	D	E	F	G	H
1	MAXIMUM SIGNIFICANT WAVE HEIGHT FREQUENCY ANALYSIS							
2	Two Parameters Log-Normal (LN2) Distribution							
3	Methods: MOM-ML							
4								
5	Estimation of Parameters and Goodness of Fit Tests							
6	Location Parameter =		1.3628					
7	Scale Parameter =		0.5287					
8	SEF =	0.46		MARD =	9.74			
9								
10	Design Values and Confidence Limits							
11				Standard	Logarithms	Two-sided	Design	Two-sided
12				Normal	Squared root	95%	Values	95%
13	Return Period			Value	Stand. Error	Lower Limit	H_T	Upper Limit
14	Tr (years)	F(x)	v	z_T	S_T		(m)	
15	2	0.5	1.1774	0.0000	0.3739	2	4	7
16	5	0.8	1.7941	0.8415	0.4886	1	6	14
17	10	0.9	2.1460	1.2817	0.6078	1	8	20
18	20	0.95	2.4477	1.6452	0.7198	0	9	29
19	50	0.98	2.7971	2.0542	0.8542	-1	12	42
20	100	0.99	3.0349	2.3268	0.9469	-3	13	55
21	Two-sided Confidence Limits:		z_α =	1.96				

Fig. 4.13 MOM and ML Estimation of the parameters, standard errors, quantiles, and confidence limits for the maximum significant wave height data sample

A graphical depiction of the empirical and theoretical frequency curves and the confidence limits of the best fit provided, in this case that of MOM-ML, is shown in Fig. 4.16.

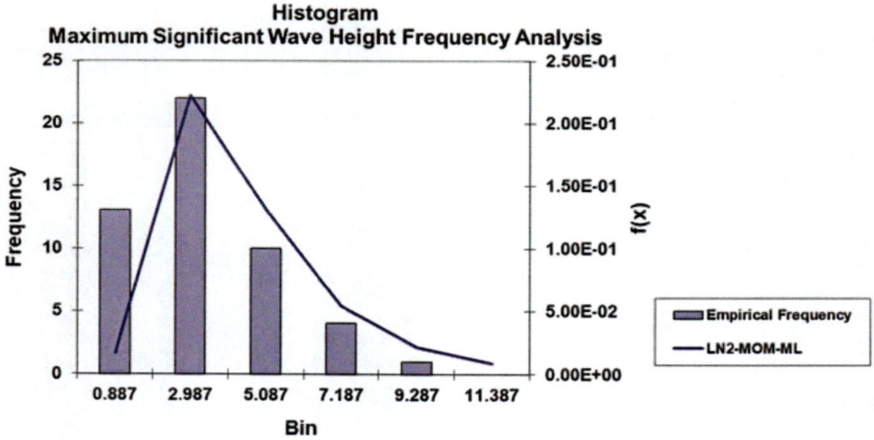

Fig. 4.14 Histogram and theoretical LN2 density for maximum significant wave height data sample

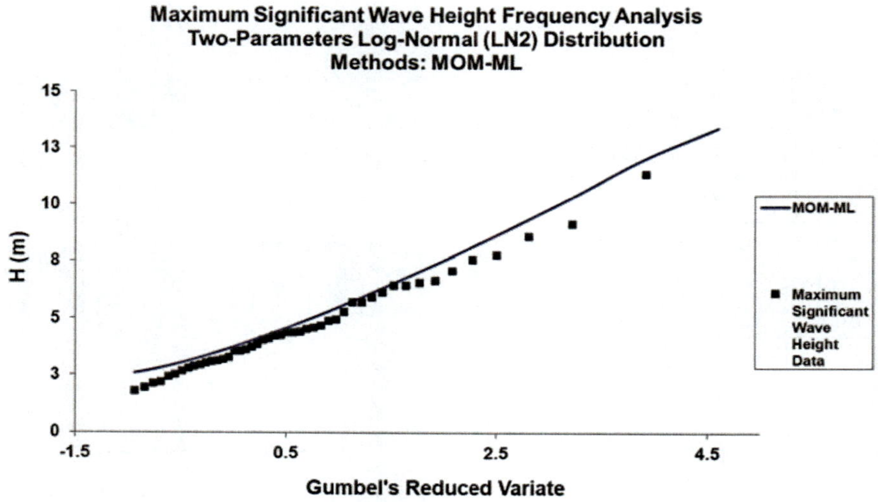

Fig. 4.15 Empirical and theoretical LN2 distributions for maximum significant wave height data sample

4.9.4 Annual Maximum Wind Speed Frequency Analysis

By using annual maximum wind speed data, Castillo (1988), the following descriptive statistics for the data and its natural logarithms were obtained and are show in Fig. 4.17.

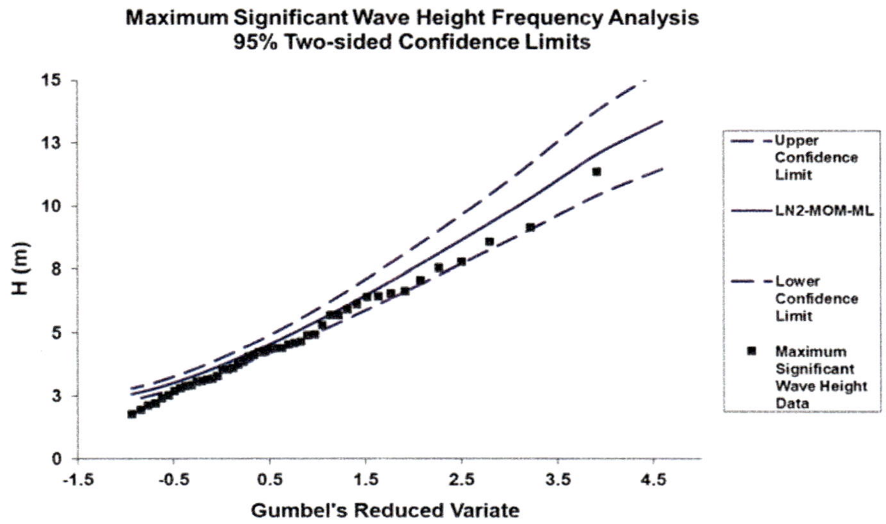

Fig. 4.16 Empirical and MOM and ML theoretical frequency curves and MOM and ML confidence limits for maximum significant wave height data sample

	A	B	C	D
1	MAXIMUM WIND SPEED FREQUENCY ANALYSIS			
2				
3	*Descriptive Statistics*		*Descriptive Statistics of the Natural Logarithms*	
4	Mean	54.7964	Mean	3.9558
5	Sample Standard Error	2.9320	Sample Standard Error	0.0408
6	Median	47.9965	Median	3.8711
7	Mode	#N/D	Mode	#N/D
8	Standard Deviation	20.7327	Standard Deviation	0.2886
9	Sample Variance	429.8439	Sample Variance	0.0833
10	Kurtosis	12.5699	Kurtosis	3.6721
11	Skewness	3.0813	Skewness	1.6052
12	Range	121.5116	Range	1.4669
13	Minimum	36.4278	Minimum	3.5953
14	Maximum	157.9394	Maximum	5.0622
15	Sum	2739.8217	Sum	197.7906
16	Count	50.0000	Count	50.0000
17	Largest(1)	157.9394	Largest(1)	5.0622
18	Smallest(1)	36.4278	Smallest(1)	3.5953
19	Mean Confidence Level(95.0%)	5.8922	Mean Confidence Level(95.0%)	0.0820

Fig. 4.17 Descriptive statistics for the annual maximum wind speed data sample

4.9.4.1 MOM-ML Methods

By using an Excel® worksheet, the results contained in Fig. 4.18, were obtained for the MOM-ML methods applied to the LN2 distribution for annual maximum wind speed data.

In Fig. 4.19 a comparison is made between the histogram and LN2 theoretical density for annual maximum wind speed data sample.

A graph containing the empirical and theoretical LN2 frequency curves is shown in Fig. 4.20.

MAXIMUM WIND SPEED FREQUENCY ANALYSIS
Two-parameters Log-Normal (LN2) Distribution
Methods: MOM-ML

Estimation of Parameters and Goodness of Fit Tests				
Location Parameter =		3.9558		
Scale Parameter =		0.2886		
SEF =	10.8		MARD =	10.4

Design Values and Confidence Limits							
Return Period Tr (years)	F(x)	v	Standard Normal Value z_T	Logarithms Squared root Stand. Error S_T	Two-sided 95% Lower Limit	Design Values V_T (m/s)	Two-sided 95% Upper Limit
2	0.5	1.1774	0.0000	0.1110	41	52	64
5	0.8	1.7941	0.8415	0.1292	51	67	85
10	0.9	2.1460	1.2817	0.1498	55	76	100
20	0.95	2.4477	1.6452	0.1703	58	84	115
50	0.98	2.7971	2.0542	0.1958	62	95	135
100	0.99	3.0349	2.3268	0.2137	64	102	150
Two-sided Confidence Limits:		z_α =	1.96				

Fig. 4.18 MOM and ML estimation of the parameters, standard errors, quantiles, and confidence limits for the annual maximum wind speed data sample

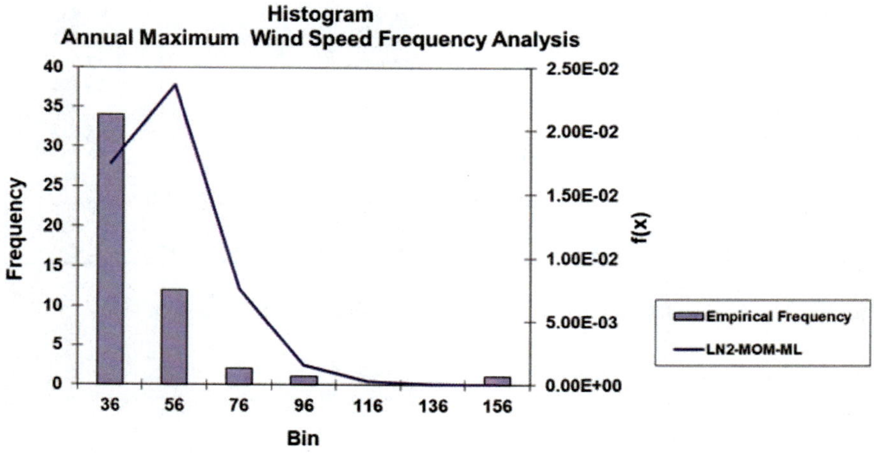

Fig. 4.19 Histogram and theoretical LN2 density for annual maximum wind speed data sample

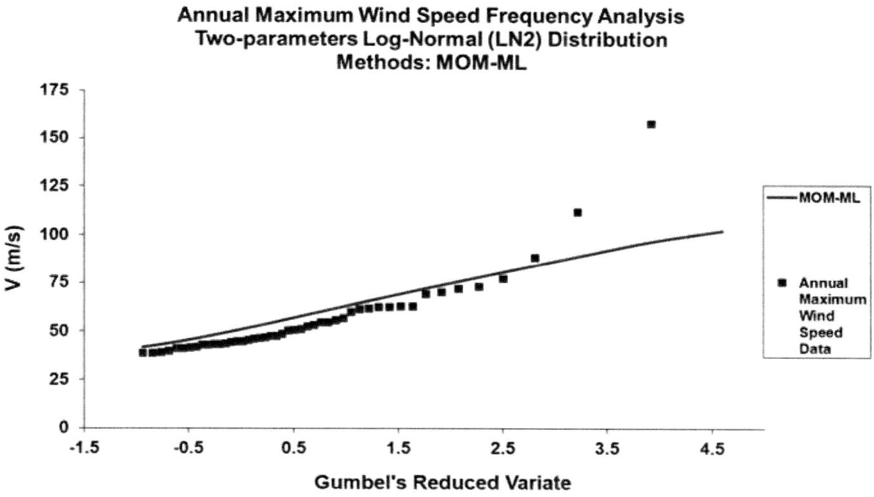

Fig. 4.20 Empirical and theoretical LN2 distributions for annual maximum wind speed data sample

A graphical depiction of the empirical and theoretical frequency curves and the confidence limits of the best fit provided, in this case that of MOM-ML, is shown in Fig. 4.21.

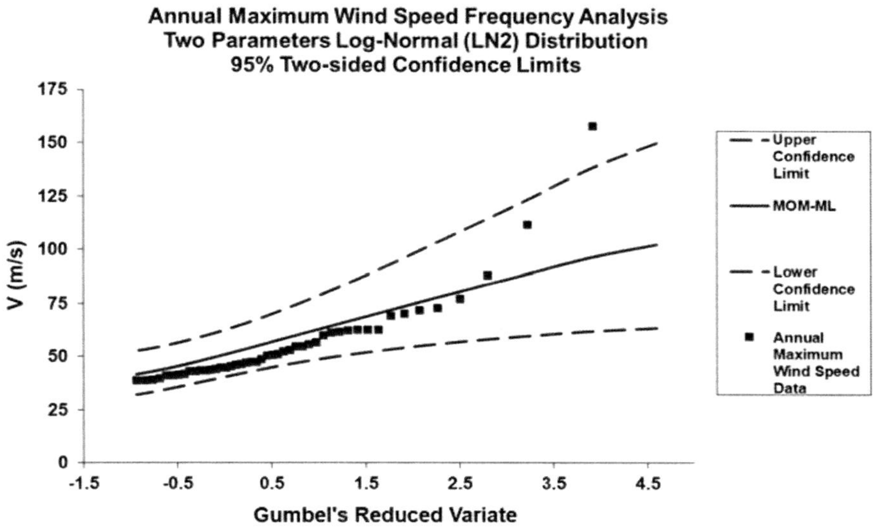

Fig. 4.21 Empirical and MOM and ML theoretical frequency curves and MOM and ML confidence limits for annual maximum wind speed data sample

5 Three-Parameters Log-Normal Distribution

Impossible is potential, impossible is temporary, impossible is nothing. M Ali

5.1 Introduction

If $y = \text{Ln}(x - x_0)$ is normally distributed, then the random variable x has a Three-parameters Log-Normal (LN3) distribution.

Chow (1954, 1959 and 1964), and Yevjevich (1972) have demonstrated the applicability of Two and Three-parameters Log-Normal distributions to solve natural extreme problems.

Sangal and Biswas (1970), proposed a method for estimating the LN3 distribution parameters, only based on the mean, median, and standard deviation of the sample data.

Burges et al. (1975), investigated the mathematical properties of the LN3 distribution, they compared two methods for estimating the location parameter of such distribution.

Hoshi et al. (1989), performed different estimation procedures for the log Normal distribution, and compared the average bias and the root mean square errors of these estimation methods.

The sample values must comply with the following restriction: $(x - x_0) > 0$, the involvement of natural logarithms in all the formulation of such probability distribution puts such a condition in its use.

J. A. Raynal Villaseñor, *Frequency Analyses of Natural Extreme Events*, Earth and Environmental Sciences Library,
https://doi.org/10.1007/978-3-030-86390-6_5

5.2 Chapter Objectives

After reading this chapter, you will know how to:

(a) Recognize the distribution and density functions of the LN3 distribution
(b) Estimate the parameters of the LN3 distribution
(c) Compute the quantiles and confidence limits of the LN3 distribution
(d) Make a graphic display of your data and the LN3 distribution
(e) Develop an application of all the above using Excel® spreadsheets.

5.3 Probability Distribution and Density Functions

The probability distribution function of the Three-parameters Log-Normal (LN3) distribution is:

$$F(x) = \int_{-\infty}^{x} \frac{1}{(x - x_0)\sigma_y\sqrt{2\pi}} \exp\left[-\frac{\left(Ln(x - x_0) - \mu_y\right)^2}{2\sigma_y^2}\right]dx \tag{5.1}$$

where x_0, σ_y and μ_y are the location, scale, and shape parameters. *F(x)* is the probability distribution function of the random variable *x* and in the case of maxima frequency analysis is equal to the non-exceedance probability, $Pr(X \leq x)$. The domain of variable *x* in this distribution is $(x - x_0) < x < \infty$, with the provision that $(x - x_0) > 0$.

The probability density function for the LN3 distribution is:

$$f(x) = \frac{1}{(x - x_0)\sigma_y\sqrt{2\pi}} \exp\left[-\frac{\left(Ln(x - x_0) - \mu_y\right)^2}{2\sigma_y^2}\right] \tag{5.2}$$

where *f(x)* is the probability density function of the random variable *x*.

5.4 Estimation of the Parameters

5.4.1 MOM Method

The MOM estimators for the parameters of the LN3 distribution are:

$$\hat{\mu}_y = \frac{1}{N}\sum_{i=1}^{N} Ln\left(x_i - \hat{x}_0\right) \tag{5.3}$$

$$\hat{\sigma}_y = \left[\frac{\sum_{i=1}^{N}\left(Ln(x_i - x_0) - \hat{\mu}_y\right)^2}{N}\right]^{1/2} \tag{5.4}$$

$$\hat{x}_0 = \mu_x\left(1 - \frac{CV_x}{CV_z}\right) \tag{5.5}$$

where,

$$CV_x = \frac{\sigma_x}{\mu_x} \tag{5.6}$$

$$CV_z = \frac{\left(1 - w^{2/3}\right)}{w^{1/3}} \tag{5.7}$$

and:

$$w = \frac{1}{2}\left[-\hat{\gamma}_x + \left(\hat{\gamma}_x^2 + 4\right)^{1/2}\right] \tag{5.8}$$

where $\hat{\gamma}_x$ is the skewness coefficient of the x's, as it was defined in Chap. 2.

5.4.1.1 Example of Application of Estimation of the Parameters of the LN3 Distribution Using the MOM Method

Find the MOM estimators for the parameters of the LN3 distribution for the flood sample data of gauging station Huites, Mexico (1942–1992), contained in Appendix A.

The following statistics can be obtained:

$$\hat{\mu} = \frac{1}{N}\sum_{i=1}^{N} x_i = 2498.9627 \text{ m}^3/\text{s}$$

$$\hat{\sigma} = \left[\frac{1}{N}\sum_{i=1}^{N}(x_i - \hat{\mu})^2\right]^{1/2} = 2199.4767 \text{ m}^3/\text{s}$$

$$\hat{\gamma} = \hat{\gamma}_x = \frac{N}{(N-1)(N-2)\hat{\sigma}^3}\sum_{i=1}^{N}(x_i - \hat{\mu})^3 = 2.0123$$

$$CV = CV_x = \frac{\hat{\sigma}}{\hat{\mu}} = \frac{2199.4767}{2498.9627} = 0.8802$$

then:

$$w = \frac{1}{2}\left[-\hat{\gamma}_x + \left(\hat{\gamma}_x^2 + 4\right)^{1/2}\right] = \frac{1}{2}\left[-2.0123 + \left((2.0123)^2 + 4\right)^{\frac{1}{2}}\right] = 0.4124$$

$$CV_z = \frac{\left(1 - w^{2/3}\right)}{w^{1/3}} = \frac{\left(1 - (0.4124)^{0.6666}\right)}{(0.4124)^{0.3333}} = 0.5990$$

so:

$$\hat{x}_0 = \mu_x\left(1 - \frac{CV_x}{CV_z}\right) = 2498.9627\left(1 - \frac{0.8802}{0.5990}\right) = -1173.1357\ \mathrm{m}^3/\mathrm{s}$$

and:

$$\hat{\mu}_y = \frac{1}{N}\sum_{i=1}^{N} Ln\left(x_i - \hat{x}_0\right) = \frac{1}{51}\sum_{i=1}^{51} Ln(x_i - (-1172.8267) = 8.0812$$

$$\hat{\sigma}_y = \left[\frac{\sum_{i=1}^{N}\left(Ln(x_i - x_0) - \hat{\mu}_y\right)^2}{N}\right]^{1/2}$$

$$= \left[\frac{\sum_{i=1}^{51}[Ln(x_i - (-1172.8267) - 8.012]^{0.5}}{51}\right] = 0.4685$$

Finally, the moment estimators of the parameters of the LN3 distribution for the flood sample data of gauging station Huites, Mexico, are:

location parameter:

$$\hat{x}_0 = -1172.8267\ \mathrm{m}^3/\mathrm{s}$$

scale parameter:

$$\hat{\sigma}_y = 0.4685$$

and:

shape parameter:

$$\hat{\mu}_y = 8.0812$$

5.4.2 ML Method

The likelihood function of the LN3 distribution is:

$$L\left(x, \mu_y, \sigma_y, x_0\right) = \prod_{i=1}^{N}\left\{\frac{1}{(x - x_0)\sigma_y\sqrt{2\pi}}\exp\left[-\frac{\left(Ln(x - x_0) - \mu_y\right)^2}{2\sigma_y^2}\right]\right\} \tag{5.9}$$

which simplifies to:

$$L\left(x, \mu_y, \sigma_y, x_0\right) = \left(\frac{1}{\sigma_y\sqrt{2\pi}}\right)^N \prod_{i=1}^{N}\left(\frac{1}{(x_i - x_0)}\right)\exp\left[-\sum_{i=1}^{N}\frac{\left(Ln(x_i - x_0) - \mu_y\right)^2}{2\sigma_y^2}\right] \tag{5.10}$$

The Log-Likelihood function of the LN3 distribution is, Kite (1988):

$$LL(x, \mu_y, \sigma_y, x_0) = -N\left(Ln(\sigma_y) + \frac{1}{2}Ln(2\pi)\right) - \sum_{i=1}^{N} Ln(x_i - x_0) - \frac{1}{\left(2\sigma_y^2\right)} \sum_{i=1}^{N} \left(Ln(x_i - x_0) - \mu_y\right)^2 \tag{5.11}$$

by taking the partial derivatives of Eq. (5.11) with respect to the parameters μ_y,σ_y and x_0 and then solving the system of equations:

$$\frac{\partial LL}{\partial \mu_y} = \frac{1}{\sigma_y^2} \sum_{i=1}^{N} \left[Ln(x_i - x_0) - \mu_y\right] = 0 \tag{5.12}$$

$$\frac{\partial LL}{\partial \sigma_y} = -\frac{N}{\sigma_y} + \frac{1}{\sigma_y^3} \sum_{i=1}^{N} \left[Ln(x_i - x_0) - \mu_y\right]^2 = 0 \tag{5.13}$$

$$\frac{\partial LL}{\partial x_0} = \left(\mu_y - \sigma_y^2\right) \sum_{i=1}^{N} \frac{1}{(x_i - x_0)} - \sum_{i=1}^{N} \frac{Ln(x_i - x_0)}{(x_i - x_0)} = 0 \tag{5.14}$$

Finally, the ML estimators of the parameters of the LN3 distribution can estimated as:

$$\hat{\mu}_y = \frac{1}{N} \sum_{i=1}^{N} Ln\left(x_i - \hat{x}_0\right) \tag{5.15}$$

$$\hat{\sigma}_y = \left[\frac{\sum_{i=1}^{N} \left(Ln\left(x_i - \hat{x}_0\right) - \hat{\mu}_y\right)^2}{N}\right]^{1/2} \tag{5.16}$$

for the case of the location parameter x_0, the following equation must be solved by a root searching procedure, the method of bisection is recommended:

$$f\left(\hat{x}_0\right) = \left[\sum_{i=1}^{N} \frac{1}{\left(x_i - \hat{x}_0\right)}\right] \left\{\frac{1}{N} \sum_{i=1}^{N} Ln^2\left(x_i - \hat{x}_0\right) - \left[\frac{1}{N} \sum_{i=1}^{N} Ln\left(x_i - \hat{x}_0\right)\right]^2 - \frac{1}{N} \sum_{i=1}^{N} Ln\left(x_i - \hat{x}_0\right)\right\} + \sum_{i=1}^{N} \frac{Ln\left(x_i - \hat{x}_0\right)}{\left(x_i - \hat{x}_0\right)} = 0 \tag{5.17}$$

5.4.2.1 Example of Application of Estimation of the Parameters of the LN3 Distribution Using the ML Method

Find the ML estimators for the parameters of the LN3 distribution for the flood sample data of gauging station Huites, Mexico (1942–1992), contained in Appendix A.

Assuming an initial value, which is less than the least value in the flood sample data, the location parameter will be:

$$\hat{x}_0 = 400 \text{ m}^3/\text{s}$$

then:

$$F(\hat{x}_0) = -1.1939 \times 10^{-1}$$

By using an adequate method of search of roots for Eq. (5.17), as the method of bisection, the following solution is found:

$$\hat{x}_0 = 243.4908 \text{ m}^3/\text{s}$$

and:

$$F(\hat{x}_0) = 4.7018 \times 10^{-7}$$

The two other parameters are found as:

$$\hat{\mu}_y = \frac{1}{N}\sum_{i=1}^{N} Ln(x_i - \hat{x}_0) = \frac{1}{51}\sum_{i=1}^{51} Ln(x_i - 243.4908) = 7.3481$$

$$\hat{\sigma}_y = \left[\frac{\sum_{i=1}^{N}\left(Ln(x_i - x_0) - \hat{\mu}_y\right)^2}{N}\right]^{1/2}$$

$$= \left[\frac{\sum_{i=1}^{51}[Ln(x_i - 243.4908) - 7.3481]^{0.5}}{51}\right] = 0.8608$$

Finally, the moment estimators of the parameters of the LN3 distribution for the flood sample data of gauging station Huites, Mexico, are:

location parameter:

$$\hat{x}_0 = 243.4908 \text{ m}^3/\text{s}$$

scale parameter:

$$\hat{\sigma}_y = 0.8608$$

and:

shape parameter:

$$\hat{\mu}_y = 7.3481$$

5.5 Estimation of Quantiles for the LN3 Distribution

The quantiles for the LN3 distribution can be obtained by inverting the Normal distribution to obtain the quantiles given by the following expression, Abramowitz and Stegun (1965):

$$z_T = w - \left(\frac{c_0 + c_1 w + c_2 w^2}{1 + d_1 w + d_2 w^2 + d_3 w^3} \right) \tag{5.18}$$

where z_T is the standard normal deviate that corresponds to a certain return period, T_r, which is linked to a certain value of the standard normal distribution function, Abramowitz and Stegun (1965):

$$F(x) = 1 - \frac{1}{T_r} \tag{5.19}$$

The coefficients c and d are as follows, Abramowitz and Stegun (1965):

$$c_0 = 2.515517; \; c_1 = 0.802853; \; c_2 = 0.010328$$
$$d_1 = 1.432788; \; d_2 = 0.189269; \; d_3 = 0.001308$$

and, Abramowitz and Stegun (1965):

$$w = \sqrt{Ln\left(\frac{1}{1 - F(x)}\right)^2} \tag{5.20}$$

where F(x) is the normal standard distribution function. If F(x) < 0.5, then put (1 – F(x)) instead F(x) in Eq. (5.17) and change the sign in the resulting values of z_T.

Now, the quantiles for the LN3 distribution are obtained as:

$$Q_T = \hat{x}_0 + \exp\left(\hat{\mu}_y + z_T \hat{\sigma}_y\right) \tag{5.21}$$

5.5.1 Examples of Estimation of MOM Quantiles for the LN3 Distribution

Find the MOM estimators of the quantiles 2, 5, 10, 20, 50, 100, 500, and 1,000 years of return period, for the LN3 distribution for the flood sample data of gauging station Huites, Mexico (1942–1992), contained in Appendix A.

The estimation of MOM quantiles for the LN3 distribution is made by using the parameters estimated in the preceding section and then inserted the parameters into Eq. (5.21). Table 5.1 summarizes these results.

Table 5.1 Estimation of MOM quantiles for the LN3 distribution for gauging station huites, Mexico

Tr (Years)	MOM
	$Q_T(m^3/s)$
2	1797
5	3448
10	4925
20	6645
50	9345
100	11,753
500	18,748
1000	22,453

5.6 Goodness of Fit Test

The standard error of fit (SEF) for the LN3 has the following form:

$$SEF = \left[\frac{\sum_{i=1}^{N}(xi - yi)^2}{(N-3)}\right]^{1/2} \tag{5.22}$$

while the mean absolute relative deviation (MARD) remains the same as it was defined in Chapter 2:

$$MARD = \frac{100}{N}\sum_{i=1}^{N}\left|\frac{(x_i - y_i)}{x_i}\right| \tag{5.23}$$

where x_i are the sample historical values, y_i are the distribution function values, corresponding to the same return periods of the historical values, N is the sample size.

5.6.1 Examples of Application of the SEF and MARD to the MOM and ML Estimators of the Parameters of the LN3 Distribution

Find the values of the SEF and MARD of the MOM and ML estimators of the parameters of the LN3 distribution for the flood sample data of gauging station Huites, Mexico (1942–1992), contained in Appendix A.

The values of SEF and MARD for the MOM and ML estimators of the parameters of the LN3 distribution for the flood sample data of gauging station Huites, Mexico have been obtained through the application of Eqs. (5.22) and (5.23) using the parameters obtained in previous example. The results are as follows:

For the MOM method:

$$SEF = \left[\frac{\sum_{i=1}^{N}(xi - yi)^2}{(N-3)}\right]^{1/2} = \left[\frac{754754.9619}{(51-3)}\right]^{\frac{1}{2}} = 869$$

$$MARD = \frac{100}{N}\sum_{i=1}^{N}\left|\frac{(x_i - y_i)}{x_i}\right| = \frac{(100)(9.8856)}{51} = 19$$

For the ML method:

$$SEF = \left[\frac{\sum_{i=1}^{N}(xi - yi)^2}{(N-3)}\right]^{1/2} = \left[\frac{305250.4119}{(51-3)}\right]^{\frac{1}{2}} = 552$$

$$MARD = \frac{100}{N}\sum_{i=1}^{N}\left|\frac{(x_i - y_i)}{x_i}\right| = \frac{(100)(4.0522)}{51} = 8$$

As a conclusion, it can be said the method of maximum likelihood showed to be the best method, when the LN3 distribution was applied to this set of flood sample data, given that the values of the SEF and MARD measures resulted to be the least in both cases when compared with the values obtained by the method of moments.

5.7 Estimation of the Confidence Limits for the LN3 Distribution

By assuming that the quantiles are normally distributed, the following form can be used to set the confidence limits on such quantiles:

$$x_l = x_T \pm z_\alpha S_T \tag{5.24}$$

where x_l is the upper or lower confidence limit, depending on the sign in the preceding formula (+) for the upper confidence limit and (–) for the lower confidence limit, z_α is the standard normal variate corresponding to a confidence level α, and S_T is the squared root of the standard error of the estimate. The evaluation procedures of the standard errors of the estimates will be described in the following subsections.

5.8 Estimation of the Standard Errors for the LN3 Distribution

5.8.1 MOM Method

The general form of the MOM estimator of the standard error of the estimate of a 3-parameter distribution is, Kite (1988):

$$S_T^2 = \left(\frac{\partial x_T}{\partial m_1^{'}}\right)^2 \mathrm{var}(m_1^{'}) + \left(\frac{\partial x_T}{\partial m_2}\right)^2 var(m_2) + \left(\frac{\partial x_T}{\partial m_3}\right)^2 var(m_3) + 2\left(\frac{\partial x_T}{\partial m_1^{'}}\right)\left(\frac{\partial x_T}{\partial m_2}\right)\mathrm{cov}(m_1^{'}, m_2) + 2\left(\frac{\partial x_T}{\partial m_1^{'}}\right)\left(\frac{\partial x_T}{\partial m_3}\right)\mathrm{cov}(m_1^{'}, m_3) + 2\left(\frac{\partial x_T}{\partial m_2}\right)\left(\frac{\partial x_T}{\partial m_3}\right)\mathrm{cov}(m_2, m_3) \quad (5.25)$$

and Eq. (5.25) can be simplified in terms of the frequency factor, K_T, as, Kite (1988):

$$S_T^2 = \frac{\mu_2}{N}\left[1 + K_T\hat{\gamma} + \frac{K_T^2}{4}(\hat{\kappa} - 1)\right] \quad (5.26)$$

and given that computing the skewness and kurtosis coefficients for the logarithms will produce values of zero and 3, respectively, also the frequency factor will be equal to z_T in the LN3 distribution, and the moment estimator of the standard error of the estimate for the LN3 distribution is, Kite (1988):

$$S_T^2 = \frac{\sigma_y^2}{N}\left(1 + \frac{z_T^2}{2}\right) \quad (5.27)$$

where z_T is the standard normal variate corresponding to the return period T. To introduce the squared root standard error of the estimate into Eq. (5.25), the following expressions must be used for the positive value associated with the upper confidence limit and the negative value associated with the lower confidence limit:

$$S_T(+) = x_T(\exp(S_T) - 1) \quad (5.28)$$

$$S_T(-) = -(x_T - x_0)(\exp(-S_T) - 1) \quad (5.29)$$

5.8.1.1 Example of Estimation of MOM Standard Errors and the Two-Sided 95% Confidence Limits for the LN3 Distribution

Find the MOM estimators of the standard errors and the two-sided 95% confidence limits for the LN3 distribution for the flood sample data of gauging station Huites, Mexico (1942–1992), contained in Appendix A.

The estimation of MOM estimators of the standard errors and the two-sided 95% confidence limits for the LN3 distribution is made by using the moment estimators of the parameters and then inserted in the Eqs. (5.26), (5.27) and (5.24), using selected values of the return intervals. Table 5.2 summarizes these results.

5.8.2 ML Method

The general form of the ML estimator of the standard error of the estimate of a three-parameter distribution is, Kite (1988):

$$\begin{aligned} S_T^2 = & \left(\frac{\partial x_T}{\partial \alpha}\right)^2 var(\alpha) + \left(\frac{\partial x_T}{\partial \beta}\right)^2 var(\beta) \\ & + \left(\frac{\partial x_T}{\partial \gamma}\right)^2 var(\gamma) + 2\left(\frac{\partial x_T}{\partial \alpha}\right)\left(\frac{\partial x_T}{\partial \beta}\right)\mathrm{cov}(\alpha, \beta) \\ & + 2\left(\frac{\partial x_T}{\partial \alpha}\right)\left(\frac{\partial x_T}{\partial \gamma}\right)\mathrm{cov}(\alpha, \gamma) + 2\left(\frac{\partial x_T}{\partial \beta}\right)\left(\frac{\partial x_T}{\partial \gamma}\right)\mathrm{cov}(\beta, \gamma) \end{aligned} \tag{5.30}$$

and for the LN3 distribution the following relationships hold, Kite (1988):

$$\frac{\partial x_T}{\partial x_0} = 1 \tag{5.31}$$

Table 5.2 Estimation of MOM standard errors and the two-sided 95% confidence limits for the LN3 distribution for gauging station huites, Mexico

Tr (Years)	Logarithmic S_T (m^3/s)	95% Lower Limit (m^3/s)	Q_T (m^3/s)	95% Upper Limit (m^3/s)
2	0.0958	1481	2060	2698
5	0.1115	2631	3623	4732
10	0.1294	3320	4721	6317
20	0.1470	3943	5816	7985
50	0.1690	4712	7292	10,347
100	0.1845	5268	8445	12,266
500	0.2174	6513	11,282	17,210
1000	0.2304	7036	12,583	19,567

Two-sided Limits: $z_T = 1.96$

$$\frac{\partial x_T}{\partial \sigma_y^2} = \frac{z_T \exp(\mu_y + z_T \sigma_y)}{2\sigma_y} \tag{5.32}$$

$$\frac{\partial x_T}{\partial \mu_y} = \exp(\mu_y + z_T \sigma_y) \tag{5.33}$$

So, a ML estimator of the standard error of the estimate for the LN3 parameters can be expressed as, Kite (1988):

$$\begin{aligned} S_T^2 = {} & var(x_0) + \frac{z_T^2 \exp[2(\mu_y + z_T\sigma_y)]}{4\sigma_y^2} var(\sigma_y^2) \\ & + \exp[2(\mu_y + z_T\sigma_y)]\mathrm{var}(\mu_y) + \frac{z_T \exp[(\mu_y + z_T\sigma_y)]}{\sigma_y}\mathrm{cov}(x_0, \sigma_y^2) \\ & + 2\exp[(\mu_y + z_T\sigma_y)]\mathrm{cov}(x_0, \mu_y) + \frac{z_T \exp[2(\mu_y + z_T\sigma_y)]}{\sigma_y}\mathrm{cov}(\sigma_y^2, \mu_y) \end{aligned} \tag{5.34}$$

where

$$\mathrm{var}(x_0) = \frac{1}{2ND} \tag{5.35}$$

$$\mathrm{var}(\mu_y) = \frac{\sigma_y^2}{ND}\left[\frac{\left(\sigma_y^2 + 1\right)}{2\sigma_y^2}\exp\left[2\left(\sigma_y^2 - \mu_y\right)\right] - \exp\left[\left(\sigma_y^2 - 2\mu_y\right)\right]\right] \tag{5.36}$$

$$\mathrm{var}(\sigma_y^2) = \frac{\sigma_y^2}{ND}\left[\left(\sigma_y^2 + 1\right)\exp\left[2\left(\sigma_y^2 - \mu_y\right)\right] - \exp\left[\left(\sigma_y^2 - 2\mu_y\right)\right]\right] \tag{5.37}$$

$$cov(x_0, \mu_y) = -\frac{\exp\left(\frac{\sigma_y^2}{2} - \mu_y\right)}{2ND} \tag{5.38}$$

$$cov(x_0, \sigma_y^2) = \frac{\sigma_y^2 \exp\left(\frac{\sigma_y^2}{2} - \mu_y\right)}{ND} \tag{5.39}$$

$$cov(\mu_y, \sigma_y^2) = -\frac{\sigma_y^2 \exp\left(\sigma_y^2 - 2\mu_y\right)}{ND} \tag{5.40}$$

and:

$$D = \frac{\left(\sigma_y^2 + 1\right)}{2\sigma_y^2}\exp\left[2\left(\sigma_y^2 - \mu_y\right)\right] - \frac{\left(2\sigma_y^2 + 1\right)}{2\sigma_y^2}\exp\left(\sigma_y^2 - 2\mu_y\right) \tag{5.41}$$

Table 5.3 Estimation of ML standard errors and the two-sided 95% confidence limits for the LN3 distribution for gauging station huites, Mexico

Tr (Years)	S_T (m^3/s)	95% lower limit (m^3/s)	Q_T (m^3/s)	95% Upper limit (m^3/s)
2	23.4703	1751	1797	1843
5	91.1637	3270	3449	3627
10	162.1414	4606	4924	5242
20	254.2953	6144	6643	7141
50	414.2338	8530	9342	10,154
100	568.6432	10,634	11,748	12,863
500	1062.2757	16,661	18,743	20,825
1000	1343.9759	19,813	22,448	25,082

Two-sided Limits: $z_T = 1.96$

5.8.2.1 Example of Estimation of ML Standard Errors and the Two-Sided 95% Confidence Limits for the LN3 Distribution

Find the ML estimators of the standard errors and the two-sided 95% confidence limits for the LN3 distribution for the flood sample data of gauging station Huites, Mexico (1942–1992), contained in Appendix A.

The estimation of ML estimators of the standard errors and the two-sided 95% confidence limits for the LN3 distribution is made by using the maximum likelihood estimators of the parameters and then inserted in the Eqs. (5.32) and (5.24), using selected values of the return intervals. Table 5.3 summarizes these results.

5.9 Examples of Application for the LN3 Distribution Using Excel® Spreadsheets

5.9.1 Flood Frequency Analysis

By using the flood data from gauging station Huites, Mexico the following descriptive statistics were obtained and are shown in Fig. 5.1.

5.9.1.1 MOM Method

By using an Excel® worksheet, the results contained in Fig. 5.2, were obtained for the MOM method applied to the LN3 distribution for gauging station Huites, Mexico.

5.9.1.2 ML Method

By using an Excel® worksheet, the results contained in Fig. 5.3, were obtained for the ML method applied to the LN3 distribution for gauging station Huites, Mexico.

	A	B	C	D
1	FLOOD FREQUENCY ANALYSIS			
2	Gauging Station: Huites, Sinaloa, Mexico (1942-1992)			
3				
4	*Descriptive Statistics*		*Descriptive Statistics of the Natural Logarithms*	
5	Mean	2498.9627	Mean	7.5383
6	Sample Standard Error	311.0530	Sample Standard Error	0.1024
7	Median	1775.7000	Median	7.4819
8	Mode	#N/D	Mode	#N/D
9	Standard Deviation	2221.3626	Standard Deviation	0.7312
10	Sample Variance	4934451.7500	Sample Variance	0.5347
11	Kurtosis	3.6701	Kurtosis	0.0081
12	Skewness	2.0123	Skewness	0.4627
13	Range	9707.5000	Range	3.1787
14	Minimum	421.8000	Minimum	6.0445
15	Maximum	10129.3000	Maximum	9.2232
16	Sum	127447.1000	Sum	384.4530
17	Count	51.0000	Count	51.0000
18	Largest(1)	10129.3000	Largest(1)	9.2232
19	Smallest(1)	421.8000	Smallest(1)	6.0445
20	Mean Confidence Level(95.0%)	624.7683	Mean Confidence Level(95.0%)	0.2057

Fig. 5.1 Descriptive statistics for the flood sample of huites, Mexico

	A	B	C	D	E	F	G	H	I
1	FLOOD FREQUENCY ANALYSIS								
2	Gauging Station: Huites, Sinaloa, Mexico (1942-1992)								
3	Three-Parameters Log-Normal (LN3) Distribution								
4	Method: MOM								
5									
6	Estimation of Parameters								
7	w =	0.7951	Skew(Ln(x-x_0)) =		1.7304				
8	n_z =	0.1530		n_x =	0.8889				
9	Location Parameter =		-12018.2351	m^3/s					
10	Scale Parameter =		0.1349						
11	Shape Parameter =		9.5733						
12	SEF =	974		MARD =	26				
13									
14	Design Values and Confidence Limits								
15–18	Return Period Tr (years)	F(x)	v	Standard Normal Variate z_T	Logarithmic Squared Root Stand.Error S_T	Squared Root Stand.Error S_T (m^3/s)	Two-sided 95% Lower Limit (m^3/s)	Design Values Q_T (m^3/s)	Two-sided 95% Upper Limit (m^3/s)
19	2	0.5	1.1774	0.0000	0.0514	739.7495	907	2357	3807
20	5	0.8	1.7941	0.8415	0.0599	964.4414	2195	4086	5976
21	10	0.9	2.1460	1.2817	0.0694	1187.2898	2744	5071	7398
22	20	0.95	2.4477	1.6452	0.0789	1417.7473	3152	5930	8709
23	50	0.98	2.7971	2.0542	0.0907	1722.8015	3572	6949	10325
24	100	0.99	3.0349	2.3268	0.0990	1951.9280	3833	7659	11485
25	500	0.998	3.5255	2.8785	0.1166	2478.3614	4322	9180	14038
26	1000	0.999	3.7169	3.0905	0.1236	2703.3794	4497	9795	15094
27	Two-sided Confidence Limits:		z_α =	1.96					

Fig. 5.2 MOM estimators for the parameters, standard errors, Quantiles, and confidence limits of the LN3 distribution for the flood sample data at huites, Mexico

A	B	C	D	E	F	G
			FLOOD FREQUENCY ANALYSIS			
			Gauging Station: Huites, Sinaloa, Mexico (1942-1992)			
			Three-Parameters Log-Normal (LN3) Distribution			
			Method: ML			
			Estimation of Parameters			
Test Location Parameter =	400.0000		Skew(Ln(x-x_0)) =		0.0571	
F(x_0) =	4.7018E-07		Log-likelihood function =	-5671.5292		
Location Parameter =	243.4908	m^3/s	Scale Parameter =	0.8608	Shape Parameter =	7.3481
SEF =	553	MARD =	7	AIC =	11349	
			Variance-Covariance Matrix			
var(x_0) =	207.3892	cov(x_0,μ_y) =	-0.1934	var(μ_y) =	0.0004	
cov(x_0,σ_y^2) =	0.2866	var(σ_y^2) =	0.0007	cov(μ_y,σ_y^2) =	0.0003	
D =	4.7273E-05					
			Design Values and Confidence Limits			
Return Period Tr (years)	F(x)	Standard Normal Variate z_T	Squared Root Stand.Error S_T (m^3/s)	Two-sided 95% Lower Limit (m^3/s)	Design Values Q_T (m^3/s)	Two-sided 95% Upper Limit (m^3/s)
2	0.5	0.0000	23.4703	1751	1797	1843
5	0.8	0.8416	91.1637	3270	3449	3627
10	0.9	1.2816	162.1414	4606	4924	5242
20	0.95	1.6449	254.2953	6144	6643	7141
50	0.98	2.0537	414.2338	8530	9342	10154
100	0.99	2.3263	568.6432	10634	11748	12863
500	0.998	2.8782	1062.2757	16661	18743	20825
1000	0.999	3.0902	1343.9759	19813	22448	25082
Two-sided Confidence Limits:	z_α =	1.96				

Fig. 5.3 ML Estimators for the parameters, standard errors, Quantiles, and confidence limits of the LN3 distribution for the flood sample data at huites, Mexico

In Fig. 5.4 a comparison is made between the histogram and LN3 theoretical density for gauging station Huites, Mexico.

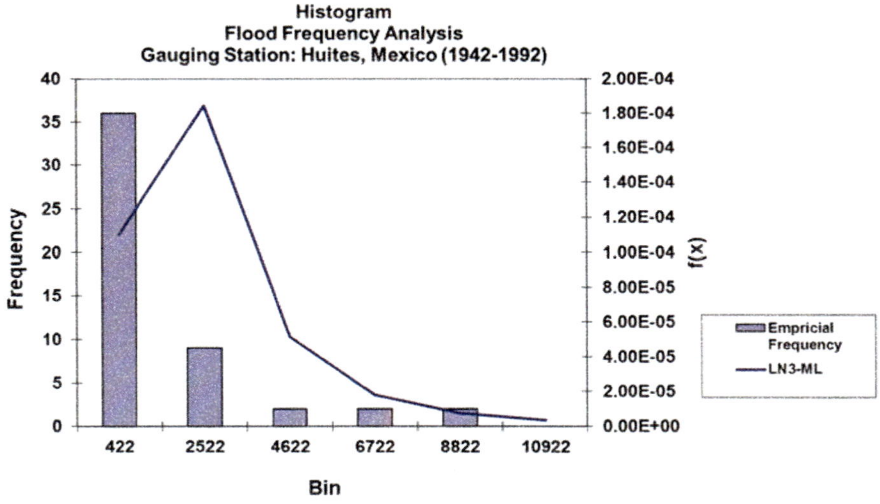

Fig. 5.4 Histogram and theoretical ML LN3 density for flood data at huites, Mexico

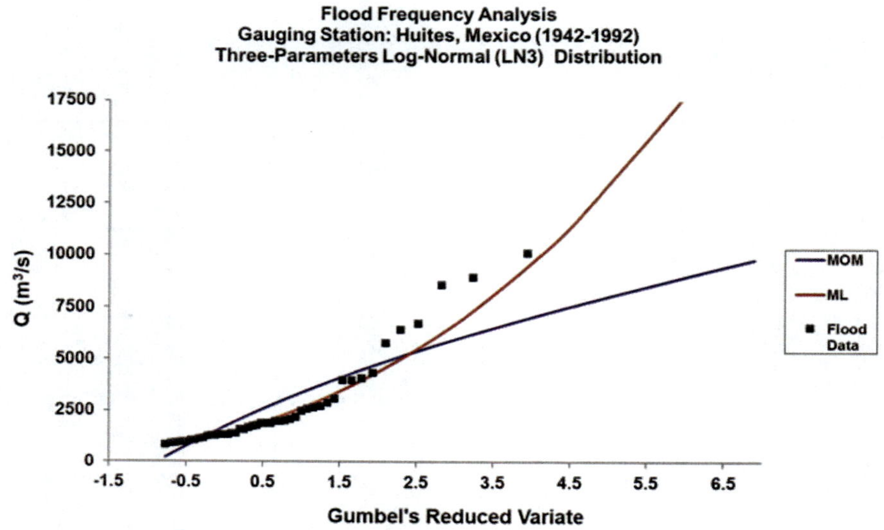

Fig. 5.5 Empirical and MOM-ML theoretical curves for the LN3 and flood data at huites, Mexico

A graph containing the empirical and theoretical LN3 frequency curves for gauging station Huites, Mexico, is shown in Fig. 5.5.

A graphical depiction of the empirical and theoretical frequency curves and the confidence limits of the best fit provided, in this case that of ML, for gauging station Huites, Mexico, is shown in Fig. 5.6.

5.9.2 Rainfall Frequency Analysis

By using the 24 h, annual maximum rainfall data from meteorological station Chihuahua, Mexico the following descriptive statistics were obtained and are displayed in Fig. 5.7.

5.9.2.1 MOM Method

By using an Excel® worksheet, the results contained in Fig. 5.8, were obtained for the MOM method applied to the LN3 distribution for meteorological station Chihuahua, Mexico.

5.9.2.2 ML Method

By using an Excel® worksheet, the results contained in Fig. 5.9, were obtained for the ML method applied to the LN3 distribution for rainfall meteorological station Chihuahua, Mexico.

In Fig. 5.10 a comparison is made between the histogram and LN3 theoretical density for rainfall meteorological station Chihuahua, Mexico.

A graph containing the empirical and theoretical LN3 frequency curves is shown in Fig. 5.11.

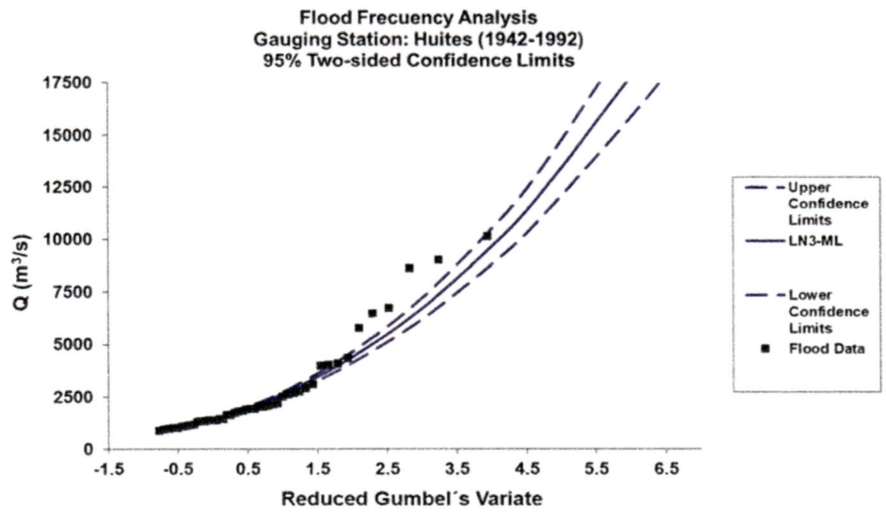

Fig. 5.6 Empirical and ML theoretical frequency curves and ML confidence limits of the LN3 distribution for the flood data at huites, Mexico

	A	B	C	D
1	**24 Hr. Annual Maximum Rainfall Frequency Analysis**			
2	**Meteorological Station: Chihuahua, Mexico (1960-2005)**			
3				
4	***Descriptive Statistics***		***Descriptive Statistics of the Natural Logarithms***	
5	Mean	42.0454	Mean	3.6871
6	Sample Standard Error	1.9784	Sample Standard Error	0.0487
7	Median	40.2400	Median	3.6948
8	Mode	#N/D	Mode	#N/D
9	Standard Deviation	13.4184	Standard Deviation	0.3302
10	Sample Variance	180.0535	Sample Variance	0.1090
11	Kurtosis	-0.7252	Kurtosis	-0.4776
12	Skewness	0.3591	Skewness	-0.2507
13	Range	54.3000	Range	1.4283
14	Minimum	17.1200	Minimum	2.8402
15	Maximum	71.4200	Maximum	4.2686
16	Sum	1934.0900	Sum	169.6058
17	Count	46.0000	Count	46.0000
18	Largest(1)	71.4200	Largest(1)	4.2686
19	Smallest(1)	17.1200	Smallest(1)	2.8402
20	Mean Confidence Level(95.0%)	3.9848	Mean Confidence Level(95.0%)	0.0981

Fig. 5.7 Descriptive statistics for the 24 h, annual maximum rainfall data at meteorological station Chihuahua, Mexico

24 Hr. Annual Maximum Rainfall Frequency Analysis
Meteorological Station: Chihuahua, Mexico (1960-2005)
Three-Parameters Log-Normal (LN3) Distribution
Method: MOM

Estimation of Parameters				
w =	0.8364		CV_x =	0.3191
CV_z =	0.1191			
Location Parameter =		-70.5801	cm	
Scale Parameter =		0.1168		
Shape Parameter =		4.7172		
SEF =	1.57		MARD =	2.61

Design Values and Confidence Limits							
Return Period Tr (years)	F(x)	Standard Normal Variate z_T	Logarithmic Squared Root Stand.Error S_T	Squared Root Stand.Error S_T (cm)	Two-sided 95% Lower Limit (cm)	Design Values P_T (cm)	Two-sided 95% Upper Limit (cm)
2	0.5	0.0000	0.0404	4.5248	32	41	50
5	0.8	0.8416	0.0471	5.8100	41	53	64
10	0.9	1.2816	0.0546	7.0939	45	59	73
20	0.95	1.6449	0.0620	8.4138	48	65	81
50	0.98	2.0537	0.0713	10.1471	52	72	91
100	0.99	2.3263	0.0779	11.4389	54	76	99
Two-sided Confidence Limits:		z_α =	1.96				

Fig. 5.8 MOM estimation of the parameters, standard errors, quantiles, and confidence limits for the 24 h, annual maximum rainfall data at meteorological station Chihuahua, Mexico

24 Hr. Annual Maximum Rainfall Frequency Analysis
Meteorological Station: Chihuahua, Mexico (1960-2005)
Three-Parameters Log-Normal (LN3) Distribution
Method: ML

Estimation of Parameters						
Test Location Parameter =	-14.4497					
$F(x_0)$ =	9.9730E-07		Log-likelihood function =	632.3470		
Location Parameter =	-14.4497	cm				
Scale Parameter =	0.2364					
Shape Parameter =	4.0064					
SEF =	1.91		MARD =	3.66	AIC =	-1259

Variance-Covariance Matrix					
$var(x_0)$ =	3.6401E-01	$cov(x_0,\mu_y)$ =	-6.8119E-03	$var(\mu_y)$ =	1.2809E-04
$cov(x_0,\sigma_y^2)$ =	7.6135E-04	$var(\sigma_y^2)$ =	1.6609E-06	$cov(\mu_y,\sigma_y^2)$ =	1.4248E-05
D =	1.9233E-02				

Design Values and Confidence Limits						
Return Period Tr (years)	F(x)	Standard Normal Variate z_T	Squared Root Stand.Error S_T (cm)	Two-sided 95% Lower Limit (cm)	Design Values P_T (cm)	Two-sided 95% Upper Limit (cm)
2	0.5	0.0000	0.0463	40	41	41
5	0.8	0.8416	0.6781	51	53	54
10	0.9	1.2816	0.9282	58	60	62
20	0.95	1.6449	1.1465	64	67	69
50	0.98	2.0537	1.4129	72	75	78
100	0.99	2.3263	1.6057	78	81	84
Two-sided Confidence Limits:	z_α =	1.96				

Fig. 5.9 ML estimation of the parameters, standard errors, Quantiles, and confidence limits for the 24 h, annual maximum rainfall data at meteorological station Chihuahua, Mexico

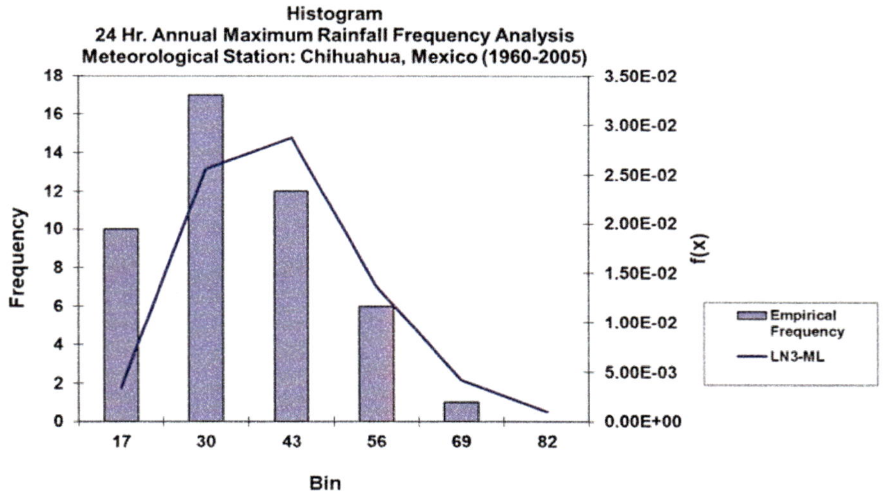

Fig. 5.10 Histogram and theoretical LN3 density for 24 h, annual maximum rainfall data at meteorological station Chihuahua, Mexico

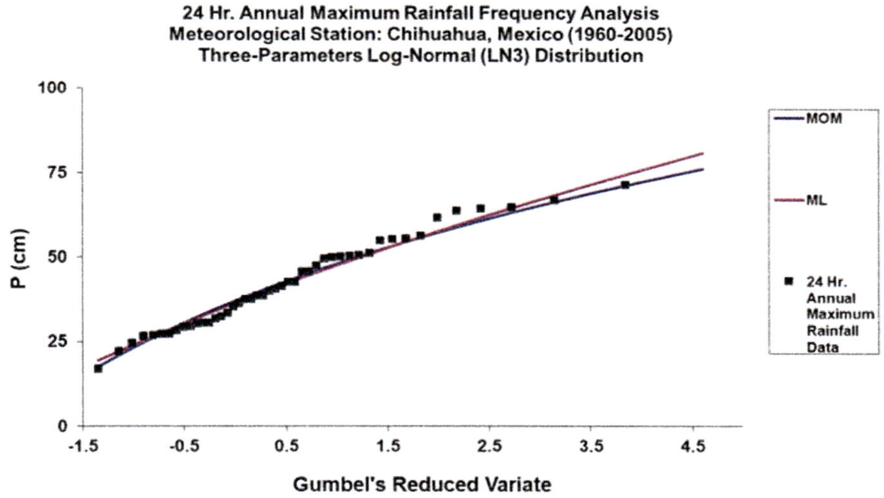

Fig. 5.11 Empirical and theoretical LN3 distributions for 24 h, annual maximum. rainfall data at meteorological station Chihuahua, Mexico

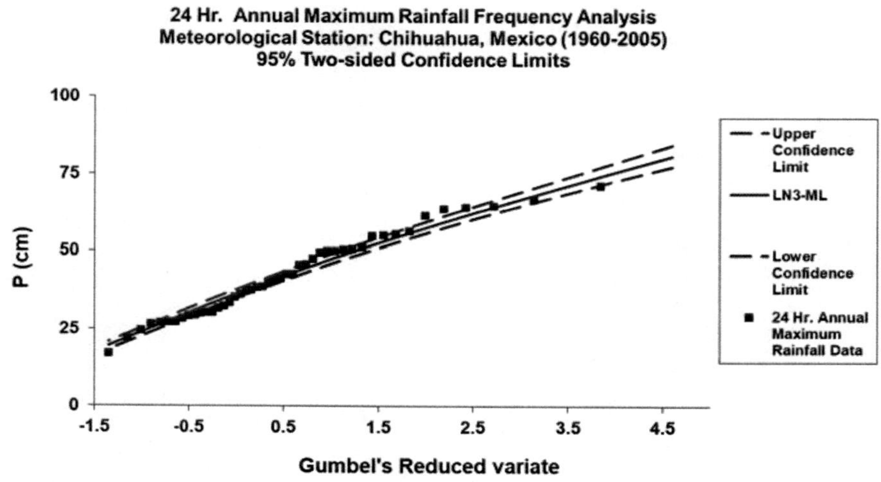

Fig. 5.12 Empirical and MOM and ML theoretical frequency curves and MOM and ML confidence limits for 24 h, annual maximum rainfall data at meteorological station Chihuahua, Mexico

A graphical depiction of the empirical and theoretical frequency curves and the confidence limits of the best fit provided, in this case that of MOM-ML, is shown in Fig. 5.12.

5.9.3 Maximum Significant Wave Height Frequency Analysis

By using the maximum significant wave height data, Castillo (1988), the following descriptive statistics for the data and its natural logarithms were obtained and are show in Fig. 5.13.

5.9.3.1 MOM Method

By using an Excel® worksheet, the results contained in Fig. 5.14, were obtained for the MOM method applied to the LN3 distribution for maximum significant wave height data.

5.9.3.2 ML Method

By using an Excel® worksheet, the results contained in Fig. 5.15, were obtained for the ML method applied to the LN3 distribution for maximum significant wave height data.

In Fig. 5.16 a comparison is made between the histogram and LN3 theoretical density for Maximum Significant Wave Height Data Sample.

A graph containing the empirical and theoretical LN3 frequency curves is shown in Fig. 5.17.

Maximum Significant Wave Height Frequency Analysis

Descriptive Statistics		*Descriptive Statistics of the Natural Logarithms*	
Mean	4.4247	Mean	1.3628
Sample Standard Error	0.3053	Sample Standard Error	0.0748
Median	4.1666	Median	1.4270
Mode	3.5509	Mode	1.2672
Standard Deviation	2.1585	Standard Deviation	0.5287
Sample Variance	4.6591	Sample Variance	0.2796
Kurtosis	1.0740	Kurtosis	0.6324
Skewness	0.9053	Skewness	-0.6102
Range	10.4485	Range	2.5478
Minimum	0.8870	Minimum	-0.1199
Maximum	11.3355	Maximum	2.4279
Sum	221.2326	Sum	68.1403
Count	50.0000	Count	50.0000
Largest(1)	11.3355	Largest(1)	2.4279
Smallest(1)	0.8870	Smallest(1)	-0.1199
Mean Confidence Level(95.0%)	0.6134	Mean Confidence Level(95.0%)	0.1503

Fig. 5.13 Descriptive statistics for the maximum significant wave height data sample

Maximum Significant Wave Height Frequency Analysis

Three-Parameters Log-Normal (LN3) Distribution

Method: MOM

Estimation of Parameters and Goodness of Fit Tests				
w =	0.6450		Skew(Ln(x-x₀)) =	-0.5022
n_z =	0.2933		n_x =	0.4878
Location Parameter =		-0.2025	m	
Scale Parameter =		0.4899		
Shape Parameter =		1.4207		
SEF =	0.21		MARD =	5.44

Design Values and Confidence Limits

Return Period Tr (years)	F(x)	v	Standard Normal Variate z_T	Logarithmic Squared Root Stand.Error S_T	Squared Root Stand.Error S_T (m)	Two-sided 95% Lower Limit (m)	Design Values H_T (m)	Two-sided 95% Upper Limit (m)
2	0.5	1.1774	0.0000	0.0980	0.4064	3	4	5
5	0.8	1.7941	0.8415	0.1140	0.7146	5	6	7
10	0.9	2.1460	1.2817	0.1323	1.0291	6	8	10
20	0.95	2.4477	1.6452	0.1504	1.3989	6	9	12
50	0.98	2.7971	2.0542	0.1728	1.9672	7	11	15
100	0.99	3.0349	2.3268	0.1887	2.4570	8	13	18
Two-sided Confidence Limits:	z_α =		1.96					

Fig. 5.14 MOM estimation of the parameters, standard errors, Quantiles, and confidence limits for the maximum significant wave height data sample

A graphical depiction of the empirical and theoretical frequency curves and the confidence limits of the best fit provided, in this case that of MOM-ML, is shown in Fig. 5.18.

Maximum Significant Wave Height
Three-Parameters Log-Normal (LN3) Distribution
Method: ML

Estimation of Parameters

Test Location Parameter =	-0.2025		Skew(Ln(x-x$_0$)) =		0.8924
F(x$_0$) =	-6.5596E-03		Log-likelihood function =	-353.0265	
Location Parameter =	-528.7500	m			
Scale Parameter =	0.0040				
Shape Parameter =	6.2788				
SEF =	0.51	MARD =	9.26	AIC =	712

Variance-Covariance Matrix

var(x$_0$) =	72.8604	cov(x$_0$,μ_y) =	-0.1367
var(μ_y) =	0.0003	cov(x$_0$,σ_y^2) =	0.0000
var(σ_y^2) =	0.0000	cov(μ_y,σ_y^2) =	0.0000
D =	1.3456E-04		

Design Values and Confidence Limits

Return Period Tr (years)	F(x)	Standard Normal Variate z_T	Squared Root Stand.Error S_T (m)	Two-sided 95% Lower Limit (m)	Design Values H_T (m)	Two-sided 95% Upper Limit (m)
2	0.5	0.0000	0.0002	4	4	4
5	0.8	0.8416	0.9940	4	6	8
10	0.9	1.2816	1.2287	5	7	10
20	0.95	1.6449	1.3940	5	8	11
50	0.98	2.0537	1.5602	6	9	12
100	0.99	2.3263	1.6624	6	9	13
Two-sided Confidence Limits:		z_α =	1.96			

Fig. 5.15 ML estimation of the parameters, standard errors, quantiles, and confidence limits for the maximum significant wave height data sample

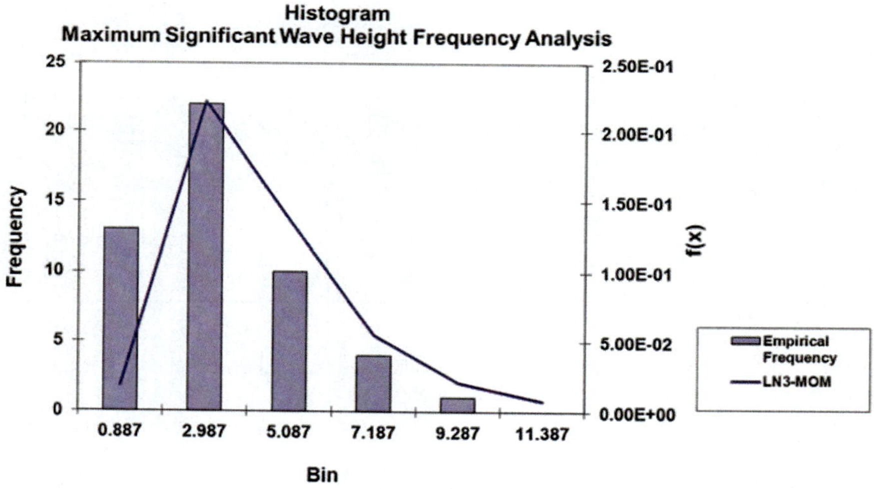

Fig. 5.16 Histogram and theoretical LN3 density for maximum significant wave height data sample

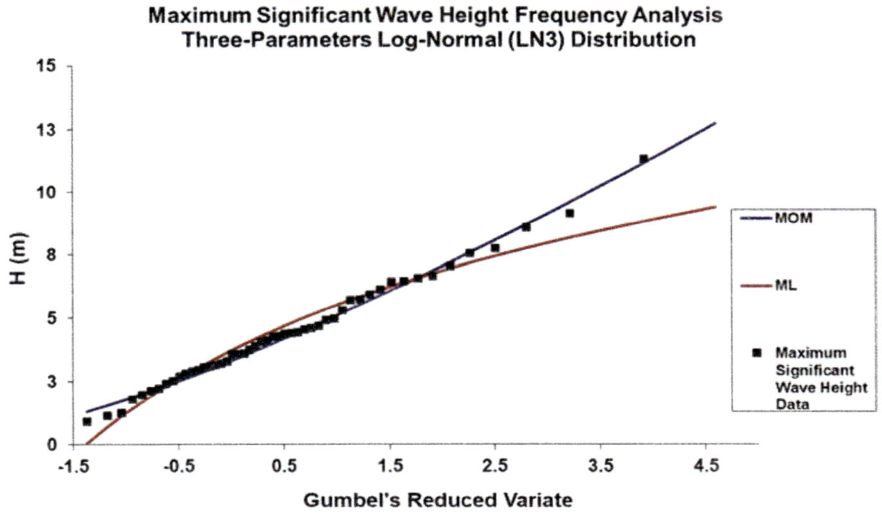

Fig. 5.17 Empirical and theoretical LN3 distributions for maximum significant wave height data sample

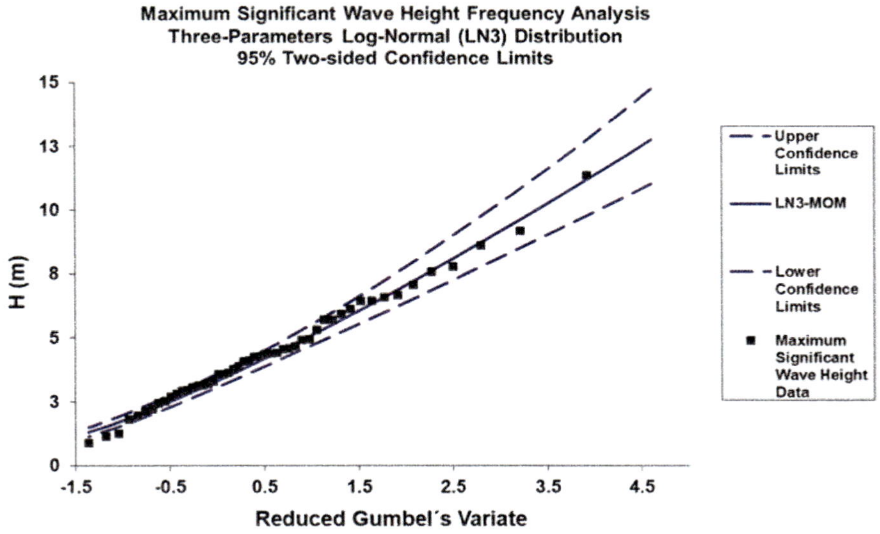

Fig. 5.18 Empirical and MOM and ML theoretical frequency curves and MOM and ML confidence limits for maximum significant wave height data sample

Maximum Annual Wind Speed Frequency Analysis			
Descriptive Statistics		*Descriptive Statistics of the Natural Logarithms*	
Mean	54.7964	Mean	3.9558
Sample Standard Error	2.9320	Sample Standard Error	0.0408
Median	47.9965	Median	3.8711
Mode	#N/D	Mode	#N/D
Standard Deviation	20.7327	Standard Deviation	0.2886
Sample Variance	429.8439	Sample Variance	0.0833
Kurtosis	12.5699	Kurtosis	3.6721
Skewness	3.0813	Skewness	1.6052
Range	121.5116	Range	1.4669
Minimum	36.4278	Minimum	3.5953
Maximum	157.9394	Maximum	5.0622
Sum	2739.8217	Sum	197.7906
Count	50.0000	Count	50.0000
Largest(1)	157.9394	Largest(1)	5.0622
Smallest(1)	36.4278	Smallest(1)	3.5953
Mean Confidence Level(95.0%)	5.8922	Mean Confidence Level(95.0%)	0.0820

Fig. 5.19 Descriptive statistics for the annual maximum wind speed data sample

5.9.4 Annual Maximum Wind Speed Frequency Analysis

By using the annual maximum wind speed data, Castillo (1988), the following descriptive statistics for the data and its natural logarithms were obtained and are show in Fig. 5.19.

5.9.4.1 MOM Method

By using an Excel® worksheet, the results contained in Fig. 5.20, were obtained for the MOM method applied to the LN3 distribution for annual maximum wind speed data.

5.9.4.2 ML Method

By using an Excel® worksheet, the results contained in Fig. 5.21, were obtained for the ML method applied to the LN3 distribution for annual maximum wind speed data.

In Fig. 5.22 a comparison is made between the histogram and LN3 theoretical density for annual maximum wind speed data sample.

A graph containing the empirical and theoretical LN3 frequency curves is shown in Fig. 5.23.

A graphical depiction of the empirical and theoretical frequency curves and the confidence limits of the best fit provided, in this case that of MOM, is shown in Fig. 5.24.

MAXIMUM ANNUAL WIND SPEED FREQUENCY ANALYSIS
Three-Parameters Log-Normal (LN3) Distribution
Method: MOM

Estimation of Parameters and Goodness of Fit Tests				
w =	0.2961		Skew(Ln(x-x$_0$)) =	0.5730
n$_z$ =	0.8338		n$_x$ =	0.3784
Location Parameter =		29.9298	m/s	
Scale Parameter =		0.6254		
Shape Parameter =		2.9943		
SEF =	8.80		MARD =	2.92

Return Period Tr (years)	F(x)	v	Standard Normal Variate z_T	Logarithmic Squared Root Stand.Error S_T	Squared Root Stand.Error S_T (m/s)	Two-sided 95% Lower Limit (m/s)	Design Values V_T (m/s)	Two-sided 95% Upper Limit (m/s)
2	0.5	1.1774	0.0000	0.1107	2.2160	46	50	54
5	0.8	1.7941	0.8415	0.1289	4.3676	55	64	72
10	0.9	2.1460	1.2817	0.1494	6.6776	61	74	88
20	0.95	2.4477	1.6452	0.1699	9.5380	67	86	105
50	0.98	2.7971	2.0542	0.1953	14.1817	74	102	130
100	0.99	3.0349	2.3268	0.2132	18.3838	79	116	152
Two-sided Confidence Limits:		z_α =	1.96					

Fig. 5.20 MOM estimation of the parameters, standard errors, Quantiles, and confidence limits for the annual maximum wind speed data sample

MAXIMUM ANNUAL WIND SPEED FREQUENCY ANALYSIS
Three-Parameters Log-Normal (LN3) Distribution
Method: ML

Estimation of Parameters					
Test Location Parameter =	30.0000		Skew(Ln(x-x$_0$)) =		2.8453
F(x$_0$) =	3.6582E-06		Log-likelihood function =	-322.7729	
Location Parameter =	-478.9432	m/s			
Scale Parameter =	0.0365				
Shape Parameter =	6.2792				
SEF =	11.93	MARD =	13.25	AIC =	652

Variance-Covariance Matrix			
var(x$_0$) =	72.7767	cov(x$_0$,μ_y) =	-0.1365
var(μ_y) =	0.0003	cov(x$_0$,σ_y^2) =	0.0004
var(σ_y^2) =	0.0000	cov(μ_y,σ_y^2) =	0.0000
D =	1.3471E-04		

Return Period Tr (years)	F(x)	Standard Normal Variate z_T	Squared Root Stand.Error S_T (m/s)	Two-sided 95% Lower Limit (m/s)	Design Values V_T (m/s)	Two-sided 95% Upper Limit (m/s)
2	0.5	0.0000	0.0151	54	54	54
5	0.8	0.8416	3.0861	65	71	77
10	0.9	1.2816	3.8699	72	80	88
20	0.95	1.6449	4.4428	79	87	96
50	0.98	2.0537	5.0391	86	96	106
100	0.99	2.3263	5.4168	91	102	112
Two-sided Confidence Limits:	z_α =		1.96			

Fig. 5.21 ML Estimation of the parameters, standard errors, Quantiles, and confidence limits for the annual maximum wind speed data sample

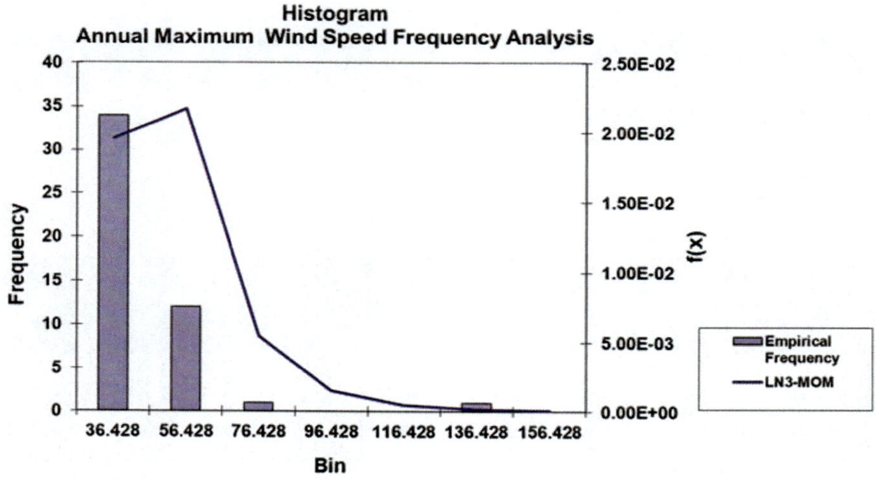

Fig. 5.22 Histogram and theoretical LN3 density for annual maximum wind speed data sample

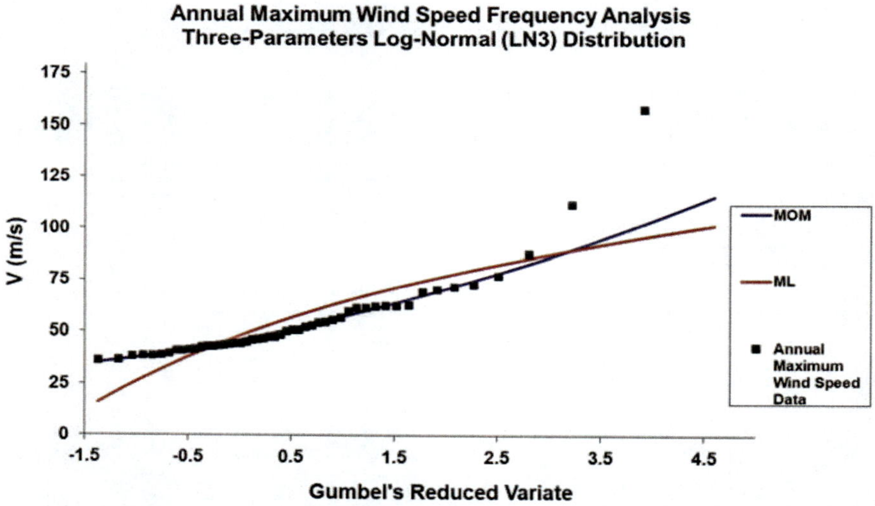

Fig. 5.23 Empirical and theoretical LN3 distributions for annual maximum wind speed data sample

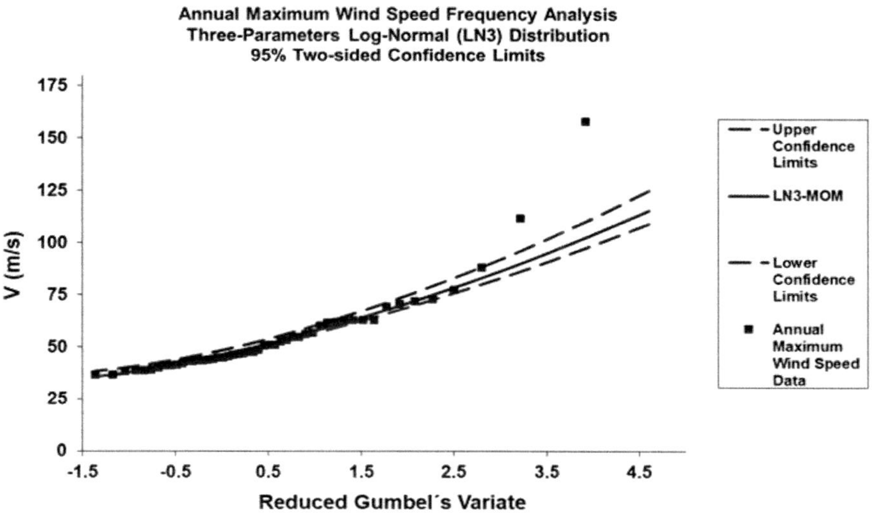

Fig. 5.24 Empirical and MOM and ML theoretical frequency curves and MOM and ML confidence limits for annual maximum wind speed data sample

Gamma Distribution 6

Today is only one day in all the days that will ever be. But what will happen in all the other days that ever come can depend on what you do today. E. Hemingway

6.1 Introduction

The Gamma (GAM) distribution is one of the members of the Gamma family of probability distribution functions, see Fig. 6.1. Other members of the GAM family are the Exponential, Pearson type III, and the Log-Pearson III distributions. The GAM distribution is a special case of the Pearson type III (PIII) distribution. In this case the location parameter x_0 has a value equal to zero.

Phien and Jivajirajah (1984) used the GAM for the computation of annual streamflow frequency analysis.

Yevjevich and Obeysekera (1984) used the GAM distribution to study the evaluation of skewness in hydrologic variables.

6.2 Chapter Objectives

After reading this chapter, you will know how to:

(a) Recognize the distribution and density functions of the GAM distribution
(b) Estimate the parameters of the GAM distribution
(c) Compute the quantiles and confidence limits of the GAM distribution
(d) Make a graphic display of your data and the GAM distribution
(e) Develop an application of all the above using Excel® spreadsheets.

J. A. Raynal Villaseñor, *Frequency Analyses of Natural Extreme Events*,
Earth and Environmental Sciences Library,
https://doi.org/10.1007/978-3-030-86390-6_6

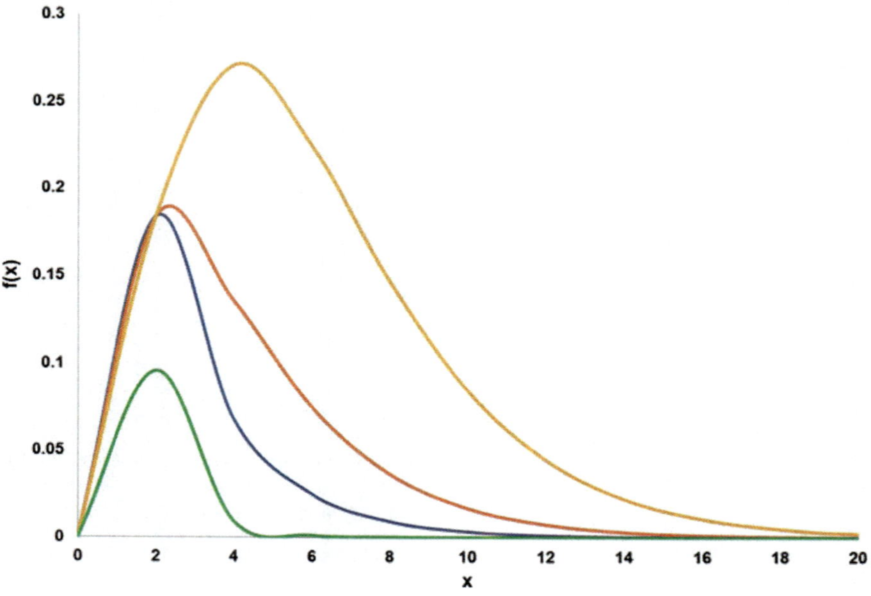

Fig. 6.1 Gamma family of distributions

6.3 Probability Distribution and Density Functions

The probability distribution function of the Gamma (GAM) distribution is:

$$F(x) = \frac{1}{\alpha^{\beta}\Gamma(\beta)}\int_{0}^{x} x^{\beta-1}\exp\left(-\frac{x}{\alpha}\right)dx \tag{6.1}$$

where α and β are the scale and shape parameters, respectively, and $\Gamma(.)$ is the complete Gamma function. $F(x)$ is the probability distribution function of the random variable x and in the case of flood frequency analysis is equal to the non-exceedance probability, $Pr(X \leq x)$. Such parameters are restricted to be greater than zero, that is $\alpha > 0$ and $\beta > 0$. The domain of variable x in this distribution is $0 < x < \infty$.

The complete Gamma function is defined as:

$$\Gamma(\beta) = \int_{0}^{\infty} z^{\beta-1}\exp(z)dz \tag{6.2}$$

The complete Gamma function has the following properties, Abramowitz and Stegun (1965):

$$\Gamma(n) = (n-1)! \tag{6.3}$$

for n being a positive integer.

$$\Gamma(n+1) = n\Gamma(n) \tag{6.4}$$

for n > 0.

$$\Gamma(n) = \frac{\Gamma(n+1)}{n} \tag{6.5}$$

for n < 1.

The complete Gamma function can be computed by using the following approximation, Abramowitz and Stegun (1965):

$$\Gamma(n+1) = 1 + b_1 n + b_2 n^2 + \ldots + b_8 n^8 + \varepsilon(n) \tag{6.6}$$

for $0 \leq n \leq 1$ and:

$b_1 = -0.57719$; $b_5 = -0.7567$; $b_2 = 0.98820$; $b_6 = 0.48219$.
$b_3 = -0.89705$; $b_7 = -0.19352$; $b_4 = 0.91820$; $b_8 = 0.03586$.

the absolute error in this approximation is $|\varepsilon(n)| \leq 3 \times 10^{-7}$.

The probability density function for the GAM distribution is:

$$f(x) = \frac{1}{\alpha^{\beta}\Gamma(\beta)} x^{\beta-1} \exp\left(-\frac{x}{\alpha}\right) \tag{6.7}$$

where $f(x)$ is the probability density function of the random variable x.

6.4 Estimation of the Parameters

6.4.1 MOM Method

The population mean, variance and coefficient of skewness of the GAM distribution are:

$$\mu = \alpha\beta \tag{6.8}$$

$$\sigma^2 = \alpha^2\beta \tag{6.9}$$

$$\gamma = \frac{2\alpha}{|\alpha|\sqrt{\beta}} \tag{6.10}$$

So, these previous equations would serve to estimate the parameters of the GAM distribution by the MOM method as:

$$\hat{\alpha} = \frac{\hat{\sigma}^2}{\hat{\mu}} \tag{6.11}$$

$$\hat{\beta} = \left(\frac{\hat{\mu}}{\hat{\sigma}}\right)^2 \tag{6.12}$$

In the above equations the mean and the variance and/or the standard deviation can be estimated by the procedures outlined in Chap. 2.

6.4.1.1 Example of Application of Estimation of the Parameters of the GAM Distribution Using the MOM Method

Find the MOM estimators for the parameters of the GAM distribution for the flood sample data of gauging station Huites, Mexico (1942–1992), contained in Appendix A.

The following statistics can be obtained:

$$\hat{\mu} = \frac{1}{N}\sum_{i=1}^{N} x_i = 2498.96 \text{ m}^3/\text{s}$$

$$\hat{\sigma} = \left[\frac{1}{N}\sum_{i=1}^{N}(x_i - \hat{\mu})^2\right]^{1/2} = 2221.36 \text{ m}^3/\text{s}$$

then:

$$\hat{\alpha} = \frac{\hat{\sigma}^2}{\hat{\mu}} = \frac{(2221.36)^2}{2498.96} = 1974.60 \text{ m}^3/\text{s}$$

$$\hat{\beta} = \left(\frac{\hat{\mu}}{\hat{\sigma}}\right)^2 = \left(\frac{2498.96}{2221.36}\right)^2 = 1.2656$$

Finally, the moment estimators of the parameters of the GAM distribution for the flood sample data of gauging station Huites, Mexico, are:

scale parameter:

$$\hat{\alpha} = 1974.60 \text{ m}^3/\text{s}$$

and:

shape parameter:

$$\hat{\beta} = 1.2656$$

6.4.2 ML Method

The likelihood function of the GAM distribution is:

$$L(x, \alpha, \beta) = \prod_{i=1}^{N}\left\{\frac{1}{\alpha^{\beta}\Gamma(\beta)}x^{\beta-1}\exp\left(-\frac{x}{\alpha}\right)\right\} \tag{6.13}$$

The Log-likelihood function of the GAM distribution is:

$$LL(x, \alpha, \beta) = -N\beta Ln(\alpha) - NLn[\Gamma(\beta)] + (\beta - 1)\sum_{i=1}^{N} Ln(x_i) - \frac{1}{\alpha}\sum_{i=1}^{N} x_i \tag{6.14}$$

Now, taking the partial derivatives of this equation with respect to parameters α and β, and set them equal to zero:

$$\frac{\partial LL}{\partial\alpha} = -\frac{N\beta}{\alpha} + \frac{1}{\alpha^2}\sum_{i=1}^{N} x_i = 0 \tag{6.15}$$

$$\frac{\partial LL}{\partial\beta} = -NLn(\alpha) - \frac{N\Gamma'(\beta)}{\Gamma(\beta)} + \sum_{i=1}^{N} Ln(x_i) = 0 \tag{6.16}$$

The solution of the system of equations conformed by Eqs. (6.15) and (6.16) provides the maximum likelihood estimators for the parameters of the GAM distribution:

$$\hat{\alpha} = \frac{\hat{\mu}}{\hat{\beta}} \tag{6.17}$$

$$F\left(\hat{\beta}\right) = \hat{\mu}_y - Ln(\hat{\mu}) + Ln\left(\hat{\beta}\right) - \psi\left(\hat{\beta}\right) = 0 \tag{6.18}$$

where $\hat{\mu}_y$ is the mean of the natural logarithms of the sample data and $\psi(\beta)$ is the so-called digamma function, that can be approximated by, Abramowitz and Stegun (1965):

$$\psi\left(\hat{\beta}\right) = \frac{\Gamma'\left(\hat{\beta}\right)}{\Gamma\left(\hat{\beta}\right)} = Ln\left(\hat{\beta}\right) - \frac{1}{2\hat{\beta}} - \frac{1}{12\hat{\beta}^2} + \frac{1}{120\hat{\beta}^4} - \frac{1}{252\hat{\beta}^6} \tag{6.19}$$

Condie and Nix (1975) found that it was necessary to include a recurrence equation to preserve accuracy at low values of β, so Eq. (6.19) it is transformed into:

$$\psi\left(\hat{\beta}\right) = \frac{\Gamma'\left(\hat{\beta}\right)}{\Gamma\left(\hat{\beta}\right)} = Ln\left(\hat{\beta}+2\right) - \frac{1}{2\left(\hat{\beta}+2\right)} - \frac{1}{12(\hat{\beta}+2)^2} + \frac{1}{120(\hat{\beta}+2)^4}$$

$$-\frac{1}{252(\hat{\beta}+2)^6}-\frac{1}{\left(\hat{\beta}+1\right)}-\frac{1}{\hat{\beta}} \tag{6.20}$$

Equation (6.18) must be solved by a procedure of finding the roots of such an equation, again the bisection method will give good results in solving this equation.

6.4.2.1 Example of Application of Estimation of the Parameters of the GAM Distribution Using the ML Method

Find the ML estimators for the parameters of the GAM distribution for the flood sample data of gauging station Huites, Mexico (1942–1992), contained in Appendix A.

The following statistics can be obtained:

$$\hat{\mu}=\frac{1}{N}\sum_{i=1}^{N}x_i=2498.96\ \mathrm{m}^3/\mathrm{s}$$

$$\hat{\sigma}=\left[\frac{1}{N}\sum_{i=1}^{N}(x_i-\hat{\mu})^2\right]^{1/2}=2221.36\ \mathrm{m}^3/\mathrm{s}$$

$$\hat{\mu}_l=\hat{\mu}_y=\frac{1}{N}\sum_{i=1}^{N}Ln(x_i)=7.5383$$

then the following equation needs to be solved:

$$\begin{aligned}F\left(\hat{\beta}\right)&=\hat{\mu}_y-Ln(\hat{\mu})+Ln\left(\hat{\beta}\right)-\psi\left(\hat{\beta}\right)\\&=7.5383-Ln(2498.96)-Ln\left(\hat{\beta}\right)-\psi\left(\hat{\beta}\right)=0\end{aligned}$$

now, using a method for finding the roots of the previous equation, the following solution is found:

$$\hat{\beta}=1.9021$$

and then:

$$\hat{\alpha}=\frac{\hat{\mu}}{\hat{\beta}}=\frac{2498.96}{1.9021}=1313.82\,\mathrm{m}^3/\mathrm{s}$$

Finally, the moment estimators of the parameters of the GAM distribution for the flood sample data of gauging station Huites, Mexico, are:

scale parameter:

$$\hat{\alpha} = 1313.82\,\text{m}^3/\text{s}$$

and:

shape parameter:

$$\hat{\beta} = 1.9021$$

6.5 Estimation of Quantiles for the GAM Distribution

By using the following reduced value, Kite (1988):

$$y = \frac{x}{\alpha} \tag{6.21}$$

into Eq. (6.1) produces the following expression, Kite (1988):

$$F(y) = \frac{1}{\Gamma(\beta)} \int_0^{y_0} y^{\beta-1} \exp(-y)dy \tag{6.22}$$

and Abramowitz and Stegun (1965) have shown that:

$$F(y) = F(\chi^2|v) \tag{6.23}$$

where $F(\chi^2|v)$ is the chi-square distribution with 2β degrees of freedom and $\chi^2 = 2y$.

So, the reduced event magnitude, y_0, may be obtained as, Kite (1988):

$$y_0 = \frac{\chi^2}{2} \tag{6.24}$$

and the expected event of magnitude associated with a given return period T_r is:

$$x_T = \frac{\alpha\chi^2}{2} \tag{6.25}$$

but using the Wilson-Hilferty approximation, Kendall and Stuart (1963):

$$\left[\left(\frac{\chi^2}{v}\right)^{\frac{1}{3}} + \frac{2}{9v} - 1\right]\left(\frac{9v}{2}\right)^{\frac{1}{2}} \sim N(0, 1) \tag{6.26}$$

and this expression is valid for $v > 30$. So, the values of χ^2 can be approximated as:

$$\chi^2 \approx v\left[1 - \frac{2}{9v} + z_T\sqrt{\frac{2}{9v}}\right]^3 \tag{6.27}$$

where z_T is the standard normal (N(0, 1)) variate corresponding to a certain return period, T_r, and from Eqs. (6.25) and (6.27) one may obtain:

$$x_T = \hat{\alpha}\hat{\beta}\left[1 - \frac{1}{9\hat{\beta}} + z_T\sqrt{\frac{1}{9\hat{\beta}}}\right]^3 \tag{6.28}$$

By using engineering terms, the following expression is widely used:

$$Q_T = \hat{\alpha}\hat{\beta}\left[1 - \frac{1}{9\hat{\beta}} + z_T\sqrt{\frac{1}{9\hat{\beta}}}\right]^3 \tag{6.29}$$

where Q_T is design value corresponding to a certain value of the return period T_r.

6.5.1 Examples of Estimation of MOM and ML Quantiles for the GAM Distribution

Find the MOM and ML estimators of the quantiles 2, 5, 10, 20, 50, 100, 500, and 1000, years of return period, for the GAM distribution for the flood sample data of gauging station Huites, Mexico (1942–1992), contained in Appendix A.

The estimation of MOM and ML quantiles for the GAM distribution is made by using the parameters estimated in the preceding sections and then inserting the parameters into Eq. (6.29). Table 6.1 summarizes these results.

Table 6.1 Estimation of MOM and ML Quantiles for the GAM distribution for Gauging station Huites, Mexico

Tr (Years)	MOM	ML
	Q_T (m^3/s)	Q_T (m^3/s)
2	1897	2086
5	3916	3751
10	5389	4897
20	6853	6002
50	8791	7432
100	10,267	8501
500	13,743	10,968
1000	15,263	12,031

6.6 Goodness of Fit Test

The standard error of fit (SEF) for the GAM has the following form:

$$SEF = \left[\frac{\sum_{i=1}^{N} (xi - yi)^2}{(N-2)} \right]^{1/2} \tag{6.30}$$

while the mean absolute relative deviation (MARD) remains the same as it was defined in Chap. 2:

$$MARD = \frac{100}{N} \sum_{i=1}^{N} \left| \frac{(x_i - y_i)}{x_i} \right| \tag{6.31}$$

where x_i are the sample historical values, y_i are the distribution function values corresponding to the same return periods of the historical values, N is the sample size.

6.6.1 Examples of Application of the SEF and MARD to the MOM and ML Estimators of the Parameters of the GAM Distribution

Find the values of the SEF and MARD of the MOM and ML estimators of the parameters of the GAM distribution for the flood sample data of gauging station Huites, Mexico (1942–1992), contained in Appendix A.

The values of SEF and MARD for the MOM and ML estimators of the parameters of the GAM distribution for the flood sample data of gauging station Huites, Mexico, have been obtained through the application of Eqs. (6.30) and (6.31) using the parameters obtained in previous examples. The results are as follows:

(a) MOM Method

$$SEF = \left[\frac{\sum_{i=1}^{N} (xi - yi)^2}{(N-2)} \right]^{1/2} = \left[\frac{344929.1969}{(51-2)} \right]^{\frac{1}{2}} = 587$$

$$MARD = \frac{100}{N} \sum_{i=1}^{N} \left| \frac{(x_i - y_i)}{x_i} \right| = \frac{(100)(11.4870)}{51} = 23$$

(b) ML Method

$$SEF = \left[\frac{\sum_{i=1}^{N} (xi - yi)^2}{(N-2)} \right]^{1/2} = \left[\frac{669041.2447}{(51-2)} \right]^{\frac{1}{2}} = 818$$

$$MARD = \frac{100}{N}\sum_{i=1}^{N}\left|\frac{(x_i - y_i)}{x_i}\right| = \frac{(100)(9.3338)}{51} = 18$$

So, in the GAM distribution the MOM estimators produce a better fitting if the SEF measure is the criterion of goodness of fit, but the maximum likelihood estimators give the MARD's least value. Given that the MOM and ML values of the MARD measure are very close, one may choose the MOM method as the best choice given that its SEF value is 40% less that the ML method SEF value and its MARD value is just 28% bigger than the corresponding value for the ML method.

6.7 Estimation of Confidence Limits for the GAM Distribution

By assuming that the quantiles are normally distributed, the following form can be used to set the confidence limits on such quantiles:

$$x_l = x_T \pm z_\alpha S_T \tag{6.32}$$

where x_l is the upper or lower confidence limit, depending on the sign in the preceding formula (+) for the upper confidence limit and (−) for the lower confidence limit, z_α is the standard normal variate corresponding to a confidence level α, and S_T is squared root of the standard error of the estimate. The evaluation procedures of the standard errors of the estimates will be described in the following subsections.

6.8 Estimation of Standard Errors for the GAM Distribution

6.8.1 MOM Method

The general form of the MOM estimator of the standard error of the estimate of a two-parameter distribution is, Kite (1988):

$$S_T^2 = \left(\frac{\partial x_T}{\partial m_1'}\right)^2 var(m_1') + \left(\frac{\partial x_T}{\partial m_2}\right)^2 var(m_2) + \left(\frac{\partial x_T}{\partial m_3}\right)^2 var(m_3) + 2\left(\frac{\partial x_T}{\partial m_1'}\right)\left(\frac{\partial x_T}{\partial m_2}\right) cov(m_1', m_2) \tag{6.33}$$

and Eq. (6.33) can be simplified in terms of the frequency factor, K_T, as, Kite (1988):

$$S_T^2 = \frac{\mu_2}{N}\left[1 + K_T\hat{\gamma} + \frac{K_T^2}{4}(\hat{\kappa} - 1)\right] \tag{6.34}$$

which further simplifies to, Bobée (1973):

$$S_T^2 = \frac{\mu_2}{N}\left[1 + \frac{K_T^2}{2}(1 + 3CV^2) + 2K_T CV\right] \tag{6.35}$$

but given that the following relationship exists for the GAM distribution:

$$CV = \frac{\hat{\gamma}}{2} \tag{6.36}$$

the standard error of the estimate of a two-parameter distribution can be expressed in terms of the skewness coefficient as:

$$S_T^2 = \frac{\alpha^2\beta}{N}\left[1 + K_T\hat{\gamma} + \frac{K_T^2}{2}\left(\frac{3\hat{\gamma}^2}{4} + 1\right)\right] \tag{6.37}$$

and the frequency factor for the GAM distribution is, Kite (1988):

$$K_T = z_T + \left(z_T^2 - 1\right)\left(\frac{\hat{\gamma}}{6}\right) + \frac{\left(z_T^3 - 6z_T\right)}{3}\left(\frac{\hat{\gamma}}{6}\right)^2 - \left(z_T^2 - 1\right)\left(\frac{\hat{\gamma}}{6}\right)^3 + z_T\left(\frac{\hat{\gamma}}{6}\right)^4 - \frac{1}{3}\left(\frac{\hat{\gamma}}{6}\right)^5 \tag{6.38}$$

6.8.1.1 Example of Estimation of MOM Standard Errors and the Two-Sided 95% Confidence Limits for the GAM Distribution

Find the MOM estimators of the standard errors and the two-sided 95% confidence limits for the GAM distribution for the flood sample data of gauging station Huites, Mexico (1942–1992), contained in Appendix A.

The estimation of MOM estimators of the standard errors and the two-sided 95% confidence limits for the GAM distribution is made by using the MOM estimators of the parameters and then inserted in the Eqs. (6.38) and (6.37), using selected values of the return intervals. Table 6.2 summarizes these results.

6.8.2 ML Method

The general form of the ML estimator of the standard error of the estimate of a two-parameter distribution is:

$$S_T^2 = \left(\frac{\partial x_T}{\partial \alpha}\right)^2 var(\alpha) + \left(\frac{\partial x_T}{\partial \beta}\right)^2 var(\beta) + 2\left(\frac{\partial x_T}{\partial \alpha}\right)\left(\frac{\partial x_T}{\partial \beta}\right) cov(\alpha, \beta) \tag{6.39}$$

Table 6.2 Estimation of MOM Standard Errors and the two-sided 95% confidence limits for the GAM distribution for Gauging Station Huites, Mexico

Return period Tr (years)	Squared root stand. error S_T (m^3/s)	Two-sided 95% lower limit (m^3/s)	Design values Q_T (m^3/s)	Two-sided 95% upper limit (m^3/s)
2	223.4520	1459	1897	2335
5	537.0002	2864	3916	4969
10	859.8045	3704	5389	7075
20	1204.7080	4491	6853	9214
50	1682.9331	5492	8791	12,089
100	2058.4212	6233	10,267	14,302
500	2969.2111	7923	13,743	19,563
1000	3376.6390	8645	15,263	21,881

Two-sided Limits: $z_\alpha = 1.96$

the first-order partial derivatives of x_T with respect to the parameters of the GAM distribution are, Kite (1988):

$$\frac{\partial x_T}{\partial \alpha} = \left[\beta^{\frac{1}{3}} - \frac{1}{9\beta^{\frac{2}{3}}} + \frac{z_T}{3\beta^{\frac{1}{6}}}\right] \tag{6.40}$$

$$\frac{\partial x_T}{\partial \beta} = \alpha\left[1 - \frac{1}{9\beta} + \frac{z_T}{3\beta^{\frac{1}{2}}}\right]^2\left[1 + \frac{2}{9\beta} - \frac{z_T}{6\beta^{\frac{1}{2}}}\right] \tag{6.41}$$

Now, the second-order partial derivatives of the log-likelihood function with respect to the parameters of the GAM distribution are, Kite (1988):

$$\frac{\partial^2 LL}{\partial \alpha^2} = -\frac{N\beta}{\alpha^2} \tag{6.42}$$

$$\frac{\partial^2 LL}{\partial \beta^2} = -N\psi'(\beta) \tag{6.43}$$

$$\frac{\partial^2 LL}{\partial \alpha \partial \beta} = -\frac{N}{\alpha} \tag{6.44}$$

Then the information matrix is:

$$[I] = N\begin{bmatrix} \frac{\beta}{\alpha^2} & \frac{1}{\alpha} \\ \frac{1}{\alpha} & \psi'(\beta) \end{bmatrix} \tag{6.45}$$

and the variances and covariance of the variance–covariance matrix of the parameters for the GAM distribution are:

$$var(\alpha) = \frac{\alpha^2 \psi'(\beta)}{N[\beta\psi'(\beta) - 1]} \tag{6.46}$$

$$var(\beta) = \frac{\beta}{N[\beta\psi'(\beta) - 1]} \tag{6.47}$$

$$cov(\alpha, \beta) = -\frac{\alpha}{N[\beta\psi'(\beta) - 1]} \tag{6.48}$$

The substitution of Eqs. (6.46) to (6.48) and (6.40) and (6.41) into Eq. (6.39) will provide the ML estimator of the standard error of the estimate for the GAM distribution.

6.8.2.1 Example of Estimation of ML Standard Errors and the Two-Sided 95% Confidence Limits for the GAM Distribution

Find the ML estimators of the standard errors and the two-sided 95% confidence limits for the GAM distribution for the flood sample data of gauging station Huites, Mexico (1942–1992), contained in Appendix A.

The estimation of ML estimators of the standard errors and the two-sided 95% confidence limits for the GAM distribution is made by using the ML estimators of the parameters and then inserted in the Eqs. (6.46) to (6.48) and (6.40) and (6.41) into Eq. (6.39), using selected values of the return intervals. Table 6.3 summarizes these results.

Table 6.3 Estimation of ML standard errors and the two-sided 95% confidence limits for the GAM distribution for Gauging Station Huites, Mexico

Return period Tr (years)	Squared root stand. error S_T (m3/s)	Two-sided 95% lower limit (m3/s)	Design values Q_T (m^3/s)	Two-sided 95% upper limit (m3/s)
2	223.4520	1459	1897	2335
5	537.0002	2864	3916	4969
10	859.8045	3704	5389	7075
20	1204.7080	4491	6853	9214
50	1682.9331	5492	8791	12,089
100	2058.4212	6233	10,267	14,302
500	2969.2111	7923	13,743	19,563
1000	3376.6390	8645	15,263	21,881

Two-sided Limits: $z_\alpha = 1.96$

6.9 Examples of Application for the GAM Distribution Using Excel® Spreadsheets

By using the flood data from gauging station Huites, Mexico the following descriptive statistics were obtained and are displayed in Fig. 6.2.

6.9.1 Flood Frequency Analysis

6.9.1.1 MOM Method

By using an Excel® worksheet, the results contained in Fig. 6.3, were obtained for the MOM method applied to the GAM distribution for gauging station Huites, Mexico.

6.9.1.2 ML Method

By using an Excel® worksheet, the results contained in Fig. 6.4, were obtained for the ML method applied to the GAM distribution for gauging station Huites, Mexico.

In Fig. 6.5 a comparison is made between the histogram and GAM theoretical density for flood sample of Huites, Mexico.

A graph containing the empirical and theoretical GAM frequency curves is shown in Fig. 6.6.

A graphical depiction of the empirical and theoretical frequency curves and the confidence limits of the best fit provided, in this case that of MOM is shown in Fig. 6.7.

	A	B	C	D
1	FLOOD FREQUENCY ANALYSIS			
2	Gauging Station: Huites, Sinaloa, Mexico (1942-1992)			
3				
4	*Descriptive Statistics*		*Descriptive Statistics of the Natural Logarithms*	
5	Mean	2498.9627	Mean	7.5383
6	Sample Standard Error	311.0530	Sample Standard Error	0.1024
7	Median	1775.7000	Median	7.4819
8	Mode	#N/D	Mode	#N/D
9	Standard Deviation	2221.3626	Standard Deviation	0.7312
10	Sample Variance	4934451.7500	Sample Variance	0.5347
11	Kurtosis	3.6701	Kurtosis	0.0081
12	Skewness	2.0123	Skewness	0.4627
13	Range	9707.5000	Range	3.1787
14	Minimum	421.8000	Minimum	6.0445
15	Maximum	10129.3000	Maximum	9.2232
16	Sum	127447.1000	Sum	384.4530
17	Count	51.0000	Count	51.0000
18	Largest(1)	10129.3000	Largest(1)	9.2232
19	Smallest(1)	421.8000	Smallest(1)	6.0445
20	Mean Confidence Level(95.0%)	624.7683	Mean Confidence Level(95.0%)	0.2057

Fig. 6.2 Descriptive statistics of the flood sample of Gauging Station Huites, Mexico

FLOOD FREQUENCY ANALYSIS
Gauging Station: Huites, Sinaloa, Mexico (1942-1992)
Gamma (GAM) Distribution
Method: MOM

Estimation of Parameters

Skewness Coef. (Gamma Correction) =	2.4194		
Scale Parameter =	1974.6000	m³/s	
Shape Parameter =	1.2656		
Γ(β) =	0.9033		
SEF =	581	MARD =	18

Design Values and Confidence Limits

Return Period Tr (years)	F(x)	Standard Normal Variate z_T	Frequency Factor K_T	Squared root Stand. Error S_T (m³/s)	Two-sided 95% Lower Limit (m³/s)	Design Values Q_T (m³/s)	Two-sided 95% Upper Limit (m³/s)
2	0.5	0.0000	-0.3412	217.3465	1471	1897	2323
5	0.8	0.8416	0.5205	537.7870	2863	3917	4971
10	0.9	1.2816	1.2261	880.7820	3662	5389	7115
20	0.95	1.6449	1.9670	1251.4295	4398	6851	9304
50	0.98	2.0537	2.9927	1769.9368	5320	8789	12258
100	0.99	2.3263	3.7999	2179.7566	5992	10265	14537
500	0.998	2.8782	5.7665	3180.7548	7506	13741	19975
1000	0.999	3.0902	6.6498	3630.9439	8144	15261	22378
Two-sided Confidence Limits:		z_α =	1.96				

Fig. 6.3 MOM estimators for the parameters, standard errors, quantiles, and the confidence limits of the GAM distribution for the flood sample of Huites, Mexico

FLOOD FREQUENCY ANALYSIS
Gauging Station: Huites, Sinaloa, Mexico (1942-1992)
Gamma Distribution
Method: ML

Estimation of Parameters

Test Shape Parameter =	1.9021	Γ(β) =	0.9624		
F(β) =	0.0000	ψ(β) =	0.3576	Log-likelihood function =	-444.8155
Scale Parameter =	1313.8189	m³/s			
Shape Parameter =	1.9021				
SEF =	817	MARD =	16	AIC =	894

Variance-Covariance Matrix

var(α) =	7.5799E+04	cov(α, β) =	-8.3975E+01	var(β) =	1.2157E-01	ψ'(β) =	0.6870

Design Values and Confidence Limits

Return Period Tr (years)	F(x)	Standard Normal Variate z_T	$\partial x_T/\partial\alpha$	$\partial x_T/\partial\beta$	Squared root Stand. Error S_T (m³/s)	Two-sided 95% Lower Limit (m³/s)	Design Values Q_T (m³/s)	Two-sided 95% Upper Limit (m³/s)
2	0.5	0.0000	1.1666	1300.8927	232.4121	1631	2086	2542
5	0.8	0.8416	1.4186	2098.8697	433.6258	2901	3751	4601
10	0.9	1.2816	1.5504	2616.1567	577.1173	3765	4896	6028
20	0.95	1.6449	1.6592	3099.6406	716.2182	4597	6001	7405
50	0.98	2.0537	1.7816	3708.2664	895.9784	5674	7430	9186
100	0.99	2.3263	1.8633	4153.7499	1029.9898	6480	8499	10518
500	0.998	2.8782	2.0285	5158.0222	1337.5716	8345	10967	13588
1000	0.999	3.0902	2.0920	5582.0347	1469.2080	9150	12029	14909
Two-sided Confidence Limits:		z_α =	1.96					

Fig. 6.4 ML estimators for the parameters, standard errors, quantiles, and the confidence limits of the GAM distribution for the flood sample of Huites, Mexico

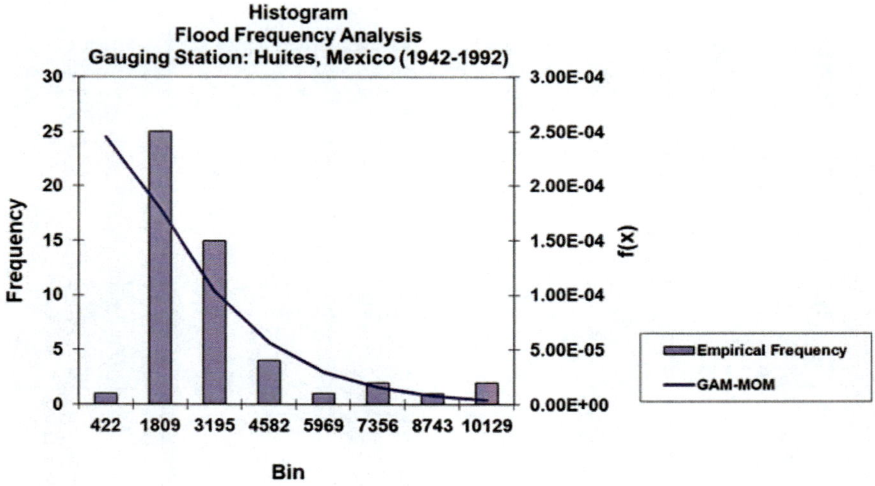

Fig. 6.5 Histogram and theoretical MOM GAM density for flood sample of Huites, Mexico

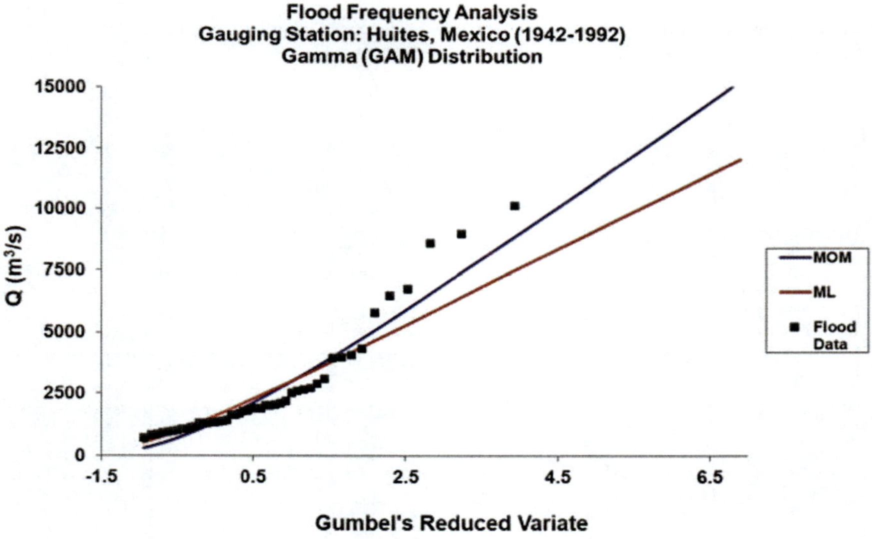

Fig. 6.6 Empirical and MOM-ML GAM theoretical curves for the flood sample of Huites, Mexico

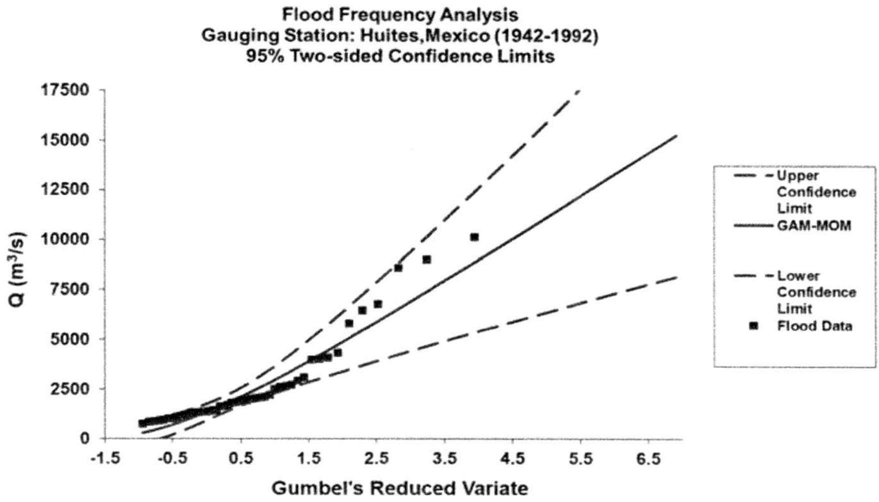

Fig. 6.7 Empirical and MOM and ML theoretical frequency curves and MOM and ML confidence limits of the GAM distribution for the flood data at Huites, Mexico

6.9.2 Rainfall Frequency Analysis

By using the 24 h, annual maximum rainfall data from rainfall gauging station Chihuahua, Mexico the following descriptive statistics were obtained and are displayed in Fig. 6.8.

6.9.2.1 MOM Method

By using an Excel® worksheet, the results contained in Fig. 6.9, were obtained for the MOM method applied to the GAM distribution for rainfall gauging station Chihuahua, Mexico.

6.9.2.2 ML Method

By using an Excel® worksheet, the results contained in Fig. 6.10, were obtained for the ML method applied to the GAM distribution for rainfall gauging station Chihuahua, Mexico.

In Fig. 6.11 a comparison is made between the histogram and GAM theoretical density for 24 h annual maximum rainfall data at Chihuahua, Mexico.

A graph containing the empirical and theoretical GAM frequency curves is shown in Fig. 6.12.

A graphical depiction of the empirical and theoretical frequency curves and the confidence limits of the best fit provided, in this case that of ML, is shown in Fig. 6.13.

24 Hr. Annual Maximum Rainfall Frequency Analysis			
Meteorological Station: Chihuahua, Mexico (1960-2005)			
Descriptive Statistics		***Descriptive Statistics of the Natural Logarithms***	
Mean	42.0454	Mean	3.6871
Sample Standard Error	1.9784	Sample Standard Error	0.0487
Median	40.2400	Median	3.6948
Mode	#N/D	Mode	#N/D
Standard Deviation	13.4184	Standard Deviation	0.3302
Sample Variance	180.0535	Sample Variance	0.1090
Kurtosis	-0.7252	Kurtosis	-0.4776
Skewness	0.3591	Skewness	-0.2507
Range	54.3000	Range	1.4283
Minimum	17.1200	Minimum	2.8402
Maximum	71.4200	Maximum	4.2686
Sum	1934.0900	Sum	169.6058
Count	46.0000	Count	46.0000
Largest(1)	71.4200	Largest(1)	4.2686
Smallest(1)	17.1200	Smallest(1)	2.8402
Mean Confidence Level(95.0%)	3.9848	Mean Confidence Level(95.0%)	0.0981

Fig. 6.8 Descriptive statistics for the 24 h, annual maximum rainfall data at meteorological station Chihuahua, Mexico

24 Hr. Annual Maximum Rainfall Frequency Analysis							
Meteorological Station: Chihuahua, Mexico (1960-2005)							
Gamma (GAM) Distribution							
Method: MOM							
Estimation of Parameters							
Skewness Coef. (Gamma Correction) =			0.4400				
Scale Parameter =		4.2824	cm				
Shape Parameter =		9.8183					
Γ(β) =	700161.3416						
SEF =	1.84		MARD =	3.63			
Design Values and Confidence Limits							
Return Period Tr (years)	F(x)	Standard Normal Variate z_T	Frequency Factor K_T	Squared root Stand. Error S_T (cm)	Two-sided 95% Lower Limit (cm)	Design Values P_T (cm)	Two-sided 95% Upper Limit (cm)
2	0.5	0.0000	-0.0729	1.9495	37	41	44
5	0.8	0.8416	0.8124	2.6062	48	53	58
10	0.9	1.2816	1.3184	3.1750	54	60	66
20	0.95	1.6449	1.7596	3.7260	59	66	74
50	0.98	2.0537	2.2819	4.4175	65	74	83
100	0.99	2.3263	2.6457	4.9151	70	79	89
Two-sided Confidence Limits:		z_α =	1.96				

Fig. 6.9 MOM estimation of the parameters, standard errors, Quantiles, and Confidence Limits for the 24 h, Annual Maximum rainfall data at meteorological station Chihuahua, Mexico

24 Hr. Annual Maximum Rainfall Frequency Analysis
Meteorological Station: Chihuahua, Mexico (1960-2005)
Gamma Distribution
Method: ML

Estimation of Parameters					
Test Shape Parameter =		9.8407		$\Gamma(\beta)$ =	715718.4846
$F(\beta)$ =	0.0000	$\psi(\beta)$ =	2.2349	Log-likelihood function =	-35.6720
Scale Parameter =		4.2726	cm		
Shape Parameter =		9.8407			
SEF =	1.85	MARD =	3.64	AIC =	75

Variance-Covariance Matrix			
var(α) =	5.2046E-01	cov(α, β) =	-1.1389E+00
var(β) =	2.6232E+00	$\psi'(\beta)$ =	0.1070

Design Values and Confidence Limits

Return Period Tr (years)	F(x)	Standard Normal Variate z_T	$\partial x_T/\partial \alpha$	$\partial x_T/\partial \beta$	Squared root Stand. Error S_T (cm)	Two-sided 0.95 Lower Limit (cm)	Design Values P_T (cm)	Two-sided 0.95 Upper Limit (cm)
2	0.5	0.0000	2.1186	4.2710	5.4383	30	41	51
5	0.8	0.8416	2.3102	5.3006	6.9704	39	53	66
10	0.9	1.2816	2.4104	5.8966	7.8648	44	60	75
20	0.95	1.6449	2.4932	6.4199	8.6539	49	66	83
50	0.98	2.0537	2.5863	7.0435	9.5983	55	74	93
100	0.99	2.3263	2.6484	7.4802	10.2616	59	79	99

Two-sided Confidence Limits: z_α = 1.96

Fig. 6.10 ML estimation of the parameters, standard errors, Quantiles, and Confidence Limits for the 24 h, Annual Maximum rainfall data at meteorological station Chihuahua, Mexico

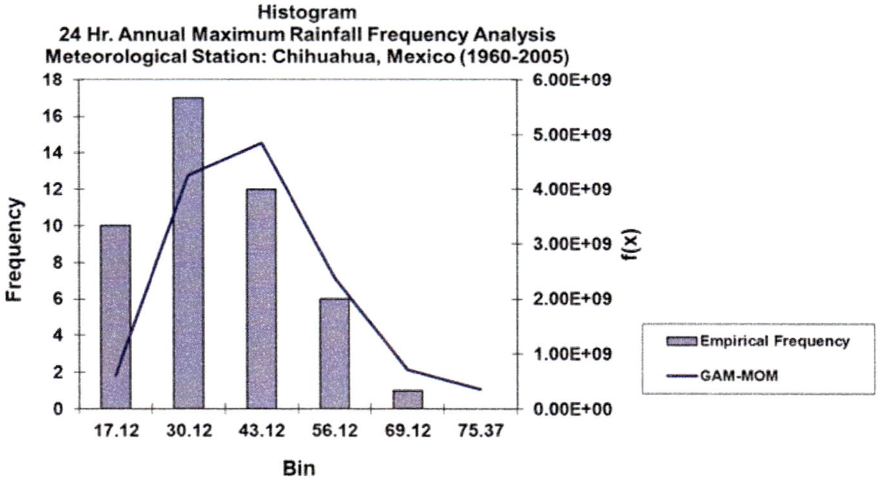

Fig. 6.11 Histogram and theoretical GAM Ddensity for 24 h, annual maximum rainfall data at meteorological station Chihuahua, Mexico

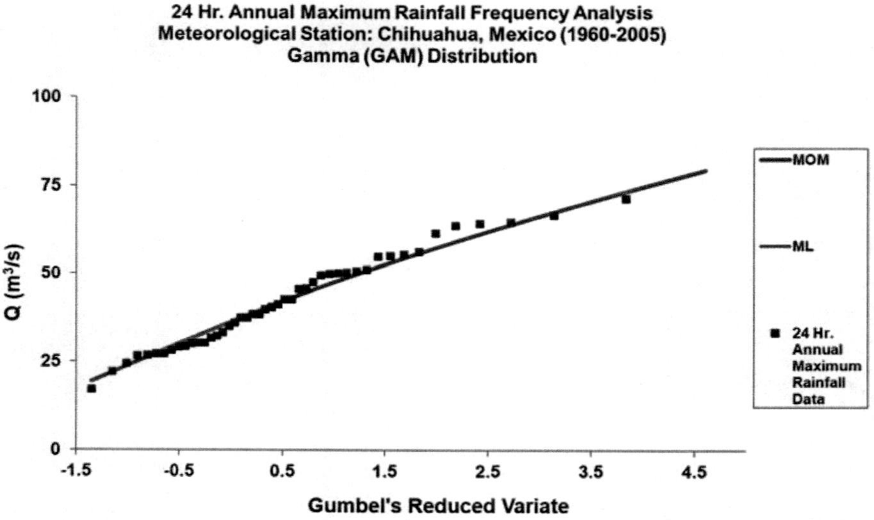

Fig. 6.12 Empirical and theoretical GAM distributions for 24 h, annual maximum rainfall data at meteorological station Chihuahua, Mexico

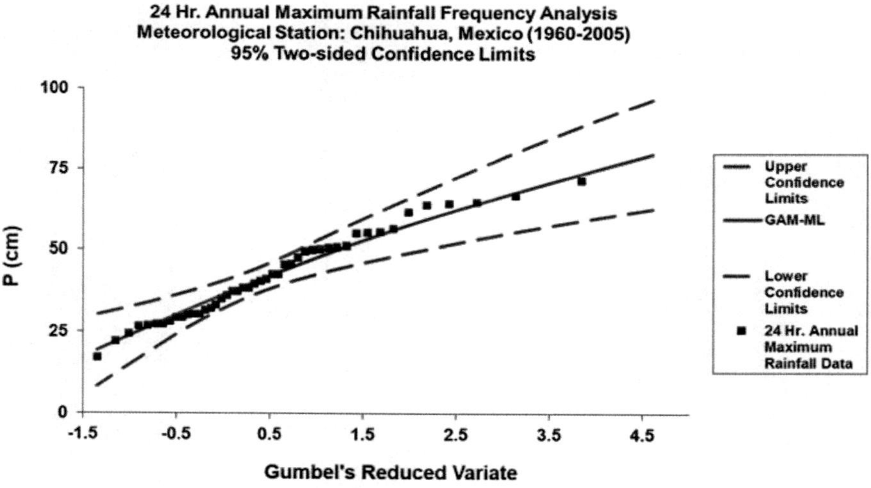

Fig. 6.13 Empirical and ML theoretical frequency curves and ML confidence limits for 24 h, annual maximum rainfall data at meteorological station Chihuahua, Mexico

6.9.3 Maximum Significant Wave Height Frequency Analysis

By using the maximum significant wave height data, Castillo (1988), the following descriptive statistics for the data and its natural logarithms were obtained and are show in Fig. 6.14.

6.9.3.1 MOM Method

By using an Excel® worksheet, the results contained in Fig. 6.15, were obtained for the MOM method applied to the GAM distribution for maximum significant wave height data.

6.9.3.2 ML Method

By using an Excel® worksheet, the results contained in Fig. 6.16, were obtained for the ML method applied to the GAM distribution for maximum significant wave height data.

In Fig. 6.17 a comparison is made between the histogram and GAM theoretical density for maximum significant wave height data sample.

A graph containing the empirical and theoretical GAM frequency curves is shown in Fig. 6.18.

A graphical depiction of the empirical and theoretical frequency curves and the confidence limits of the best fit provided, in this case that of MOM, is shown in Fig. 6.19.

	A	B	C	D
1	**Maximum Significant Wave Height Frequency Analysis**			
2				
3	*Descriptive Statistics*		*Descriptive Statistics of the Natural Logarithms*	
4	Mean	4.4247	Mean	1.3628
5	Sample Standard Error	0.3053	Sample Standard Error	0.0748
6	Median	4.1666	Median	1.4270
7	Mode	3.5509	Mode	1.2672
8	Standard Deviation	2.1585	Standard Deviation	0.5287
9	Sample Variance	4.6591	Sample Variance	0.2796
10	Kurtosis	1.0740	Kurtosis	0.6324
11	Skewness	0.9053	Skewness	-0.6102
12	Range	10.4485	Range	2.5478
13	Minimum	0.8870	Minimum	-0.1199
14	Maximum	11.3355	Maximum	2.4279
15	Sum	221.2326	Sum	68.1403
16	Count	50.0000	Count	50.0000
17	Largest(1)	11.3355	Largest(1)	2.4279
18	Smallest(1)	0.8870	Smallest(1)	-0.1199
19	Mean Confidence Level(95.0%)	0.6134	Mean Confidence Level(95.0%)	0.1503

Fig. 6.14 Descriptive statistics for the maximum significant wave height data sample

MAXIMUM SIGNIFICANT WAVE HEIGHT FREQUENCY ANALYSIS
Gamma (GAM) Distribution
Method: MOM

Estimation of Parameters				
Skewness Coef. (Gamma Correction) =	0.4400			
Scale Parameter =	1.0530	m		
Shape Parameter =	4.2020			
$\Gamma(\beta)$ =	69.6678			
SEF =	0.40		MARD =	7.97

Return Period Tr (years)	F(x)	Standard Normal Variate z_T	Frequency Factor K_T	Squared root Stand. Error S_T (m)	Two-sided 95% Lower Limit (m)	Design Values H_T (m)	Two-sided 95% Upper Limit (m)
2	0.5	0.0000	-0.0729	0.3136	3	4	5
5	0.8	0.8416	0.8124	0.4192	5	6	7
10	0.9	1.2816	1.3184	0.5107	6	7	8
20	0.95	1.6449	1.7596	0.5994	7	8	10
50	0.98	2.0537	2.2819	0.7106	8	10	11
100	0.99	2.3263	2.6457	0.7907	9	11	12

Two-sided Confidence Limits: z_α = 1.96

Fig. 6.15 MOM estimation of the parameters, standard errors, Quantiles, and Confidence Limits for the Maximum Significant wave height data sample

MAXIMUM SIGNIFICANT WAVE HEIGHT FREQUENCY ANALYSIS
Gamma (GAM) Distribution
Method: ML

Estimation of Parameters					
Test Shape Parameter =	4.2020		$\Gamma(\beta)$ =	70.2716	
$F(\beta)$ = 0.0000	$\psi(\beta)$ =	1.3119	Log-likelihood function =		-429.2059
Scale Parameter =	1.0522	m			
Shape Parameter =	4.2049				
SEF = 0.40		MARD =	7.96	AIC =	862

Variance-Covariance Matrix			
var(α) =	3.2439E-02	cov(α,β) =	-1.1490E-01
var(β) =	4.5914E-01	$\psi'(\beta)$ =	0.2683

Return Period Tr (years)	F(x)	Standard Normal Variate z_T	$\partial x_T/\partial\alpha$	$\partial x_T/\partial\beta$	Squared root Stand. Error S_T (m)	Two-sided 95% Lower Limit (m)	Design Values H_T (m)	Two-sided 95% Upper Limit (m)
2	0.5	0.0000	1.5713	1.0501	0.4552	3	4	5
5	0.8	0.8416	1.7922	1.4547	0.6904	5	6	7
10	0.9	1.2816	1.9076	1.7007	0.8370	6	7	9
20	0.95	1.6449	2.0029	1.9227	0.9708	7	8	10
50	0.98	2.0537	2.1102	2.1940	1.1361	8	10	12
100	0.99	2.3263	2.1818	2.3878	1.2550	8	11	13

Two-sided Confidence Limits: z_α = 1.96

Fig. 6.16 ML estimation of the parameters, standard errors, quantiles, and confidence limits for the maximum significant wave height data sample

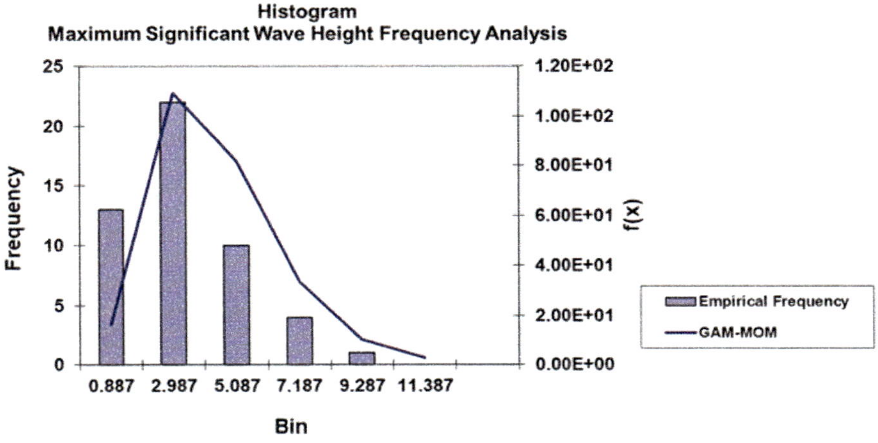

Fig. 6.17 Histogram and theoretical GAM density for maximum significant wave height data sample

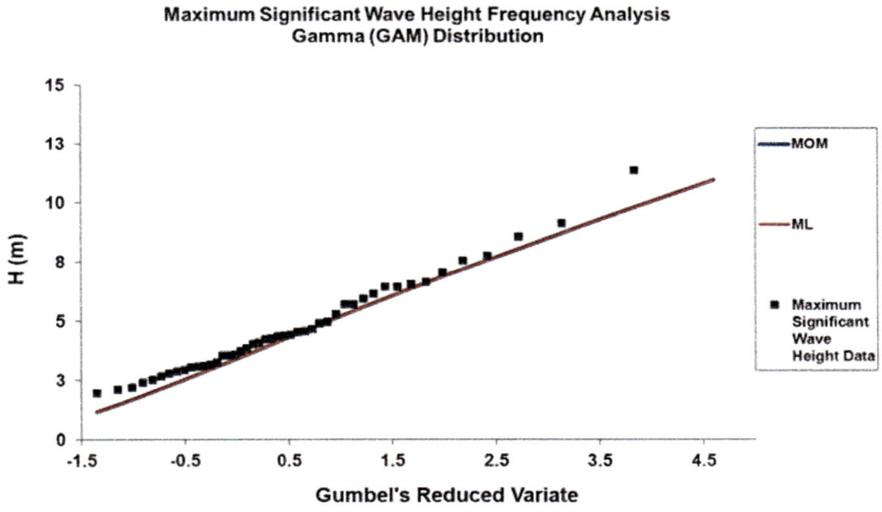

Fig. 6.18 Empirical and theoretical GAM distributions for maximum significant wave height data sample

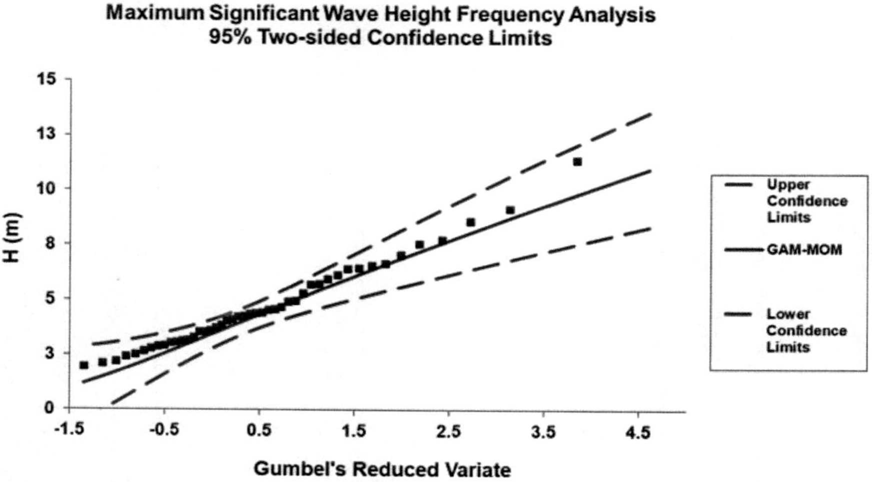

Fig. 6.19 Empirical and MOM theoretical frequency curves and MOM confidence limits for maximum significant wave height data sample

6.9.4 Annual Maximum Wind Speed Frequency Analysis

By using the annual maximum wind speed data, Castillo (1988), the following descriptive statistics for the data and its natural logarithms were obtained and are show in Fig. 6.20.

6.9.4.1 MOM Method

By using an Excel® worksheet, the results contained in Fig. 6.21, were obtained for the MOM method applied to the GAM distribution for annual maximum wind speed data.

6.9.4.2 ML Method

By using an Excel® worksheet, the results contained in Fig. 6.22, were obtained for the ML method applied to the GAM distribution for annual maximum wind speed data.

In Fig. 6.23 a comparison is made between the histogram and GAM theoretical density for annual maximum wind speed data sample.

A graph containing the empirical and theoretical GAM frequency curves is shown in Fig. 6.24.

A graphical depiction of the empirical and theoretical frequency curves and the confidence limits of the best fit provided, in this case that of MOM, is shown in Fig. 6.25.

	A	B	C	D
1	**MAXIMUM WIND SPEED FREQUENCY ANALYSIS**			
2				
3	*Descriptive Statistics*		*Descriptive Statistics of the Natural Logarithms*	
4	Mean	54.7964	Mean	3.9558
5	Sample Standard Error	2.9320	Sample Standard Error	0.0408
6	Median	47.9965	Median	3.8711
7	Mode	#N/D	Mode	#N/D
8	Standard Deviation	20.7327	Standard Deviation	0.2886
9	Sample Variance	429.8439	Sample Variance	0.0833
10	Kurtosis	12.5699	Kurtosis	3.6721
11	Skewness	3.0813	Skewness	1.6052
12	Range	121.5116	Range	1.4669
13	Minimum	36.4278	Minimum	3.5953
14	Maximum	157.9394	Maximum	5.0622
15	Sum	2739.8217	Sum	197.7906
16	Count	50.0000	Count	50.0000
17	Largest(1)	157.9394	Largest(1)	5.0622
18	Smallest(1)	36.4278	Smallest(1)	3.5953
19	Mean Confidence Level(95.0%)	5.8922	Mean Confidence Level(95.0%)	0.0820

Fig. 6.20 Descriptive statistics for the annual maximum wind speed data sample

	A	B	C	D	E	F	G	H
1	**ANNUAL MAXIMUM WIND SPEED FREQUENCY ANALYSIS**							
2	Gamma (GAM) Distribution							
3	Method: MOM							
4								
5	Estimation of Parameters							
6	Skewness Coef. (Gamma Correction) =		3.7176					
7	Scale Parameter =	7.8444	m/s					
8	Shape Parameter =	6.9854						
9	Γ(β) =	23185.3739						
10	SEF =	10.11		MARD =	10.05			
11								
12	Design Values and Confidence Limits							
13–16	Return Period Tr (years)	F(x)	Standard Normal Variate z_T	Frequency Factor K_T	Squared root Stand. Error S_T (m/s)	Two-sided 95% Lower Limit (m/s)	Design Values V_T (m/s)	Two-sided 95% Upper Limit (m/s)
17	2	0.5	0.0000	-0.4122	2.0118	48	52	56
18	5	0.8	0.8416	0.2540	4.6467	62	71	80
19	10	0.9	1.2816	0.9706	9.6479	64	82	101
20	20	0.95	1.6449	1.8145	15.7225	62	93	124
21	50	0.98	2.0537	3.0860	24.9450	56	105	154
22	100	0.99	2.3263	4.1479	32.6653	50	114	178
23	Two-sided Confidence Limits:		z_α =	1.96				

Fig. 6.21 MOM estimation of the parameters, standard errors, quantiles, and confidence limits for the annual maximum wind speed data sample

ANNUAL MAXIMUM WIND SPEED FREQUENCY ANALYSIS

Gamma (GAM) Distribution

Method: ML

Estimation of Parameters

Test Shape Parameter =	6.6615		Γ(β) =	23183.9355	
F(β) = 0.0000	ψ(β) =	1.8194	Log-likelihood function =		-1547.5810
Scale Parameter =	7.8444	m/s			
Shape Parameter =	6.9854				
SEF = 10.11		MARD =	10.05	AIC =	3099

Variance-Covariance Matrix

var(α) =	1.7683E+00	cov(α,β) =	-1.4649E+00
var(β) =	1.3045E+00	ψ'(β) =	0.1539

Design Values and Confidence Limits

Return Period Tr (years)	F(x)	Standard Normal Variate z_T	$\partial x_T/\partial\alpha$	$\partial x_T/\partial\beta$	Squared root Stand. Error S_T (m/s)	Two-sided 95% Lower Limit (m/s)	Design Values V_T (m/s)	Two-sided 95% Upper Limit (m/s)
2	0.5	0.0000	1.8811	7.8385	6.5733	39	52	65
5	0.8	0.8416	2.0840	10.1155	8.9104	54	71	88
10	0.9	1.2816	2.1901	11.4570	10.3051	62	82	103
20	0.95	1.6449	2.2777	12.6468	11.5507	70	93	115
50	0.98	2.0537	2.3763	14.0778	13.0575	80	105	131
100	0.99	2.3263	2.4420	15.0873	14.1256	87	114	142
Two-sided Confidence Limits:		z_α =	1.96					

Fig. 6.22 ML estimation of the parameters, standard errors, quantiles, and confidence limits for the annual maximum wind speed data sample

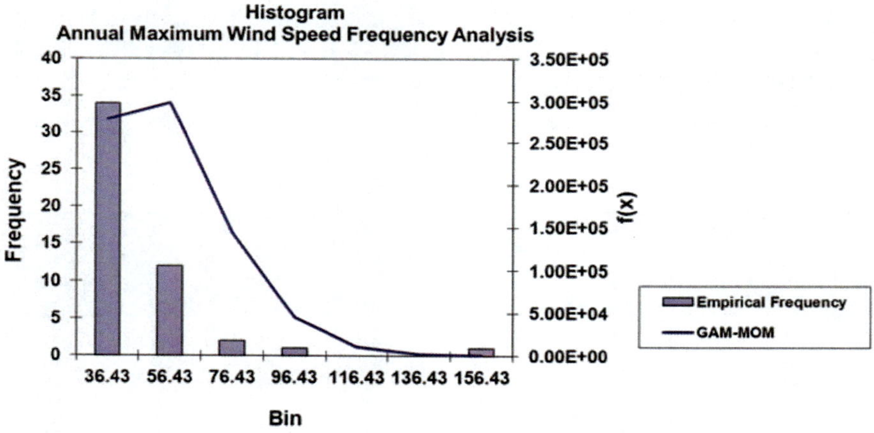

Fig. 6.23 Histogram and theoretical GAM density for annual maximum wind speed data sample

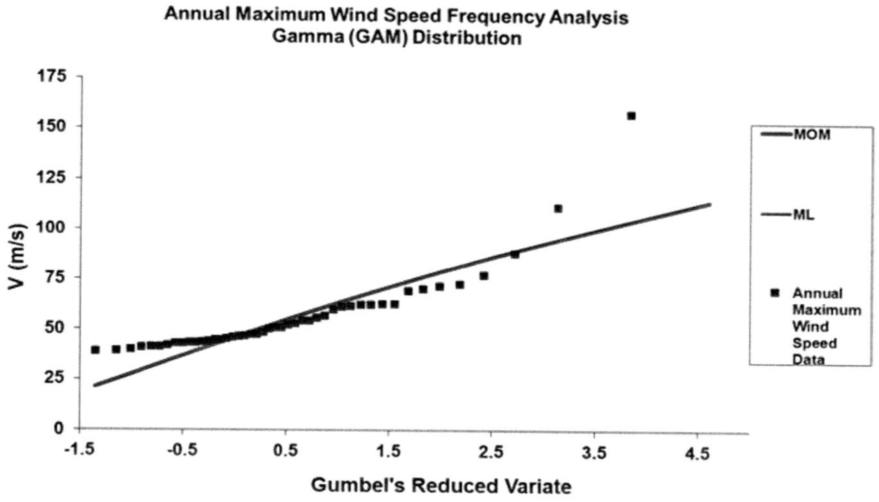

Fig. 6.24 Empirical and theoretical GAM distributions for annual maximum wind speed data sample

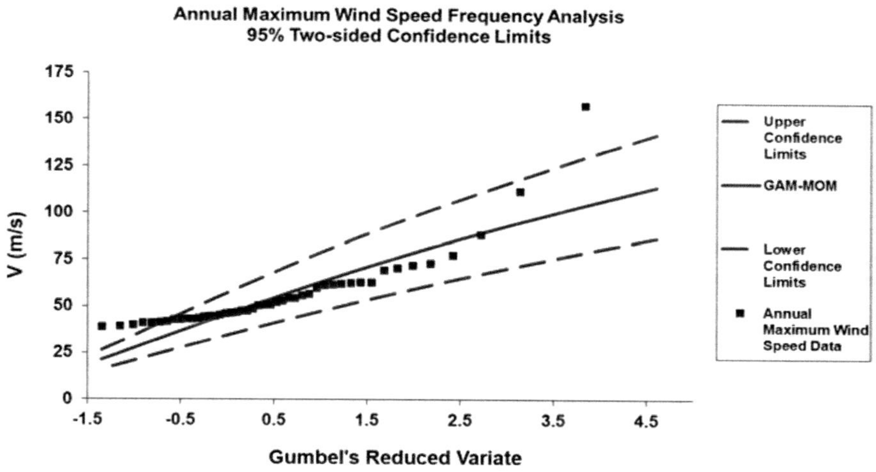

Fig. 6.25 Empirical and MOM and ML theoretical frequency curves and MOM and ML confidence limits for annual maximum wind speed data sample

7 Pearson Type III Distribution

Don't stir out all the warmth of your coffee; drink it. K. Chopin

7.1 Introduction

The probability distribution function Pearson type III (PIII) is one of the members of the Pearson family of distributions. These distributions were named in such a way after the great British mathematician Karl Pearson (1857–1936), the father of Mathematical Statistics. The most popular members of this family of distributions, are Pearson types I, III, IV, V, and VI. All these probability distributions are part of the Gamma family of distributions, in addition to those mentioned distributions there are other widely used probability distributions within this family of distributions, like the one-parameter and two-parameters Gamma distributions, the Log-Pearson type III, and the generalized Gamma distribution. The PIII distribution it is a Gamma distribution with three parameters.

Matalas and Wallis (1973) compared MOM and ML estimates produced by using the PIII distribution variates. They found a less biased and less variable results for ML estimates than those estimates produced by the MOM method. The differences between the MOM and ML estimates were bigger in small samples. They also noted that for very small values of the sample coefficient of skewness, in the case of the maximum likelihood method a solution may be not possible.

The frequency factors for the PIII distribution were developed by Haktanir (1991) using a practical method for their computation. Shaligram and Lele (1978) showed that the PIII distribution confidence limits were larger than those of the EVI distribution, when they used hydrologic data.

J. A. Raynal Villaseñor, *Frequency Analyses of Natural Extreme Events*,
Earth and Environmental Sciences Library,
https://doi.org/10.1007/978-3-030-86390-6_7

7.2 Chapter Objectives

After reading this chapter, you will know how to:

(a) Recognize the distribution and density functions of the PIII distribution
(b) Estimate the parameters of the PIII distribution
(c) Compute the quantiles and confidence limits of the PIII distribution
(d) Make a graphic display of your data and the PIII distribution
(e) Develop an application of all the above using Excel® spreadsheets.

7.3 Probability Distribution and Density Functions

The probability distribution function of the Pearson type III (PIII) distribution is:

$$F(x) = \frac{1}{\alpha\Gamma(\beta)}\int_{x_0}^{x}\left(\frac{x - x_0}{\alpha}\right)^{\beta-1}\exp\left(-\frac{(x - x_0)}{\alpha}\right)dx \tag{7.1}$$

where x_0, α and β are the location, scale, and shape parameters, respectively, and $\Gamma(.)$ is the complete gamma function with argument (.), as it was defined in Chap. 6. $F(x)$ is the probability distribution function of the random variable x and in the case of maxima natural extreme frequency analysis is equal to the non-exceedance probability, $Pr(X \leq x)$. Such parameters are restricted to be greater than zero, that is $\alpha > 0$ and $\beta > 0$. The domain of variable x in this distribution is $x_0 \leq x < \infty$.

The probability density function for the PIII distribution is:

$$f(x) = \frac{1}{\alpha\Gamma(\beta)}\left(\frac{x - x_0}{\alpha}\right)^{\beta-1}\exp\left(-\frac{(x - x_0)}{\alpha}\right) \tag{7.2}$$

where $f(x)$ is the probability density function of the random variable x.

7.4 Estimation of the Parameters

7.4.1 MOM Method

The mean, variance and coefficients of skewness and kurtosis of the PIII distribution are, Kite(1988):

$$\mu = \alpha\beta + x_0 \tag{7.3}$$

$$\sigma^2 = \alpha^2\beta \tag{7.4}$$

$$\gamma = \frac{2\alpha}{|\alpha|\sqrt{\beta}} \tag{7.5}$$

$$\kappa = 3\left(1 + \frac{\gamma^2}{2}\right) \tag{7.6}$$

Another set of useful relationships are, Kite (1988):

$$\lambda_1 = \frac{\mu_5}{\mu_2^{2.5}} = \gamma\left(10 + 3\gamma^2\right) \tag{7.7}$$

$$\lambda_2 = \frac{\mu_6}{\mu_2^3} = 5\left(3 + \frac{13\gamma^2}{2} + \frac{3\gamma^4}{2}\right) \tag{7.8}$$

So, these previous equations would serve to estimate the parameters of the PIII distribution by the MOM method as:

$$\hat{x}_0 = \hat{\mu} - \hat{\sigma}\sqrt{\hat{\beta}} \tag{7.9}$$

$$\hat{\alpha} = \frac{\hat{\sigma}}{\sqrt{\hat{\beta}}} \tag{7.10}$$

$$\hat{\beta} = \left(\frac{2}{\gamma}\right)^2 \tag{7.11}$$

In the above equations the mean, the variance and/or the standard deviation and the coefficient of skewness can be estimated by the procedures outlined in Chap. 2. In the case of the skewness coefficient, the following formula allows to correct the bias in such coefficient:

$$\hat{\gamma}_{PIII} = \frac{\frac{1}{N}\sum_{i=1}^{N}\left(x_i - \hat{\mu}\right)^3}{\hat{\sigma}^3}\left[\frac{\left(N(N-1)\right)^{\frac{1}{2}}}{(N-2)}\right]\left[1 + \frac{8.5}{N}\right] \tag{7.12}$$

7.4.1.1 Example of Application of Estimation of the Parameters of the PIII Distribution Using the MOM Method

Find the MOM estimators for the parameters of the PIII distribution for the flood sample data of gauging station Huites, Mexico (1942–1992), contained in Appendix A.

The following statistics can be obtained:

$$\hat{\mu} = \frac{1}{N}\sum_{i=1}^{N} x_i = 2498.96\,\mathrm{m}^3/\mathrm{s}$$

$$\hat{\sigma} = \left[\frac{1}{N}\sum_{i=1}^{N}(x_i - \hat{\mu})^2\right]^{1/2} = 2221.36\,\text{m}^3/\text{s}$$

$$\hat{\gamma} = \hat{\gamma}_x = \frac{N}{(N-1)(N-2)\hat{\sigma}^3}\sum_{i=1}^{N}(x_i - \hat{\mu})^3 = 2.0123$$

then:

$$\hat{\beta} = \left(\frac{2}{\gamma^2}\right)^2 = \left(\frac{2}{2.0123}\right)^2 = 0.7702$$

$$\hat{x}_0 = \hat{\mu} - \hat{\sigma}\sqrt{\hat{\beta}} = 2498.96 - (2221.36)(0.7702)^{0.5} = 549.4789\ \text{m}^3/\text{s}$$

$$\hat{\alpha} = \frac{\hat{\sigma}}{\sqrt{\hat{\beta}}} = \frac{2221.36}{(0.7702)^{0.5}} = 2531.158\,\text{m}^3/\text{s}$$

Finally, the MOM estimators of the parameters of the PIII distribution for the flood sample data of gauging station Huites, Mexico are:

location parameter:

$$\hat{x}_0 = 549.4789\,\text{m}^3/\text{s}$$

scale parameter:

$$\hat{\alpha} = 2531.158\,\text{m}^3/\text{s}$$

and:

shape parameter:

$$\hat{\beta} = 0.7702$$

7.4.2 ML Method

The likelihood function of the PIII distribution is:

$$L(\alpha, \beta, x_0) = \prod_{i=1}^{N}\left\{\frac{1}{\alpha\Gamma(\beta)}\left(\frac{x - x_0}{\alpha}\right)^{\beta-1}\exp\left(-\frac{(x - x_0)}{\alpha}\right)\right\} \tag{7.13}$$

So, the Log-likelihood function of the PIII distribution is:

$$LL(\alpha, \beta, x_0) = -NLn(\Gamma(\beta)) - \frac{1}{\alpha}\sum_{i=1}^{N}(x_i - x_0) + (\beta - 1)\sum_{i=1}^{N} Ln(x_i - x_0) - N\beta Ln(\alpha) \quad (7.14)$$

Now, taking the first-order partial derivatives of this equation with respect to parameters α, β and x_0 and set them equal to zero:

$$\frac{\partial LL}{\partial \alpha} = -\frac{N\beta}{\alpha} + \frac{1}{\alpha^2}\sum_{i=1}^{N}(x_i - x_0) = 0 \quad (7.15)$$

$$\frac{\partial LL}{\partial \beta} = -NLn(\alpha) - \frac{N\Gamma'(\beta)}{\Gamma(\beta)} + \sum_{i=1}^{N} Ln(x_i - x_0) = 0 \quad (7.16)$$

$$\frac{\partial LL}{\partial x_0} = \frac{N}{\alpha} - (\beta - 1)\sum_{i=1}^{N}\left[\frac{1}{(x_i - x_0)}\right] = 0 \quad (7.17)$$

The solution of the system of equations conformed by Eqs. (7.15)–(7.17) provides the ML estimators for the parameters of the PIII distribution:

$$\hat{\alpha} = \frac{1}{N}\sum_{i=1}^{N}(x_i - \hat{x}_0) - N\left[\sum_{i=1}^{N}\frac{1}{(x_i - \hat{x}_0)}\right]^{-1} \quad (7.18)$$

$$\hat{\beta} = \left\{1 - N^2\left[\sum_{i=1}^{N}(x_i - \hat{x}_0)\sum_{i=1}^{N}\frac{1}{(x_i - \hat{x}_0)}\right]^{-1}\right\}^{-1} \quad (7.19)$$

$$F(\hat{x}_0) = -N\psi(\hat{\beta}) + \sum_{i=1}^{N} Ln(x_i - \hat{x}_0) - NLn(\hat{\alpha}) = 0 \quad (7.20)$$

Equation (7.20) must be solved by a procedure of finding the roots of such an equation, again the bisection method will give good results in solving this equation.

Kite (1988) proposed the following procedure to solve the system of equation formed by Eqs. (7.18)–(7.20), the results are:

$$A = \frac{1}{N}\sum_{i=1}^{N}(x_i - \hat{x}_0) \quad (7.21)$$

$$B = \left[\prod_{i=1}^{N} (x_i - \hat{x}_0) \right]^{\frac{1}{N}} \tag{7.22}$$

and if:

$$C = Ln(A) - Ln(B) \tag{7.23}$$

and then:

$$C = Ln\left(\hat{\beta}\right) - \frac{\Gamma'\left(\hat{\beta}\right)}{\Gamma(\beta)} \tag{7.24}$$

An approximation for the solution for $\hat{\beta}$ is given by, Greenwood and Durand (1960):

$$\hat{\beta} = \frac{1}{C}\left(0.500876 + 0.1648852C - 0.054427C^2\right) \tag{7.25}$$

for $0 \leq C \leq 0.5772$, and:

$$\hat{\beta} = \frac{8.898919 + 9.05995C + 0.9775373C^2}{C\left(17.7928 + 11.966847C + C^2\right)} \tag{7.26}$$

for $0.5772 \leq C \leq 17.0$.

As a word of warning, it must be said that the ML method may not always be applicable for the PIII distribution, this was specifically observed for small values of the sample skewness coefficient, Matalas and Wallis (1973). Furthermore, if $\beta < 1$ the ML solution is not possible, Kite (1988). This situation is coupled with the fact that the sample skewness coefficient must be less than two ($\hat{\gamma} < 2$), see Eq. (7.11). If the sample skewness coefficient is negative, then the PIII distribution becomes upper-bounded and it is no longer suitable for maximum events frequency analysis.

7.5 Estimation of Quantiles for the PIII Distribution

By using the following reduced value, Kite (1988):

$$y = \frac{x - x_0}{\alpha} \tag{7.27}$$

into Eq. (7.1) produces the following expression, Kite (1988):

$$F(y) = \frac{1}{\Gamma(\beta)} \int_0^{y_0} y^{\beta-1} \exp(-y) dy \tag{7.28}$$

and Abramowitz and Stegun (1965) have shown that:

$$F(y) = F(\chi^2|v) \tag{7.29}$$

where $F(\chi^2/v)$ is the chi-square distribution with 2β degrees of freedom and $\chi^2 = 2\ y$. So, the reduced event magnitude, y_0, may be obtained as, Kite (1988):

$$y_0 = \frac{\chi^2}{2} \tag{7.30}$$

and the expected event of magnitude associated with a given return period T_r is, Kite (1988):

$$x_T = \frac{\alpha\chi^2}{2} + x_0 \tag{7.31}$$

but using the Wilson-Hilferty approximation, Kendall and Stuart (1963):

$$\left[\left(\frac{\chi^2}{v}\right)^{\frac{1}{3}} + \frac{2}{9v} - 1\right]\left(\frac{9v}{2}\right)^{\frac{1}{2}} \sim N(0, 1) \tag{7.32}$$

and this expression is valid for $v > 30$. So, the values of χ^2 can be approximated as:

$$\chi^2 \approx v\left[1 - \frac{2}{9v} + z_T\sqrt{\frac{2}{9v}}\right]^3 \tag{7.33}$$

where z_T is the standard normal (N(0,1)) variate corresponding a certain return period, T_r, and from Eqs. (7.31) and (7.33) one may obtain:

$$x_T = \hat{\alpha}\hat{\beta}\left[1 - \frac{1}{9\hat{\beta}} + z_T\sqrt{\frac{1}{9\hat{\beta}}}\right]^3 + \hat{x}_0 \tag{7.34}$$

7.5.1 Examples of Estimation of MOM and ML Quantiles for the PIII Distribution

Find the MOM and ML estimators of the quantiles 2, 5, 10, 20, 50, 100, 500, and 1000 years of return period, for the PIII distribution for the flood sample data of gauging station Huites, Mexico (1942–1992), contained in Appendix A.

The estimation of MOM quantiles for the PIII distribution is made by using the parameters estimated in the preceding sections and then inserting the parameters into Eq. (7.34). The ML was not applicable for the PIII distribution for this set of data. Table 7.1 summarizes these results.

Table 7.1 Estimation of MOM and ML quantiles for the PIII distribution for gauging station Huites, Mexico

Tr (Years)	MOM
	Q_T (m^3/s)
2	1771
5	3715
10	5267
20	6877
50	9085
100	10,811
500	14,984
1000	16,848

7.6 Goodness of Fit Test

The standard error of fit (SEF) for the PIII has the following form:

$$SEF = \left[\frac{\sum_{i=1}^{N} (xi - yi)^2}{(N - 3)} \right]^{1/2} \tag{7.35}$$

while the mean absolute relative deviation (MARD) remains the same as it was defined in Chap. 2:

$$MARD = \frac{100}{N} \sum_{i=1}^{N} \left| \frac{(x_i - y_i)}{x_i} \right| \tag{7.36}$$

where x_i are the sample historical values, y_i are the distribution function values, corresponding to the same return periods of the historical values, N is the sample size.

7.6.1 Examples of Application of the SEF and MARD to the MOM and ML Estimators of the Parameters of the PIII Distribution

Find the values of the SEF and MARD of the MOM and ML estimators of the parameters of the PIII distribution for the flood sample data of gauging station Huites, Mexico (1942–1992), contained in Appendix A.

The values of SEF and MARD for the MOM estimators of the parameters of the PIII distribution for the flood sample data of gauging station Huites, Mexico, have been obtained through the application of Eqs. (7.35) and (7.36) using the parameters obtained in previous examples. The results are as follows:

(a) MOM Method

$$SEF = \left[\frac{\sum_{i=1}^{N}(xi - yi)^2}{(N-3)}\right]^{1/2} = \left[\frac{261937.3412}{(51-3)}\right]^{\frac{1}{2}} = 512$$

$$MARD = \frac{100}{N}\sum_{i=1}^{N}\left|\frac{(x_i - y_i)}{x_i}\right| = \frac{(100)(6.2858)}{51} = 12$$

(b) ML Method

This option was not obtainable.

7.7 Estimation of Confidence Limits for the PIII Distribution

By assuming that the quantiles are normally distributed, the following form can be used to set the confidence limits on such quantiles:

$$x_l = x_T \pm z_\alpha S_T \tag{7.37}$$

where x_l is the upper or lower confidence limit, depending on the sign in the preceding formula (+) for the upper confidence limit and (−) for the lower confidence limit, z_α is the standard normal variate corresponding to a confidence level α, and S_T is squared root of the standard error of the estimate. The evaluation procedures of the standard errors of the estimates will be described in the following subsections.

7.8 Estimation of Standard Errors for the PIII Distribution

7.8.1 MOM Method

The general form of the MOM estimator of the standard error of the estimate of a three-parameter distribution is, Kite (1988):

$$\begin{aligned} S_T^2 = {} & \left(\frac{\partial x_T}{\partial m_1'}\right)^2 \mathrm{var}(m_1') + \left(\frac{\partial x_T}{\partial m_2}\right)^2 \mathrm{var}(m_2) + \left(\frac{\partial x_T}{\partial m_3}\right)^2 \mathrm{var}(m_3) \\ & + 2\left(\frac{\partial x_T}{\partial m_1'}\right)\left(\frac{\partial x_T}{\partial m_2}\right)\mathrm{cov}(m_1', m_2) + 2\left(\frac{\partial x_T}{\partial m_1'}\right)\left(\frac{\partial x_T}{\partial m_3}\right)\mathrm{cov}(m_1', m_3) \\ & + 2\left(\frac{\partial x_T}{\partial m_2}\right)\left(\frac{\partial x_T}{\partial m_3}\right)\mathrm{cov}(m_2, m_3) \end{aligned} \tag{7.38}$$

and Eq. (7.36) can be simplified in terms of the frequency factor, K_T, as, Kite (1988):

$$S_T^2 = \frac{\mu_2}{N}\Bigg\{1 + K_T\gamma + \frac{K_T^2}{4}(\kappa - 1) + \frac{\partial K_T}{\partial \gamma}\left[2\kappa - 3\gamma^2 - 6 + K_T\left(\lambda_1 - \frac{6\gamma\kappa}{4} - \frac{10\gamma}{4}\right)\right] + \left(\frac{\partial K_T}{\partial \gamma}\right)^2\left[\lambda_2 - 3\gamma\lambda_1 - 6\gamma + \frac{9\gamma^2\kappa}{4} + \frac{35\gamma^2}{4} + 9\right]\Bigg\} \tag{7.39}$$

and:

$$\hat{\lambda}_1 = \frac{\left[\sum_{i=1}^{N}(x_i - \hat{\mu})^5\right] / N}{\left(\sigma^2\right)^{2.5}} \tag{7.40}$$

$$\hat{\lambda}_2 = \frac{\left[\sum_{i=1}^{N}(x_i - \hat{\mu})^6\right] / N}{\left(\sigma^2\right)^{3}} \tag{7.41}$$

The previous equation may be further simplified by using the relationships given in Eqs. (7.6)–(7.8), Kite (1988):

$$S_T^2 = \frac{\mu_2}{N}\Bigg\{1 + K_T\gamma + \frac{K_T^2}{2}\left(\frac{3\gamma^2}{4} + 1\right) + 3K_T\frac{\partial K_T}{\partial \gamma}\left(\gamma + \frac{\gamma^3}{4}\right) + 3\left(\frac{\partial K_T}{\partial \gamma}\right)^2\left(2 + 3\gamma^2 + \frac{5\gamma^4}{8}\right)\Bigg\} \tag{7.42}$$

The partial derivative of the frequency factor with respect to skewness coefficient may be evaluated as, Kite (1988):

$$\frac{\partial K_T}{\partial \gamma} \approx \frac{z_T^2 - 1}{6} + \frac{4\left(z_T^3 - 6z_T\right)\gamma}{6^3} - \frac{3\left(z_T^2 - 1\right)\gamma^2}{6^3} + \frac{4z_T\gamma^3}{6^4} - \frac{10\gamma^4}{6^6} \tag{7.43}$$

and the frequency factor for the PIII distribution is, Kite (1988):

$$K_T = z_T + \left(z_T^2 - 1\right)\left(\frac{\hat{\gamma}}{6}\right) + \frac{\left(z_T^3 - 6z_T\right)}{3}\left(\frac{\hat{\gamma}}{6}\right)^2 - \left(z_T^2 - 1\right)\left(\frac{\hat{\gamma}}{6}\right)^3 + z_T\left(\frac{\hat{\gamma}}{6}\right)^4 - \frac{1}{3}\left(\frac{\hat{\gamma}}{6}\right)^5 \tag{7.44}$$

By inserting these results into Eq. (7.42), the MOM estimator of the standard error of the estimate for the PIII distribution may be obtained.

7.8.1.1 Example of Estimation of MOM Standard Errors and the Two-Sided 95% Confidence Limits for the PIII Distribution

Find the MOM estimators of the standard errors and the two-sided 95% confidence limits for the PIII distribution for the flood sample data of gauging station Huites, Mexico (1942–1992), contained in Appendix A.

The estimation of MOM estimators of the standard errors and the two-sided 95% confidence limits for the PIII distribution is made by using the MOM estimators of the parameters and then inserted in the Eqs. (7.42)–(7.44), using selected values of the return intervals. Table 7.2 summarizes these results.

7.8.2 ML Method

The general equation for the ML estimator of the standard error of the estimate of a three-parameters distribution is, Kite (1988):

$$\begin{aligned} S_T^2 = & \left(\frac{\partial x_T}{\partial \alpha}\right)^2 \operatorname{var}(\alpha) + \left(\frac{\partial x_T}{\partial \beta}\right)^2 \operatorname{var}(\beta) \\ & + \left(\frac{\partial x_T}{\partial \gamma}\right)^2 \operatorname{var}(\gamma) + 2\left(\frac{\partial x_T}{\partial \alpha}\right)\left(\frac{\partial x_T}{\partial \beta}\right)\operatorname{cov}(\alpha, \beta) \\ & + 2\left(\frac{\partial x_T}{\partial \alpha}\right)\left(\frac{\partial x_T}{\partial \gamma}\right)\operatorname{cov}(\alpha, \gamma) + 2\left(\frac{\partial x_T}{\partial \beta}\right)\left(\frac{\partial x_T}{\partial \gamma}\right)\operatorname{cov}(\beta, \gamma) \end{aligned} \tag{7.45}$$

and the elements of the Fisher's information matrix, are:

$$-\frac{\partial^2 LL}{\partial \alpha^2} = \frac{2\sum_{i=1}^{N}(x_i - x_0)}{\alpha^3} - \frac{N\beta}{\alpha^2} \tag{7.46}$$

Table 7.2 Estimation of MOM standard errors and the two-sided 95% confidence limits for the PIII distribution for gauging station Huites, Mexico

Return period Tr (years)	Squared root Stand. error S_T (m^3/s)	Two-sided 95% lower limit (m^3/s)	Design values Q_T (m^3/s)	Two-sided 95% upper limit (m^3/s)
2	447.7693	893	1771	2649
5	702.0084	2339	3715	5091
10	815.3443	3669	5267	6865
20	1189.1455	4547	6877	9208
50	2073.9995	5020	9085	13,150
100	2942.7722	5043	10,811	16,578
500	5410.9402	4378	14,984	25,589
1000	6628.8745	3855	16,848	29,840

Two-sided Limits: $z_\alpha = 1.96$

$$-\frac{\partial^2 LL}{\partial \beta^2} = N\psi'(\beta) \tag{7.47}$$

$$-\frac{\partial^2 LL}{\partial x_0^2} = (\beta - 1)\sum_{i=1}^{N} \frac{1}{(x_i - x_0)^2} \tag{7.48}$$

$$-\frac{\partial^2 LL}{\partial \alpha \partial \beta} = \frac{N}{\alpha} \tag{7.49}$$

$$-\frac{\partial^2 LL}{\partial \alpha \partial x_0} = \frac{N}{\alpha^2} \tag{7.50}$$

$$-\frac{\partial^2 LL}{\partial \beta \partial x_0} = \frac{N}{\alpha(\beta - 1)} \tag{7.51}$$

but the Eqs. (7.44) and (7.46) can be simplified by using the following result, Kite (1988):

$$\sum_{i=1}^{N} (x_i - x_0)^r = \frac{N\alpha^r \Gamma(\beta + r)}{\Gamma(\beta)} \tag{7.52}$$

so:

$$-\frac{\partial^2 LL}{\partial \alpha^2} = \frac{N\beta}{\alpha^2} \tag{7.53}$$

$$-\frac{\partial^2 LL}{\partial x_0^2} = \frac{N}{\alpha^2(\beta - 2)} \tag{7.54}$$

and the Fisher's information matrix is constructed as:

$$[I] = N\begin{bmatrix} \frac{\beta}{\alpha^2} & \frac{1}{\alpha} & \frac{1}{\alpha^2} \\ \frac{1}{\alpha} & \psi'(\beta) & \frac{1}{\alpha(\beta-1)} \\ \frac{1}{\alpha^2} & \frac{1}{\alpha(\beta-1)} & \frac{1}{\alpha^2(\beta-2)} \end{bmatrix} \tag{7.55}$$

Finally, the variances and covariances of the parameters of the PIII distribution are found as:

$$\text{var}(\alpha) = \frac{1}{ND\alpha^2}\left[\frac{\psi'(\beta)}{(\beta - 2)} - \frac{1}{(\beta - 1)^2}\right] \tag{7.56}$$

$$\text{var}(\beta) = \frac{2}{ND\alpha^4(\beta - 2)} \tag{7.57}$$

$$\text{var}(x_0) = \frac{1}{ND\alpha^2}\left[\beta\psi'(\beta) - 1\right] \tag{7.58}$$

$$\mathrm{cov}(\alpha,\beta) = -\frac{1}{ND\alpha^3}\left[\frac{1}{(\beta-2)} - \frac{1}{(\beta-1)}\right] \tag{7.59}$$

$$\mathrm{cov}(\alpha,x_0) = \frac{1}{ND\alpha^2}\left[\frac{1}{(\beta-1)} - \psi'(\beta)\right] \tag{7.60}$$

$$\mathrm{cov}(\beta,x_0) = -\frac{1}{ND\alpha^3}\left[\frac{\beta}{(\beta-1)} - 1\right] \tag{7.61}$$

and D is the determinant of the Fisher's information matrix with the following form:

$$D = \frac{1}{(\beta-2)\alpha^4}\left[2\psi'(\beta) - \frac{(2\beta-3)}{(\beta-1)^2}\right] \tag{7.62}$$

The first order partial derivatives of x_T with respect to the parameters are evaluated as, Kite (1988):

$$\frac{\partial x_T}{\partial \alpha} = \left[\beta^{\frac{1}{3}} - \frac{1}{9\beta^{\frac{2}{3}}} + \frac{z_T}{3\beta^{\frac{1}{6}}}\right]^3 \tag{7.63}$$

$$\frac{\partial x_T}{\partial \beta} = 3\alpha\left[\beta^{\frac{1}{3}} - \frac{1}{9\beta^{\frac{2}{3}}} + \frac{z_T}{3\beta^{\frac{1}{6}}}\right]^2\left[\frac{1}{3\beta^{\frac{2}{3}}} + \frac{2}{27\beta^{\frac{5}{3}}} - \frac{z_T}{18\beta^{\frac{7}{6}}}\right] \tag{7.64}$$

$$\frac{\partial x_T}{\partial x_0} = 1 \tag{7.65}$$

The substitution of the results of Eqs. (7.56)–(7.65) into Eq. (7.45) will provide the value of the ML estimator of the standard error of the estimate of a PIII distribution.

7.8.2.1 Example of Estimation of ML Standard Errors and the Two-Sided 95% Confidence Limits for the PIII Distribution

Find the ML estimators of the standard errors and the two-sided 95% confidence limits for the PIII distribution for the flood sample data of gauging station Huites, Mexico (1942–1992), contained in Appendix A.

The estimation of ML estimators of the standard errors and the two-sided 95% confidence limits for the PIII distribution is made by using the MOM estimators of the parameters and then inserted in the Eqs. (7.43)–(7.63), using selected values of the return intervals. For this set of data there was not a feasible solution for the PIII distribution.

7.9 Examples of Application for the PIII Distribution Using Excel® Spreadsheets

7.9.1 Flood Frequency Analysis

By using the flood data from gauging station Huites, Mexico the following descriptive statistics were obtained, and are displayed in Fig. 7.1.

7.9.1.1 MOM Method

By using an Excel® spreadsheet, the results contained in Fig. 7.2, were obtained for the MOM method applied to the PIII distribution for gauging station Huites, Mexico.

7.9.1.2 ML Method

This method produced a non-feasible solution for this set of data, as is shown in Fig. 7.3.

In Fig. 7.4 a comparison is made between the histogram and PIII theoretical density for flood sample of Huites, Mexico.

A graph containing the empirical and theoretical PIII frequency curves is shown in Fig. 7.5.

A graphical depiction of the empirical and theoretical frequency curves and the confidence limits of the best fit provided, in this case that of MOM, is shown in Fig. 7.6.

	A	B	C	D
1	FLOOD FREQUENCY ANALYSIS			
2	Gauging Station: Huites, Sinaloa, Mexico (1942-1992)			
3				
4	*Descriptive Statistics*		*Descriptive Statistics of the Natural Logarithms*	
5	Mean	2498.9627	Mean	7.5383
6	Sample Standard Error	311.0530	Sample Standard Error	0.1024
7	Median	1775.7000	Median	7.4819
8	Mode	#N/D	Mode	#N/D
9	Standard Deviation	2221.3626	Standard Deviation	0.7312
10	Sample Variance	4934451.7500	Sample Variance	0.5347
11	Kurtosis	3.6701	Kurtosis	0.0081
12	Skewness	2.0123	Skewness	0.4627
13	Range	9707.5000	Range	3.1787
14	Minimum	421.8000	Minimum	6.0445
15	Maximum	10129.3000	Maximum	9.2232
16	Sum	127447.1000	Sum	384.4530
17	Count	51.0000	Count	51.0000
18	Largest(1)	10129.3000	Largest(1)	9.2232
19	Smallest(1)	421.8000	Smallest(1)	6.0445
20	Mean Confidence Level(95.0%)	624.7683	Mean Confidence Level(95.0%)	0.2057

Fig. 7.1 Descriptive statistics for the flood sample of Huites, Mexico

FLOOD FREQUENCY ANALYSIS
Gauging Station: Huites, Mexico (1942-1992)
Pearson Type III (PIII) Distribution
Method: MOM

Estimation of Parameters and Goodness of Fit Tests

Skewness Coef. (Gamma Correction) =		2.4194	$\Gamma(\beta)$ =	1.3252	
Location Parameter =	662.6554	m³/s			
Scale Parameter =	2687.1600	m³/s			
Shape Parameter =	0.6834				
SEF =	502		MARD =	9	

Design Values and Confidence Limits

Return Period Tr (years)	F(x)	Standard Normal Variate z_T	$\partial K_T/\partial \gamma$	Frequency Factor K_T	Squared root Stand. Error S_T (m³/s)	Two-sided 95% Lower Limit (m³/s)	Design Values Q_T (m³/s)	Two-sided 95% Upper Limit (m³/s)
2	0.5	0.0000	-0.0927	-0.3412	451.8811	855	1741	2627
5	0.8	0.8416	-0.1950	0.5205	752.2394	2181	3655	5129
10	0.9	1.2816	-0.1467	1.2261	848.8603	3559	5223	6886
20	0.95	1.6449	-0.0326	1.9670	1211.4873	4494	6868	9243
50	0.98	2.0537	0.1932	2.9927	2139.2370	4954	9147	13340
100	0.99	2.3263	0.4097	3.7999	3072.5399	4918	10940	16962
500	0.998	2.8782	1.0348	5.7665	5759.7033	4019	15308	26597
1000	0.999	3.0902	1.3490	6.6498	7095.5492	3363	17271	31178
Two-sided Confidence Limits:		z_α =	1.96					

Fig. 7.2 MOM estimators of the parameters, standard errors, quantiles, and confidence limits of the PIII distribution for the flood sample of Huites, Mexico

FLOOD FREQUENCY ANALYSIS
Gauging Station: Huites, Mexico (1942-1992)
Pearson Type III (PIII) Distribution
Method: ML

Estimation of Parameters and Goodness of Fit Tests

Test Location Parameter =	421.5000	$F(x_o)$ =	-9.9038E-01	$\psi(\beta)$ =	-0.5653	
Skewness Coef. (Gamma Correction) =		2.4194	>	2	No solution exists	
Log-Likelihood function =	-455.1613	AIC =	916			
Location Parameter =	421.5000	m³/s	Scale Parameter =		2062.4624	m³/s
Shape Parameter =	1.0073	SEF =	614	MARD =	13	

Variance-Covariance Matrix

$\psi'(\beta)$ =	1.6276	$var(x_0)$ =	-2.8416E+00	$cov(\alpha, x_0)$ =	-6.0376E+02	$cov(\beta, x_0)$ =	2.9624E-01
$var(\alpha)$ =	8.4015E+04	$cov(\alpha, \beta)$ =	-2.9841E-01	$var(\beta)$ =	2.1046E-06	Determinant =	-1.0373E-09

Design Values and Confidence Limits

Return Period Tr (years)	F(x)	Standard Normal Variate z_T	$\partial x_T/\partial \alpha$	$\partial x_T/\partial \beta$	Squared root Stand. Error S_T (m³/s)	Two-sided 95% Lower Limit (m³/s)	Design Values Q_T (m³/s)	Two-sided 95% Upper Limit (m³/s)
2	0.5	0.0000	0.7094	1992.7102	204.3557	1484	1885	2285
5	0.8	0.8416	1.6100	3047.4996	463.3915	2834	3742	4650
10	0.9	1.2816	2.2922	3596.0792	660.2238	3855	5149	6443
20	0.95	1.6449	2.9827	4029.5334	859.6885	4888	6573	8258
50	0.98	2.0537	3.9114	4481.7475	1128.2249	6277	8489	10700
100	0.99	2.3263	4.6270	4754.9252	1335.2404	7348	9965	12582
500	0.998	2.8782	6.3324	5217.4107	1828.8495	9897	13482	17066
1000	0.999	3.0902	7.0854	5356.5805	2046.8998	11023	15035	19047
Two-sided Confidence Limits:			z_α =	1.96				

Fig. 7.3 ML estimators for the parameters, standard errors, quantiles, and confidence limits of the PIII distribution for the flood sample of Huites, Mexico

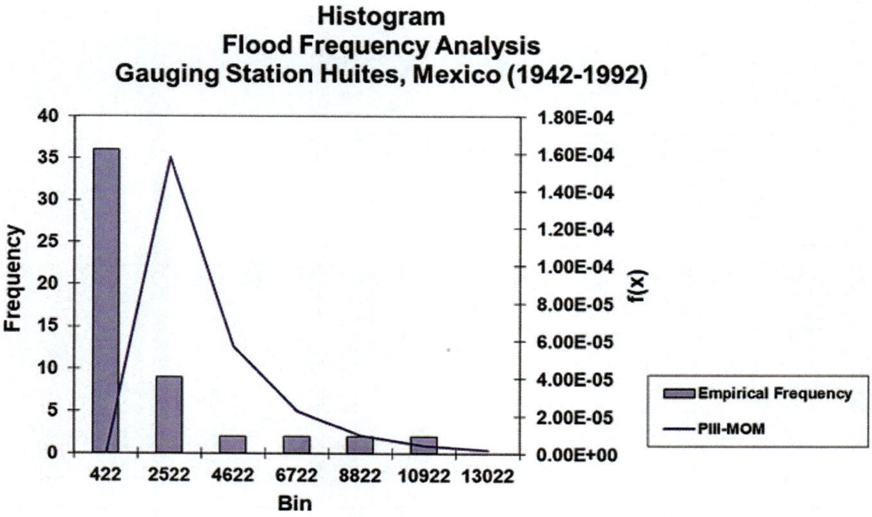

Fig. 7.4 Histogram and theoretical ML PIII density for flood data at Huites, Mexico

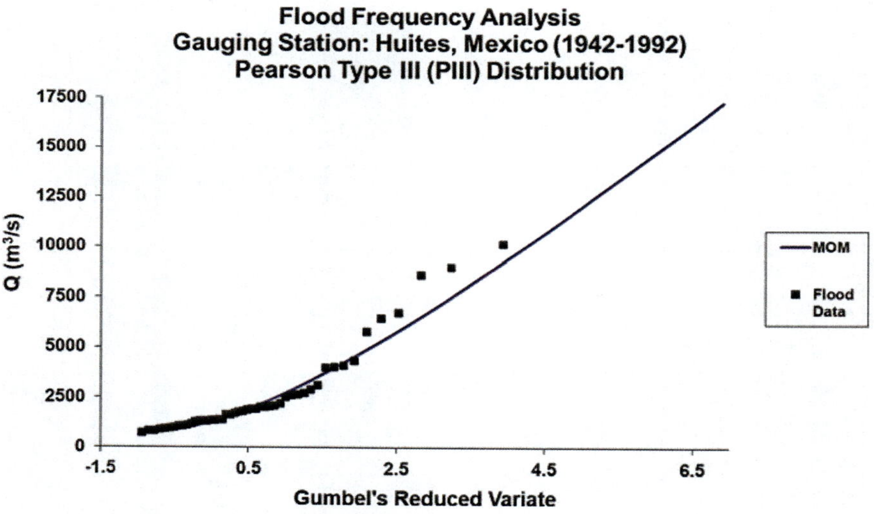

Fig. 7.5 Empirical and MOM PIII theoretical curves for the flood sample of Huites, Mexico

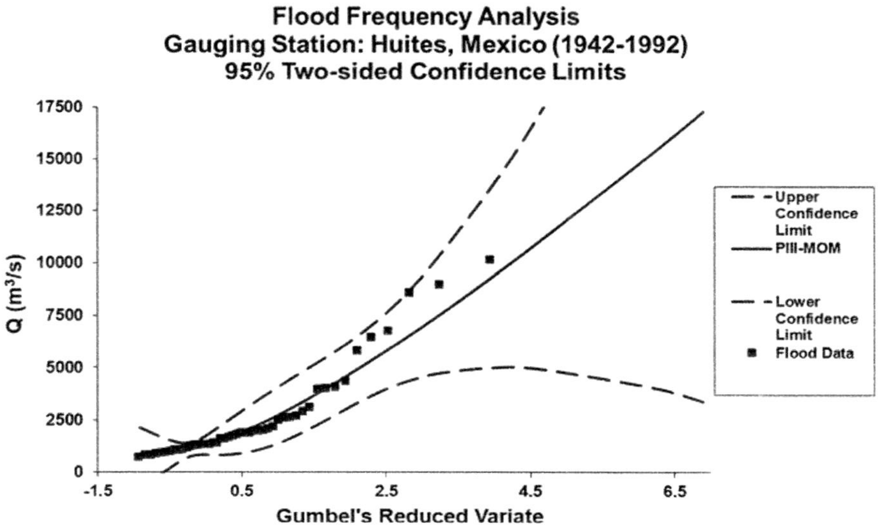

Fig. 7.6 Empirical and MOM theoretical frequency curves and MOM confidence limits of the PIII distribution for the flood data at Huites, Mexico

7.9.2 Rainfall Frequency Analysis

By using the 24 h annual rainfall data from rainfall gauging station Chihuahua, Mexico the following descriptive statistics were obtained and are shown in Fig. 7.7.

7.9.2.1 MOM Method

By using an Excel® spreadsheet, the results contained in Figs. 7.8 and 7.9, were obtained for the MOM Method applied to the PIII distribution for rainfall gauging station Chihuahua, Mexico.

7.9.2.2 ML Method

By using an Excel® spreadsheet, the results contained in Fig. 7.10, were obtained for the MOM Method applied to the PIII distribution for rainfall gauging station Chihuahua, Mexico.

In Fig. 7.10 a comparison is made between the histogram and PIII theoretical density for 24 h annual maximum rainfall data at Chihuahua, Mexico.

A graph containing the empirical and theoretical MOM and ML PIII frequency curves is shown in Fig. 7.11.

A graphical depiction of the empirical and theoretical frequency curves and the confidence limits of the best fit provided, in this case that of MOM, is shown in Fig. 7.12.

24 Hr. Annual Maximum Rainfall Frequency Analysis
Meteorological Station: Chihuahua, Mexico (1960-2005)

Descriptive Statistics		*Descriptive Statistics of the Natural Logarithms*	
Mean	42.0454	Mean	3.6871
Sample Standard Error	1.9784	Sample Standard Error	0.0487
Median	40.2400	Median	3.6948
Mode	#N/D	Mode	#N/D
Standard Deviation	13.4184	Standard Deviation	0.3302
Sample Variance	180.0535	Sample Variance	0.1090
Kurtosis	-0.7252	Kurtosis	-0.4776
Skewness	0.3591	Skewness	-0.2507
Range	54.3000	Range	1.4283
Minimum	17.1200	Minimum	2.8402
Maximum	71.4200	Maximum	4.2686
Sum	1934.0900	Sum	169.6058
Count	46.0000	Count	46.0000
Largest(1)	71.4200	Largest(1)	4.2686
Smallest(1)	17.1200	Smallest(1)	2.8402
Mean Confidence Level(95.0%)	3.9848	Mean Confidence Level(95.0%)	0.0981

Fig. 7.7 Descriptive statistics for the 24 h, annual maximum rainfall data at Chihuahua, Mexico

24 Hr. Annual Maximum Rainfall Frequency Analysis
Meteorological Station: Chihuahua, Mexico (1960-2005)
Pearson Type III (PIII) Distribution
Method: MOM

Estimation of Parameters and Goodness of Fit Tests				
Skewness Coef. (Gamma Correction) =		0.4400	$\Gamma(\beta)$ =	44335415.7246
Location Parameter =	-18.9538	cm		
Scale Parameter =	2.9517	cm		
Shape Parameter =	20.6656			
SEF = 1.84			MARD =	3.73

Design Values and Confidence Limits

Return Period Tr (years)	F(x)	Standard Normal Variate z_T	$\partial K_T/\partial\gamma$	Frequency Factor K_T	Squared root Stand. Error S_T (cm)	Two-sided 95% Lower Limit (cm)	Design Values P_T (cm)	Two-sided 95% Upper Limit (cm)
2	0.5	0.0000	-0.1640	-0.0729	2.1651	37	41	45
5	0.8	0.8416	-0.0839	0.8124	2.5765	48	53	58
10	0.9	1.2816	0.0602	1.3184	3.2590	53	60	66
20	0.95	1.6449	0.2359	1.7596	4.2231	57	66	74
50	0.98	2.0537	0.4984	2.2819	5.7679	61	73	84
100	0.99	2.3263	0.7129	2.6457	7.0651	64	78	91
Two-sided Confidence Limits:		z_α =	1.96					

Fig. 7.8 MOM estimation of the parameters, standard errors, quantiles, and confidence limits for the 24 h, annual maximum rainfall data at Chihuahua, Mexico

	A	B	C	D	E	F	G	H	I
1	24 Hr. Annual Maximum Rainfall Frequency Analysis								
2	Meteorological Station: Chihuahua, Mexico (1960-2005)								
3	Pearson Type III (PIII) Distribution								
4	Method: ML								
5									
6	Estimation of Parameters and Goodness of Fit Tests								
7	Test Location Parameter =		5.5500	$F(x_o)$ =	-4.0027E-05	$\Gamma(\beta)$ =	17285.2342	$\psi(\beta)$ =	1.9085
8	Skewness Coef. (Gamma Correction) =		0.4400		<	2	Log-Likelihood function =		-854.3038
9	Location Parameter =		5.5500	cm	Scale Parameter =	5.0429	cm		
10	Shape Parameter =		7.2371						
11	SEF =	1.87		MARD =	3.54	AIC =	1715		
12									
13	Variance-Covariance Matrix								
14	$var(x_0)$ =	9.8576E+01	$cov(\alpha,x_0)$ =	1.6602E+01	$cov(\beta,x_0)$ =	-4.3373E+01	$var(\alpha)$ =	3.5262E+00	
15	$var(\beta)$ =	2.0487E+01	$cov(\alpha,\beta)$ =	-8.2820E+00	$\psi'(\beta)$ =	0.1482	Determinant =	4.0360E-07	
16									
17	Design Values and Confidence Limits								
18–21	Return Period Tr (years)	F(x)	Standard Normal Variate z_T	$\partial x_T/\partial\alpha$	$\partial x_T/\partial\beta$	Squared root Stand. Error S_T (cm)	Two-sided 95% Lower Limit (cm)	Design Values P_T (cm)	Two-sided 95% Upper Limit (cm)
22	2	0.5	0.0000	6.9074	5.0393	1.6450	37	40	44
23	5	0.8	0.8416	9.3432	5.8517	2.2144	48	53	57
24	10	0.9	1.2816	10.8181	6.2726	2.8953	54	60	66
25	20	0.95	1.6449	12.1469	6.6160	3.7199	60	67	74
26	50	0.98	2.0537	13.7679	6.9961	4.9486	65	75	85
27	100	0.99	2.3263	14.9250	7.2448	5.9490	69	81	92
28	Two-sided Confidence Limits:		z_α =	1.96					

Fig. 7.9 ML estimation of the parameters, standard errors, quantiles, and confidence limits for the 24 h, annual maximum rainfall data at Chihuahua, Mexico

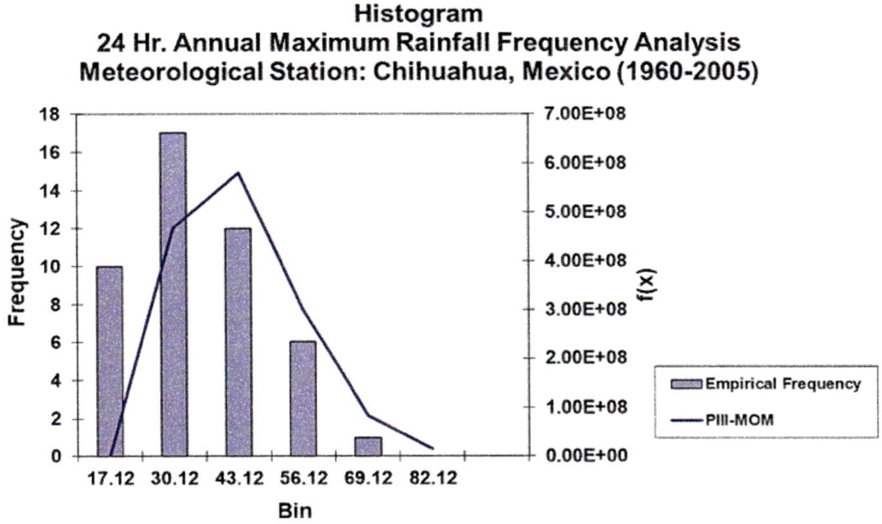

Fig. 7.10 Histogram and theoretical PIII density for 24 h, annual maximum rainfall data at Chihuahua, Mexico

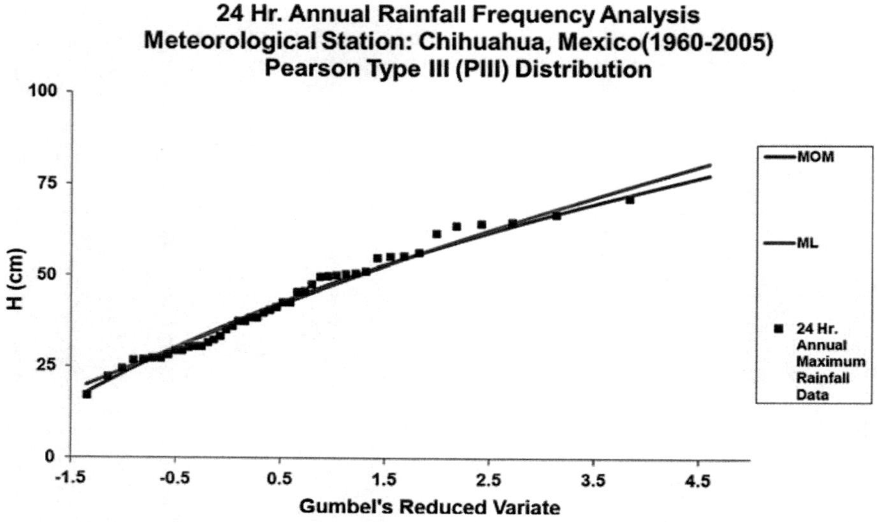

Fig. 7.11 Empirical and theoretical PIII distributions for 24 h, annual maximum rainfall data at Chihuahua, Mexico

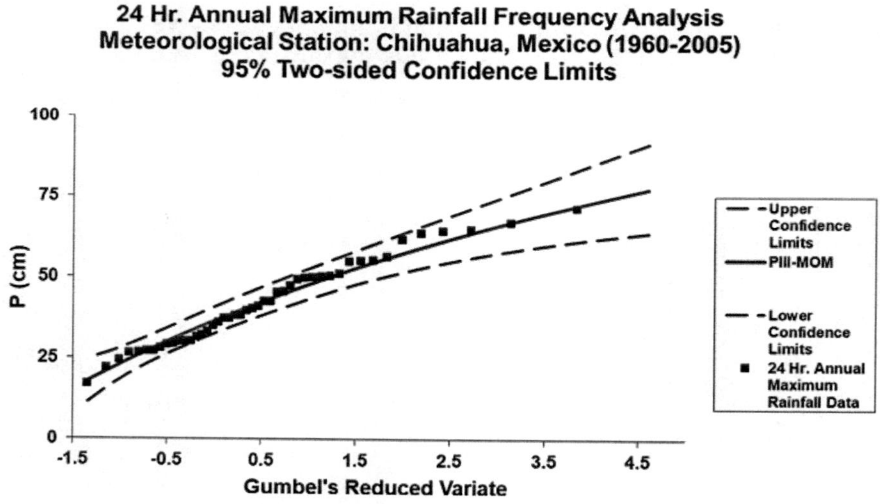

Fig. 7.12 Empirical and MOM theoretical frequency curves and MOM confidence limits for 24 h, annual maximum rainfall data at Chihuahua, Mexico

7.9.3 Maximum Significant Wave Height Frequency Analysis

By using the maximum significant wave height data, Castillo (1988), the following descriptive statistics were obtained and are shown in Fig. 7.13.

7.9.3.1 MOM Method

By using an Excel® spreadsheet, the results contained in Fig. 7.14, were obtained for the MOM Method applied to the PIII distribution for maximum significant wave height data.

7.9.3.2 ML Method

By using an Excel® spreadsheet, the results contained in Fig. 7.15, were obtained for the ML Method applied to the PIII distribution for maximum significant wave height data.

In Fig. 7.16 a comparison is made between the histogram and PIII theoretical density for maximum significant wave height data.

A graph containing the empirical and theoretical MOM and ML PIII frequency curves is shown in Fig. 7.17.

A graphical depiction of the empirical and theoretical frequency curves and the confidence limits of the best fit provided, in this case that of MOM, is shown in Fig. 7.18.

	A	B	C	D
1	**Maximum Significant Wave Height Frequency Analysis**			
2				
3	*Descriptive Statistics*		*Descriptive Statistics of the Natural Logarithms*	
4	Mean	4.4247	Mean	1.3628
5	Sample Standard Error	0.3053	Sample Standard Error	0.0748
6	Median	4.1666	Median	1.4270
7	Mode	3.5509	Mode	1.2672
8	Standard Deviation	2.1585	Standard Deviation	0.5287
9	Sample Variance	4.6591	Sample Variance	0.2796
10	Kurtosis	1.0740	Kurtosis	0.6324
11	Skewness	0.9053	Skewness	-0.6102
12	Range	10.4485	Range	2.5478
13	Minimum	0.8870	Minimum	-0.1199
14	Maximum	11.3355	Maximum	2.4279
15	Sum	221.2326	Sum	68.1403
16	Count	50.0000	Count	50.0000
17	Largest(1)	11.3355	Largest(1)	2.4279
18	Smallest(1)	0.8870	Smallest(1)	-0.1199
19	Mean Confidence Level(95.0%)	0.6134	Mean Confidence Level(95.0%)	0.1503

Fig. 7.13 Descriptive statistics for the maximum significant wave height data

	A	B	C	D	E	F	G	H	I
1	MAXIMUM SIGNIFICANT WAVE HEIGHT								
2	Pearson Type III (PIII) Distribution								
3	Method: MOM								
4									
5	Estimation of Parameters and Goodness of Fit Tests								
6	Skewness Coef. (Gamma Correction) =			1.0922	$\Gamma(\beta)$ =	32.1209			
7	Location Parameter =		0.4720	m					
8	Scale Parameter =		1.1787	m					
9	Shape Parameter =		3.3532						
10	SEF =	0.39		MARD =	7.79				
11									
12	Design Values and Confidence Limits								
13–16	Return Period Tr (years)	F(x)	Standard Normal Variate z_T	$\partial K_T/\partial\gamma$	Frequency Factor K_T	Squared root Stand. Error S_T (m)	Two-sided 95% Lower Limit (m)	Design Values H_T (m)	Two-sided 95% Upper Limi (m)
17	2	0.5	0.0000	-0.1504	-0.1761	0.3597	3	4	5
18	5	0.8	0.8416	-0.1308	0.7420	0.4580	5	6	7
19	10	0.9	1.2816	-0.0117	1.3343	0.6166	6	7	9
20	20	0.95	1.6449	0.1527	1.8869	0.8694	7	8	10
21	50	0.98	2.0537	0.4169	2.5819	1.2982	7	10	13
22	100	0.99	2.3263	0.6436	3.0902	1.6697	8	11	14
23	Two-sided Confidence Limits:		z_α =	1.96					

Fig. 7.14 MOM estimation of the parameters, standard errors, quantiles, and confidence limits for the maximum significant wave height data

	A	B	C	D	E	F	G	H	I
1	MAXIMUM SIGNIFICANT WAVE HEIGHT								
2	Pearson Type III (PIII) Distribution								
3	Method: ML								
4									
5	Estimation of Parameters and Goodness of Fit Tests								
6	Test Location Parameter =		1.8715	$F(x_o)$ =	5.5631E-05	$\Gamma(\beta)$ =	1.0431	$\psi(\beta)$ =	0.3409
7	Skewness Coef. (Gamma Correction) =			1.0922	<	2	Log-Likelihood function =		-97.3380
8	Location Parameter =		1.8715	m	Scale Parameter =	1.3853	m		
9	Shape Parameter =		1.8780						
10	SEF =	0.44		MARD =	5.11	AIC =	201		
11									
12	Variance-Covariance Matrix								
13	$var(x_0)$ =	-3.5058E-03	$cov(\alpha, x_0)$ =	-4.9650E-03	$cov(\beta, x_0)$ =	9.2617E-03	$var(\alpha)$ =	7.9071E-02	
14	$var(\beta)$ =	9.6216E-02	$cov(\alpha, \beta)$ =	-7.5905E-02	$\psi'(\beta)$ =	0.6982	Determinant =	-9.2522E-01	
15									
16	Design Values and Confidence Limits								
17–20	Return Period Tr (years)	F(x)	Standard Normal Variate z_T	$\partial x_T/\partial\alpha$	$\partial x_T/\partial\beta$	Squared root Stand. Error S_T (m)	Two-sided 95% Lower Limit (m)	Design Values H_T (m)	Two-sided 95% Upper Limit (m)
21	2	0.5	0.0000	1.5639	1.3713	0.2348	4	4	4
22	5	0.8	0.8416	2.8230	1.8469	0.4117	5	6	7
23	10	0.9	1.2816	3.6904	2.0919	0.5697	6	7	8
24	20	0.95	1.6449	4.5279	2.2874	0.7390	7	8	10
25	50	0.98	2.0537	5.6121	2.4964	0.9748	8	10	12
26	100	0.99	2.3263	6.4234	2.6273	1.1602	8	11	13
27	Two-sided Confidence Limits:		z_α =	1.96					

Fig. 7.15 ML estimation of the parameters, standard errors, quantiles, and confidence limits for the maximum significant wave height data

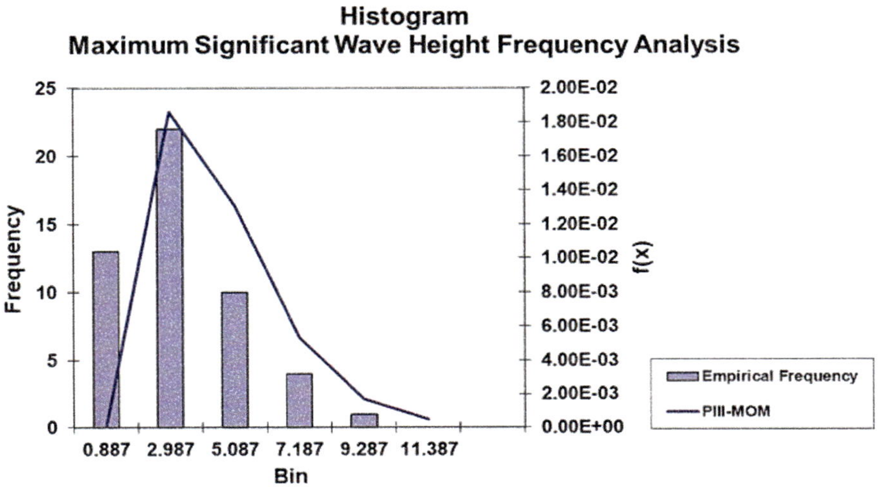

Fig. 7.16 Histogram and theoretical PIII density for maximum significant wave height data

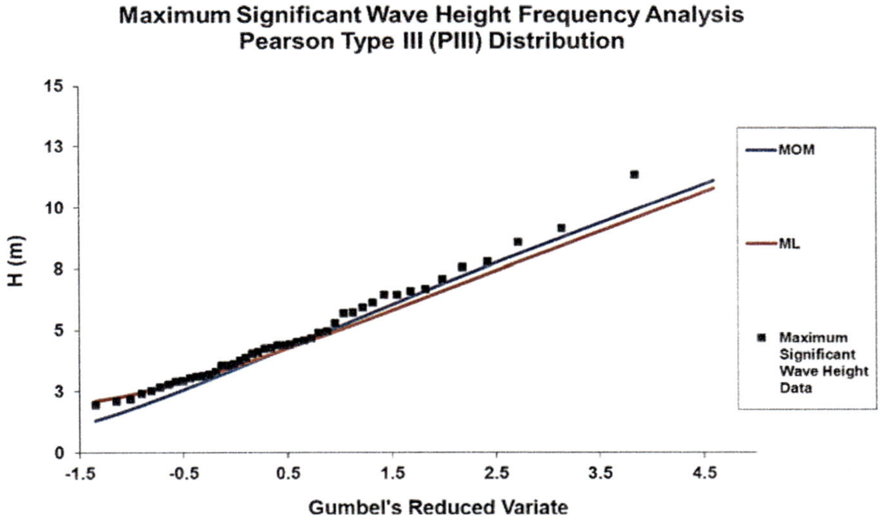

Fig. 7.17 Empirical and theoretical PIII distributions for maximum significant wave height data

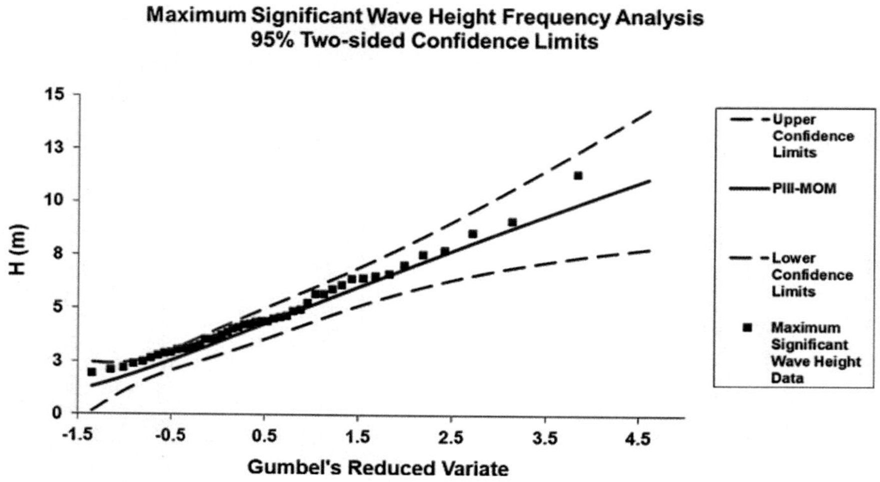

Fig. 7.18 Empirical and MOM theoretical frequency curves and MOM confidence limits for maximum significant wave height data

7.9.4 Annual Maximum Wind Speed Frequency Analysis

By using the annual maximum wind speed data, Castillo (1988), the following descriptive statistics were obtained and are shown in Fig. 7.19.

7.9.4.1 MOM Method

By using an Excel® spreadsheet, the results contained in Fig. 7.20, were obtained for the MOM Method applied to the PIII distribution for annual maximum wind speed data.

7.9.4.2 ML Method

By using an Excel® spreadsheet, the results contained in Fig. 7.21, were obtained for the ML Method applied to the PIII distribution for annual maximum wind speed data.

In Fig. 7.22 a comparison is made between the histogram and PIII theoretical density for annual maximum wind speed data.

In Fig. 7.23 are shown the empirical and mom theoretical frequency curves for annual maximum wind speed data.

In Fig. 7.24 are shown the empirical and mom theoretical frequency curves and mom confidence limits for annual maximum wind speed data.

Annual Maximum Wind Speed Frequency Analysis

Descriptive Statistics		*Descriptive Statistics of the Natural Logarithms*	
Mean	54.7964	Mean	3.9558
Sample Standard Error	2.9320	Sample Standard Error	0.0408
Median	47.9965	Median	3.8711
Mode	#N/D	Mode	#N/D
Standard Deviation	20.7327	Standard Deviation	0.2886
Sample Variance	429.8439	Sample Variance	0.0833
Kurtosis	12.5699	Kurtosis	3.6721
Skewness	3.0813	Skewness	1.6052
Range	121.5116	Range	1.4669
Minimum	36.4278	Minimum	3.5953
Maximum	157.9394	Maximum	5.0622
Sum	2739.8217	Sum	197.7906
Count	50.0000	Count	50.0000
Largest(1)	157.9394	Largest(1)	5.0622
Smallest(1)	36.4278	Smallest(1)	3.5953
Mean Confidence Level(95.0%)	5.8922	Mean Confidence Level(95.0%)	0.0820

Fig. 7.19 Descriptive statistics for annual maximum wind speed data

ANNUAL MAXIMUM WIND SPEED FREQUENCY ANALYSIS
Pearson Type III (PIII) Distribution
Method: MOM

Estimation of Parameters and Goodness of Fit Tests				
Skewness Coef. (Gamma Correction) =		3.7176	$\Gamma(\beta)$ =	3.1066
Location Parameter =	43.6426	m		
Scale Parameter =	38.5377	m		
Shape Parameter =	0.2894			
SEF = 6.69		MARD =	5.14	

Return Period Tr (years)	F(x)	Standard Normal Variate z_T	$\partial K_T/\partial\gamma$	Frequency Factor K_T	Squared root Stand. Error S_T (m)	Two-sided 95% Lower Limit (m)	Design Values H_T (m)	Two-sided 95% Upper Limit (m)
2	0.5	0.0000	-0.0157	-0.4122	0.2853	46	46	47
5	0.8	0.8416	-0.2067	0.2540	1.3846	57	60	63
10	0.9	1.2816	-0.2384	0.9706	1.5635	72	75	78
20	0.95	1.6449	-0.1963	1.8145	1.6031	89	92	96
50	0.98	2.0537	-0.0486	3.0860	2.3703	114	119	123
100	0.99	2.3263	0.1222	4.1479	3.6970	134	141	148
Two-sided Confidence Limits:		z_α =	1.96					

Fig. 7.20 MOM estimation of the parameters, standard errors, quantiles, and confidence limits for annual maximum wind speed data

	A	B	C	D	E	F	G	H	I
1	ANNUAL MAXIMUM WIND SPEED FREQUENCY ANALYSIS								
2	Pearson Type III (PIII) Distribution								
3	Method: ML								
4									
5	Estimation of Parameters and Goodness of Fit Tests								
6	Test Location Parameter =		36.4150	$F(x_o)$ =	-4.0344E+00				
7	Skewness Coef. (Gamma Correction) =		3.7176		>	2			
8	$\Gamma(\beta)$ =	0.9775	$\psi(\beta)$ =	-0.5108	No solution exists				
9	Location Parameter =		36.4150	m/s	Scale Parameter =	13.7463	m/s		
10	Shape Parameter =		1.0416		Log-Likelihood function =		-183.2113		
11	SEF =	11.21		MARD =	6.03	AIC =	372		
12									
13	Variance-Covariance Matrix								
14	$var(x_0)$ =	-4.1772E-03	$cov(\alpha,x_0)$ =	-1.5283E-01	$cov(\beta,x_0)$ =	1.1884E-02	$var(\alpha)$ =	3.9387E+00	
15	$var(\beta)$ =	7.5041E-05	$cov(\alpha,\beta)$ =	-1.2400E-02	$\psi'(\beta)$ =	1.5503	Determinant =	-1.5577E-02	
16									
17	Design Values and Confidence Limits								
18			Standard			Squared root	Two-sided	Design	Two-sided
19			Normal			Stand. Error	95%	Values	95%
20	Return Period		Variate			S_T	Lower Limit	V_T	Upper Limit
21	Tr (years)	F(x)	z_T	$\partial x_T/\partial\alpha$	$\partial x_T/\partial\beta$	(m/s)	(m/s)	(m/s)	(m/s)
22	2	0.5	0.0000	0.7425	13.3103	1.4231	44	47	49
23	5	0.8	0.8416	1.6606	20.1837	3.1667	53	59	65
24	10	0.9	1.2816	2.3518	23.7544	4.5036	60	69	78
25	20	0.95	1.6449	3.0494	26.5768	5.8620	67	78	90
26	50	0.98	2.0537	3.9856	29.5250	7.6936	76	91	106
27	100	0.99	2.3263	4.7057	31.3097	9.1070	83	101	119
28	Two-sided Confidence Limits:		z_α =	1.96					

Fig. 7.21 ML estimation of the parameters, standard errors, quantiles, and confidence limits for annual maximum wind speed data

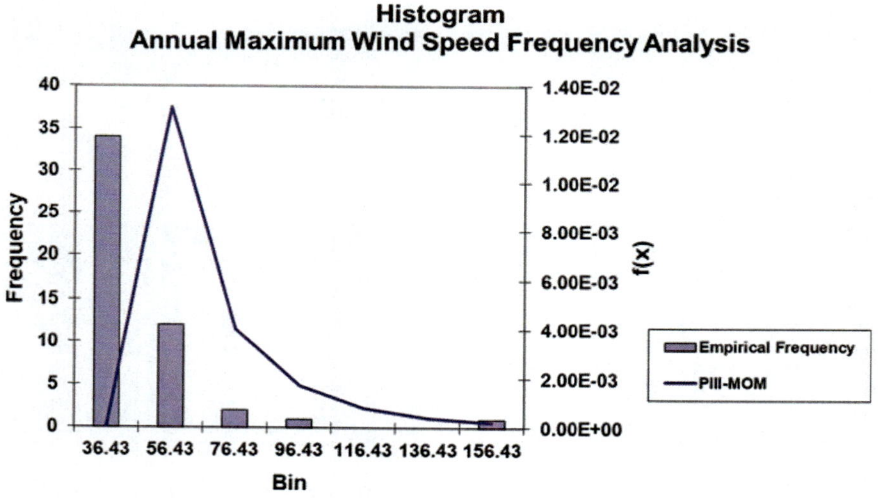

Fig. 7.22 Histogram and theoretical PIII density for annual maximum wind speed data

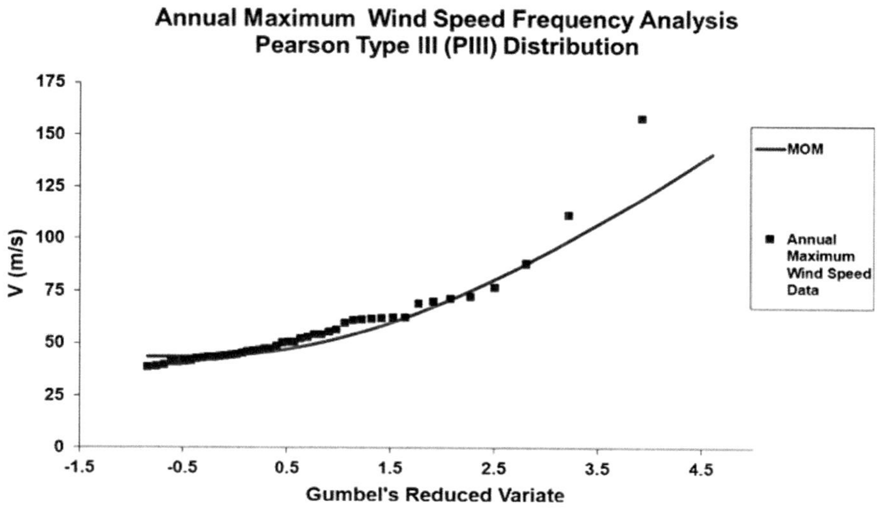

Fig. 7.23 Empirical and theoretical MOM PIII distributions for annual maximum wind speed data

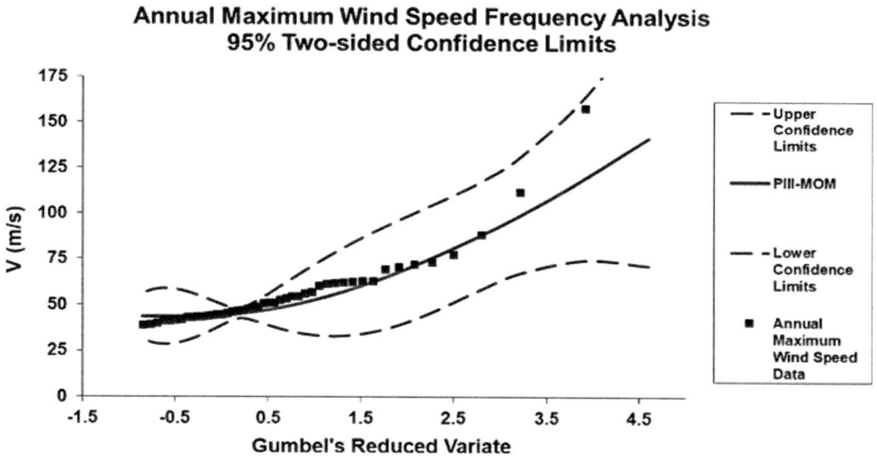

Fig. 7.24 Empirical and MOM theoretical frequency curves and MOM confidence limits for annual maximum wind speed data

8 Log-Pearson Type III Distribution

Happiness makes up in height for what it lacks in length.
R. Frost.

8.1 Introduction

The probability distribution function Log-Pearson type III (PIII) is one of the members of the Pearson family of distributions. These distributions were named in such a way after the great British mathematician Karl Pearson (1857–1936), the father of Mathematical Statistics.

The Log-Pearson type III (LPIII) distribution was declared in 1967, by the U. S. Federal Water Resources Council, as the official distribution to be used by all U. S. government agencies. Bulletins 17, 17A, 17B, and 17C contained the official accepted procedures to be used by such government agencies.

There has been a lot of research done in this distribution trying to find the best methods of estimation of its parameters, standard errors of the variates and some other features of the LPIII distribution.

The LPIII distribution behaves as Pearson type III distribution in the log domain, that is, if y = Ln(x) is Pearson type III distributed, then x is LPIII distributed. The LPIII distribution may assume many shapes, Bobée (1975): J-shaped, reverse J-shaped, U-shaped, inverted U-shaped, inverted U-shaped with inflexions, bell shaped with an upper bound, bell shaped with a lower bound, etc. In flood frequency analysis, the only form of interest is the unimodal, continuous from zero to infinity, has either an infinitely high order ($\partial f(x)/\partial x = \infty$) or smooth contact ($\partial f(x)/\partial x = 0$) with the lower limit and is unbounded at the upper limit, Kite (1988). The LPIII distribution complies with these criteria only when $\beta > 1$ and when $1/\alpha > 0$. When the coefficient of skewness, γ, is negative, this means that the

J. A. Raynal Villaseñor, *Frequency Analyses of Natural Extreme Events*,
Earth and Environmental Sciences Library,
https://doi.org/10.1007/978-3-030-86390-6_8

scale parameter, α, is also negative. This condition it is not suitable for flood frequency analysis. Reich (1972) has shown that maximum observed events exceeded the computed upper bounds when fitting the LPIII distribution to samples with negative values in their coefficients of skewness and with scale parameters negative, too.

8.2 Chapter Objectives

After reading this chapter, you will know how to:

(a) Recognize the distribution and density functions of the LPIII distribution
(b) Estimate the parameters of the LPIII distribution
(c) Compute the quantiles and confidence limits of the LPIII distribution
(d) Make a graphic display of your data and the LPIII distribution
(e) Develop an application of all the above using Excel® spreadsheets

8.3 Probability Distribution and Density Functions

The probability distribution function of the Log-Pearson type III (LPIII) distribution is:

$$F(x) = \frac{1}{|\alpha|\Gamma(\beta)x} \int_0^x \left(\frac{Ln(x) - y_0}{\alpha}\right)^{\beta-1} \exp\left(-\frac{(Ln(x) - y_0)}{\alpha}\right) dx \tag{8.1}$$

where y_0, α and β are the location, scale, and shape parameters, respectively, and $\Gamma(.)$ is the complete gamma function, as it was defined in Chap. 6. $F(x)$ is the probability distribution function of the random variable x and in the case of maxima natural extreme frequency analysis is equal to the non-exceedance probability, $Pr(X \leq x)$. The scale and shape parameters are restricted to be $1/\alpha > 0$ and $\beta > 1$, respectively. Additionally, if the coefficient of skewness, γ, is negative this means that the scale parameter, α, is negative, and such condition it not suitable for maxima natural extreme frequency analysis. The domain of variable x in this distribution is $0 < x < \infty$.

The probability density function for the LPIII distribution is:

$$f(x) = \frac{1}{|\alpha|\Gamma(\beta)x} \left(\frac{Ln(x) - y_0}{\alpha}\right)^{\beta-1} \exp\left(-\frac{(Ln(x) - y_0}{\alpha}\right) \tag{8.2}$$

where $f(x)$ is the probability density function of the random variable x.

8.4 Estimation of the Parameters

8.4.1 MOM Method

There are three options in MOM the method to estimate the parameters of the LPIII distribution:

(a) Application of the PIII distribution to the natural logarithms of x. This option will be designated as the MOM1 method
(b) Direct application of the LPIII distribution. This option will be designated as the MOM2 method
(c) U.S. Water Resources Council Method (WRCM). This option will be designated as the WRC method. In this option the logarithms base 10 of x are used.

8.4.1.1 MOM1 Method

In this option, the method of moments is applied to the natural logarithms of x using the PIII distribution. So, the methodology contained in Sect. 7.3.1 applies and Eqs. (7.9–7.11) are needed to estimate the parameters of the LPIII distribution by the MOM1 method.

8.4.1.2 MOM2 Method

The r-th population moments with respect to the origin of the LPIII distribution can be computed by the following general expression, Bobée (1975):

$$\mu_r^{'} = \frac{\exp(r y_0)}{(1 - r\alpha)^{\beta}} \tag{8.3}$$

The natural logarithms of the three first moments with respect to the origin are, Bobée (1975):

$$Ln\left(\mu_1^{'}\right) = y_0 - \beta Ln(1 - \alpha) \tag{8.4}$$

$$Ln\left(\mu_2^{'}\right) = 2y_0 - \beta Ln(1 - 2\alpha) \tag{8.5}$$

$$Ln\left(\mu_3^{'}\right) = 3y_0 - \beta Ln(1 - 3\alpha) \tag{8.6}$$

these three equations produce the following, Bobée (1975):

$$\frac{\left[Ln(\mu_3^{'}) - 3Ln(\mu_1^{'})\right]}{\left[Ln(\mu_2^{'}) - 2Ln(\mu_1^{'})\right]} = \frac{Ln\left[\frac{(1-\alpha)^3}{(1-3\alpha)}\right]}{Ln\left[\frac{(1-\alpha)^2}{(1-2\alpha)}\right]} \tag{8.7}$$

The left-hand side of the previous equation may be designated as B, Bobée (1975):

$$B = \frac{\left[Ln\left(\mu_3'\right) - 3Ln\left(\mu_1'\right)\right]}{\left[Ln\left(\mu_2'\right) - 2Ln\left(\mu_1'\right)\right]} \tag{8.8}$$

and can be computed directly from the sample moments about the origin. The right-hand side of the Eq. (8.7) may be designated as A:

$$A = \frac{Ln\left[\frac{(1-\alpha)^3}{(1-3\alpha)}\right]}{Ln\left[\frac{(1-\alpha)^2}{(1-2\alpha)}\right]} \tag{8.9}$$

So, Eq. (8.7) becomes to:

$$F\left(\hat{\alpha}\right) = B - A = 0 \tag{8.10}$$

Equation (8.10) must be solved by a procedure of finding the roots of such an equation, again the bisection method will give good results in solving this equation. The solution of this equation provides the value of the estimator of the scale parameter obtained by this option of the method of moments. The other two parameters are calculated as, Kite (1988):

$$\hat{\beta} = \frac{\left[Ln\left(\hat{\mu}_2'\right) - 2Ln\left(\hat{\mu}_1'\right)\right]}{\left[Ln(1-\hat{\alpha})^2 - Ln\left(1-2\hat{\alpha}\right)\right]} \tag{8.11}$$

$$\hat{y}_0 = Ln\left(\hat{\mu}_1'\right) + \hat{\beta}Ln\left(1-\hat{\alpha}\right) \tag{8.12}$$

and the parameters are used to compute the mean, standard deviation, and the coefficient of skewness of the logarithms of x, Kite (1988):

$$\hat{\mu}_y = \hat{y}_0 + \hat{\alpha}\hat{\beta} \tag{8.13}$$

$$\hat{\sigma}_y = \hat{\alpha}\sqrt{\hat{\beta}} \tag{8.14}$$

$$\hat{\gamma}_y = \frac{2}{\sqrt{\hat{\beta}}} \tag{8.15}$$

The coefficient of skewness must be corrected for bias using the following expression, Kite (1988):

$$\hat{\gamma}_{LPIII} = \frac{\frac{1}{N}\sum_{i=1}^{N}\left(Ln(x_i) - \hat{\mu}_y\right)^3}{\hat{\sigma}_y^3}\left[\frac{(N(N-1))^{\frac{1}{2}}}{(N-2)}\right]\left[1 + \frac{8.5}{N}\right] \tag{8.16}$$

Another way to compute the scale parameter, α, is that proposed by Kite (1988):

$$A = \frac{1}{\alpha} - 3 \tag{8.17}$$

$$C = \frac{1}{(B - 3)} \tag{8.18}$$

a series of polynomial approximation can be derived, Kite (1988), as:

(a) for $3.5 < B \leq 6.0$:

$$A = -0.23019 + 1.65262C + 0.20911C^2 - 0.04557C^3 \tag{8.19}$$

(b) for $3.0 < B \leq 3.5$:

$$A = -0.47157 + 1.99955C \tag{8.20}$$

So, the scale parameter, α, can be evaluated as:

$$\hat{\alpha} = \frac{1}{(A + 3)} \tag{8.21}$$

the other two parameters are computed as described before.

8.4.1.3 WRC Method

The methodology proposed by the U. S. Water Resources Council, is that contained in USGS (2019) and consist in the following procedures. The data must be transformed to its logarithms base 10, $y = \text{Log}_{10}(x)$. The evaluation of parameters of the LPIII distribution is performed by applying the PIII distribution to the logarithms base 10, USGS (2019), as follows:

$$\hat{y}_0 = \hat{\mu}_y - \hat{\alpha}\hat{\beta} \tag{8.22}$$

$$\hat{\alpha} = \text{sign}\left(\hat{\gamma}_y\right)\left(\frac{\widehat{\sigma_y}^2}{\hat{\beta}}\right)^{\frac{1}{2}} \tag{8.23}$$

$$\hat{\beta} = \frac{4}{\left(\hat{\gamma}_y\right)^2} \tag{8.24}$$

where

$\widehat{\mu_y}$ is the mean of the logarithms base 10 of the x's.

$\widehat{\sigma_y}^2$ is the variance of the logarithms base 10 of the x's.

$\widehat{\gamma_y}$ is the skewness coeficient of the logarithms base 10 of the x's.

8.4.2 ML Method

The likelihood function of the LPIII distribution is:

$$L(\alpha,\beta,y_0)=\prod_{i=1}^{N}\left\{\frac{1}{|\alpha|\Gamma(\beta)x_i}\left(\frac{Ln(x_i)-y_0}{\alpha}\right)^{\beta-1}\exp\left(-\frac{(Ln(x_i)-y_0}{\alpha}\right)\right\} \tag{8.25}$$

and the Log-Likelihood function of the LPIII distribution is:

$$LL(\alpha,\beta,y_0)=-N\beta Ln(|\alpha|)-NLn(\Gamma(\beta))-\sum_{i=1}^{N}Ln(x_i)$$
$$+(\beta-1)\sum_{i=1}^{N}Ln(Ln(x_i)-y_0)-\frac{1}{\alpha}\sum_{i=1}^{N}(Ln(x_i)-y_0) \tag{8.26}$$

Now, taking the first-order partial derivatives of this equation with respect to parameters α, β and y_0 and set them equal to zero:

$$\frac{\partial LL}{\partial\alpha}=-\frac{N\beta}{\alpha}+\frac{1}{\alpha^2}\sum_{i=1}^{N}(Ln(x_i-y_0)=0 \tag{8.27}$$

$$\frac{\partial LL}{\partial\beta}=-NLn(\alpha)-N\psi(\beta)+\sum_{i=1}^{N}Ln(Ln(x_i)-y_0)=0 \tag{8.28}$$

$$\frac{\partial LL}{\partial y_0}=\frac{N}{\alpha}-(\beta-1)\sum_{i=1}^{N}\left[\frac{1}{(Ln(x_i)-y_0)}\right]=0 \tag{8.29}$$

The parameters of the LPIII distribution can be estimated by the ML method by solving the following three equations:

$$\hat{\alpha}=\frac{1}{N}\sum_{i=1}^{N}[Ln(x_i)-y_0]-\frac{N}{\sum_{i=1}^{N}[Ln(x_i)-y_0]^{-1}} \tag{8.30}$$

$$\hat{\beta}=\left[1-\frac{N^2}{\sum_{i=1}^{N}[Ln(x_i)-y_0]\sum_{i=1}^{N}[Ln(x_i)-y_0]^{-1}}\right]^{-1} \tag{8.31}$$

$$F(\hat{y}_0)=-N\ln(\alpha)-N\psi(\beta)+\sum_{i=1}^{N}Ln[Ln(x_i)-y_0] \tag{8.32}$$

Equation (8.32) must be solved by a procedure of finding the roots of such an equation, again the bisection method will give good results in solving this equation.

8.5 Estimation of Quantiles for the LPIII Distribution

8.5.1 Estimation of MOM1, MOM2 and ML Quantiles for the LPIII Distribution

The MOM1, MOM2, and ML quantiles of the LPIII distribution can be estimated using the following expressions derived for the PIII distribution, but using the LPIII estimated parameters:

$$y_T = Ln(x_T) = \hat{\alpha}\hat{\beta}\left[1 - \frac{1}{9\hat{\beta}} + z_T\sqrt{\frac{1}{9\hat{\beta}}}\right]^3 + \hat{y}_0 \tag{8.33}$$

$$x_T = e^{y_T} \tag{8.34}$$

8.5.2 Estimation of WRC Quantiles for the LPIII Distribution

The WRC quantiles of the LPIII distribution can be estimated using the following expressions derived for the PIII distribution, but using the LPIII estimated parameters:

$$y_T = Log_{10}(x_T) = \hat{\alpha}\hat{\beta}\left[1 - \frac{1}{9\hat{\beta}} + z_T\sqrt{\frac{1}{9\hat{\beta}}}\right]^3 + \hat{y}_0 \tag{8.35}$$

and the quantiles in the real domain are obtained as:

$$x_T = 10^{y_T} \tag{8.36}$$

8.5.3 Examples of Estimation of MOM1, MOM2, WRC and ML Quantiles for the LPIII Distribution

Find the MOM1, and ML estimators of the quantiles 2, 5, 10, 20, 50, 100, 500, and 1000 years of return period, for the LPIII distribution for the flood data of gauging station Huites, Mexico, contained in Appendix A.

The estimation of MOM1, and ML quantiles, the MOM2 quantiles were not obtainable for this flood sample data, for the LPIII distribution is made by using the parameters estimated in the preceding sections and then inserting the parameters into Eqs. (8.33) to (8.36). Table 8.1 summarizes these results.

Table 8.1 Estimation of MOM1, WRC, and ML quantiles for the LPIII distribution for gauging station Huites, Mexico

Tr (Years)	MOM1	WRC	ML
	Q_T (m^3/s)	Q_T (m^3/s)	Q_T (m^3/s)
2	1464	1776	1774
5	2749	3398	3374
10	4605	4931	4889
20	7916	6828	6763
50	16,758	10,048	9946
100	30,238	13,157	13,022
500	127,370	23,406	23,178
1000	242,983	29,529	29,257

8.6 Goodness of Fit Test

The standard error of fit (SEF) for the LPIII has the following form:

$$SEF = \left[\frac{\sum_{i=1}^{N} (xi - yi)^2}{(N-3)} \right]^{1/2} \tag{8.37}$$

while the mean absolute relative deviation (MARD) remains the same as it was defined in Chap. 2:

$$MARD = \frac{100}{N} \sum_{i=1}^{N} \left| \frac{(x_i - y_i)}{x_i} \right| \tag{8.38}$$

where x_i are the sample historical values, y_i are the distribution function values, corresponding to the same return periods of the historical values, N is the sample size.

8.6.1 Examples of Application of the SEF and MARD to the MOM1, WRC and ML Estimators of the Parameters of the LPIII Distribution

Find the values of the SEF and MARD of the MOM1, and ML estimators of the parameters of the LPIII distribution for the flood sample data of gauging station Huites, Mexico (1942–1992), contained in Appendix A.

The values of SEF and MARD for the MOM1, and ML estimators of the parameters of the LPIII distribution for the flood sample data of gauging station Huites, Mexico, have been obtained through the application of Eqs. (8.32) and (8.33) using the parameters obtained in previous examples. The results are as follows:

(a) MOM1 Method

$$SEF = \left[\frac{\sum_{i=1}^{N} (xi - yi)^2}{(N - 3)} \right]^{1/2} = \left[\frac{259575.7998}{(51 - 3)} \right]^{\frac{1}{2}} = 509$$

$$MARD = \frac{100}{N} \sum_{i=1}^{N} \left| \frac{(x_i - y_i)}{x_i} \right| = \frac{(100)(3.5019)}{51} = 7$$

(b) MOM2 Method

This option was not suitable.

(c) Maximum Likelihood

$$SEF = \left[\frac{\sum_{i=1}^{N} (xi - yi)^2}{(N - 3)} \right]^{1/2} = \left[\frac{276570.7784}{(51 - 3)} \right]^{\frac{1}{2}} = 526$$

$$MARD = \frac{100}{N} \sum_{i=1}^{N} \left| \frac{(x_i - y_i)}{x_i} \right| = \frac{(100)(3.5373)}{51} = 7$$

(d) WRC Method

$$SEF = \left[\frac{\sum_{i=1}^{N} (xi - yi)^2}{(N - 3)} \right]^{1/2} = \left[\frac{259575.7998}{(51 - 3)} \right]^{\frac{1}{2}} = 509$$

$$MARD = \frac{100}{N} \sum_{i=1}^{N} \left| \frac{(x_i - y_i)}{x_i} \right| = \frac{(100)(3.5737)}{51} = 7$$

In this case all the methods, MOM1, WRC, and ML, the results are comparable. Their SEF values are very close one to each other and their MARD values are the same. So, any of these methods will give good results.

8.7 Estimation of Confidence Limits for the LPIII Distribution

8.7.1 Estimation of Confidence Limits for the LPIII Distribution for MOM1, MOM2, and ML Methods

By assuming that the quantiles are normally distributed, the following form can be used to set the confidence limits on such quantiles, for the MOM1, MOM2 and ML methods:

$$x_l = x_T \pm z_\alpha S_T \tag{8.39}$$

where x_l is the upper or lower confidence limit, depending on the sign in the preceding formula (+) for the upper confidence limit and (−) for the lower confidence limit, z_α is the standard normal variate corresponding to a confidence level α, and S_T is squared root of the standard error of the estimate. The evaluation procedures of the standard errors of the estimates will be described in the following subsections.

8.7.2 Estimation of Confidence Limits for the LPIII Distribution for WRC Method

By assuming that the quantiles are normally distributed, the following form can be used to set the confidence limits on such quantiles, for the WRC method, Rao and Hamed (2000):

$$x_u = \overline{y} + K_u \hat{\sigma}_y \tag{8.40}$$

$$x_l = \overline{y} + K_l \hat{\sigma}_y \tag{8.41}$$

where x_u is the upper confidence limit, x_l is the lower confidence limit, $\overline{y}$ is the mean of the logarithms base 10 of the sample of data, K_u and K_l are coefficients associated with the upper confidence limit and the lower confidence limit, and $\hat{\sigma}$ is the standard deviation of the logarithms base 10 of the sample of data.

The procedure to obtain K_u and K_l are as follows, Rao and Hamed (2000):

(a) First obtain a normal standard variate z_a with exceedance probability α, where:

$$\alpha = \frac{1-\beta}{2} \tag{8.42}$$

(b) The confidence limits factors K_u and K_l are calculated as, Rao and Hamed (2000):

$$K_u = \frac{K_T + \sqrt{K_T^2 - pq}}{p} \tag{8.43}$$

$$K_l = \frac{K_l - \sqrt{K_T^2 - pq}}{p} \tag{8.44}$$

and:

$$p = 1 - \frac{z_\alpha^2}{2(N-1)} \tag{8.45}$$

$$q = K_T^2 - \frac{Z_\alpha^2}{N} \tag{8.46}$$

and K_T is, Kite (1988):

$$\begin{aligned} K_T &= z_T + \left(z_T^2 - 1\right)\left(\frac{\hat{\gamma}_y}{6}\right) + \frac{\left(z_T^3 - 6z_T\right)}{3}\left(\frac{\hat{\gamma}_y}{6}\right)^2 \\ &\quad - \left(z_T^2 - 1\right)\left(\frac{\hat{\gamma}_y}{6}\right)^3 + z_T\left(\frac{\hat{\gamma}_y}{6}\right)^4 - \frac{1}{3}\left(\frac{\hat{\gamma}_y}{6}\right)^5 \end{aligned} \tag{8.47}$$

8.8 Estimation of Standard Errors for the LPIII Distribution

8.8.1 MOM Method

8.8.1.1 MOM-1 Method

The general form of the MOM estimator of the standard error of the estimate of a three- parameters distribution is, Kite (1988):

$$\begin{aligned} S_T^2 &= \left(\frac{\partial x_T}{\partial m_1'}\right)^2 \mathrm{var}(m_1') + \left(\frac{\partial x_T}{\partial m_2}\right)^2 \mathrm{var}(m_2) + \left(\frac{\partial x_T}{\partial m_3}\right)^2 \mathrm{var}(m_3) \\ &\quad + 2\left(\frac{\partial x_T}{\partial m_1'}\right)\left(\frac{\partial x_T}{\partial m_2}\right)\mathrm{cov}(m_1', m_2) + 2\left(\frac{\partial x_T}{\partial m_1'}\right)\left(\frac{\partial x_T}{\partial m_3}\right)\mathrm{cov}(m_1', m_3) \\ &\quad + 2\left(\frac{\partial x_T}{\partial m_2}\right)\left(\frac{\partial x_T}{\partial m_3}\right)\mathrm{cov}(m_2, m_3) \end{aligned} \tag{8.48}$$

and Eq. (8.35) can be applied to the natural logarithms of the natural extreme data and simplified in terms of the frequency factor, K_T, as, Kite (1988):

$$\begin{aligned} S_{Ty}^2 = \frac{\sigma_y^2}{N}\Bigg\{ & 1 + K_T\gamma_y + \frac{K_T^2}{4}\left(\kappa_y - 1\right) \\ & + \frac{\partial K_T}{\partial \gamma_y}\left[2\kappa_y - 3\gamma_y^2 - 6 + K_T\left(\lambda_{1y} - \frac{6\gamma_y\kappa_y}{4} - \frac{10\gamma_y}{4}\right)\right] \end{aligned} \tag{8.49}$$

The previous equation may be further simplified by using the relationships given in Eqs. (7.6–7.8) in the logarithmic domain:

$$\begin{aligned} S_{Ty}^2 = \frac{\sigma_y^2}{N}\Bigg\{ & 1 + K_T\gamma_y + \frac{K_T^2}{2}\left(\frac{3\gamma_y^2}{4} + 1\right) + 3K_T\left(\frac{\partial K_T}{\partial \gamma_y}\right)\left(\gamma_y + \frac{\gamma_y^3}{4}\right) \\ & + 3\left(\frac{\partial K_T}{\partial \gamma_y}\right)^2\left(2 + 3\gamma_y^2 + \frac{5\gamma_y^4}{8}\right)\Bigg\} \end{aligned} \tag{8.50}$$

Table 8.2 Estimation of MOM1 standard errors and the two-sided 95% confidence limits for the LPIII distribution for gauging station Huites, Mexico

Return period Tr (years)	Squared root Standard Error S_T (m³/s)	Two-sided 95% lower (m³/s)	Design values Q_T (m³/s)	Two-sided 95% upper (m³/s)
2	218.54	1035	1464	1892
5	687.59	1401	2749	4096
10	1303.51	2050	4605	7160
20	3241.08	1563	7916	14,268
50	12,800.40	−8331	16,758	41,847
100	36,070.67	−40,460	30,238	100,937
500	414,512.44	−685,074	127,370	939,815
1000	1,244,058.29	−2,195,371	242,983	2,681,337

Two-sided Confidence Limits: $z_\alpha = 1.96$.

The partial derivative of the frequency factor with respect to skewness coefficient may be evaluated as:

$$\frac{\partial K_T}{\partial \gamma_y} \approx \frac{z_T^2 - 1}{6} + \frac{4\left(z_T^3 - 6z_T\right)\gamma_y}{6^3} - \frac{3\left(z_T^2 - 1\right)\gamma_y^2}{6^3} + \frac{4z_T\gamma_y^3}{6^4} - \frac{10\gamma_y^4}{6^6} \tag{8.51}$$

and the frequency factor for the LPIII distribution was shown in Eq. (8.47)

By inserting these results into Eq. (8.37), the moment estimator of the standard error of the estimate for the LPIII distribution, in the logarithmic domain, may be obtained. Now, to transform the standard error of the estimate for the LPIII distribution into the real domain, the following transformation has been suggested, Kite (1988):

$$S_T = \frac{x_T\left(\exp(S_{Ty}) - \exp(-S_{Ty})\right)}{2} \tag{8.52}$$

Example of Estimation of MOM1 Standard Errors and the Two-Sided 95% Confidence Limits for the LPIII Distribution

Find the MOM1 estimators of the standard errors and the two-sided 95% confidence limits for the LPIII distribution for the flood sample data of gauging station Huites, Mexico (1942–1992), contained in Appendix A.

The computation of MOM1 estimators of the standard errors and the two-sided 95% confidence limits for the LPIII distribution is made by using the MOM1 estimators of the parameters and then inserted in the Eqs. (8.37) to (8.40), using selected values of the return intervals. Table 8.2 summarizes these results.

8.8.1.2 MOM2 Method

This option was not obtainable.

Table 8.3 Estimation of WRC two-sided 95% confidence limits for the LPIII distribution for gauging station Huites, Mexico

Return Period Tr (years)	Two-sided 95% Lower limit (m^3/s)	Design values Q_T (m^3/s)	Two-sided 95% upper limit (m^3/s)
2	1427	1776	2151
5	2728	3398	4382
10	3871	4931	6836
20	5223	6828	10,223
50	7427	10,048	16,670
100	9487	13,157	23,578
500	16,009	23,406	50,021
1000	19,782	29,529	67,997

Two-sided Confidence Limits: $z_\alpha = 1.96$

8.8.1.3 WRC Method

For the WRC method the standard errors for the LPIII distribution are evaluated as the standard deviation of the logarithms base 10 of the sample, as it may be seen in Eqs. (8.40) and (8.41), Rao and Hamed (2000).

Example of Estimation of WRC Standard Errors and the Two-Sided 95% Confidence Limits for the LPIII Distribution

Find the WRC estimators of the standard errors and the two-sided 95% confidence limits for the LPIII distribution for the flood sample data of gauging station Huites, Mexico (1942–1992), contained in Appendix A.

The computation of WRC estimators of the standard errors and the two-sided 95% confidence limits for the LPIII distribution is made by using the WRC estimators of the parameters and then inserted in the Eqs. (8.37) to (8.40), using selected values of the return intervals. Table 8.3 summarizes these results.

8.8.2 ML Method

The general equation for the ML estimator of the standard error of the estimate of a three-parameters distribution is, Kite (1988):

$$\begin{aligned} S_T^2 = {} & \left(\frac{\partial x_T}{\partial \alpha}\right)^2 \operatorname{var}(\alpha) + \left(\frac{\partial x_T}{\partial \beta}\right)^2 \operatorname{var}(\beta) \\ & + \left(\frac{\partial x_T}{\partial \gamma}\right)^2 \operatorname{var}(\gamma) + 2\left(\frac{\partial x_T}{\partial \alpha}\right)\left(\frac{\partial x_T}{\partial \beta}\right)\operatorname{cov}(\alpha, \beta) \\ & + 2\left(\frac{\partial x_T}{\partial \alpha}\right)\left(\frac{\partial x_T}{\partial \gamma}\right)\operatorname{cov}(\alpha, \gamma) + 2\left(\frac{\partial x_T}{\partial \beta}\right)\left(\frac{\partial x_T}{\partial \gamma}\right)\operatorname{cov}(\beta, \gamma) \end{aligned} \tag{8.53}$$

and the elements of the Fisher's information matrix are, Condie (1977, 1979):

$$-\frac{\partial^2 LL}{\partial \alpha^2} = \frac{2\sum_{i=1}^{N}(Ln(x_i) - y_0)}{\alpha^3} - \frac{N\beta}{\alpha^2} \tag{8.54}$$

$$-\frac{\partial^2 LL}{\partial \beta^2} = N\psi'(\beta) \tag{8.55}$$

$$-\frac{\partial^2 LL}{\partial x_0^2} = (\beta - 1)\sum_{i=1}^{N} \frac{1}{(Ln(x_i) - y_0)^2} \tag{8.56}$$

$$-\frac{\partial^2 LL}{\partial \alpha \partial \beta} = \frac{N}{\alpha} \tag{8.57}$$

$$-\frac{\partial^2 LL}{\partial \alpha \partial x_0} = \frac{N}{\alpha^2} \tag{8.58}$$

$$-\frac{\partial^2 LL}{\partial \beta \partial x_0} = \sum_{i=1}^{N}[Ln(x_i) - y_0]^{-1} \tag{8.59}$$

but the Eqs. (8.42), (8.44) and (8.47) can be simplified by using the following result, Condie (1977, 1979):

$$\sum_{i=1}^{N}[Ln(x_i) - y_0]^r = \frac{N\alpha^r\Gamma(\beta + r)}{\Gamma(\beta)} \tag{8.60}$$

so:

$$-\frac{\partial^2 LL}{\partial \alpha^2} = \frac{N\beta}{\alpha^2} \tag{8.61}$$

$$-\frac{\partial^2 LL}{\partial x_0^2} = \frac{N}{\alpha^2(\beta - 2)} \tag{8.62}$$

$$-\frac{\partial^2 LL}{\partial \beta \partial x_0} = \frac{N}{[\alpha(\beta - 1)]} \tag{8.63}$$

and the Fisher's information matrix is constructed as, Condie (1977, 1979):

$$[I] = N\begin{bmatrix} \frac{\beta}{\alpha^2} & \frac{1}{\alpha} & \frac{1}{\alpha^2} \\ \frac{1}{\alpha} & \psi'(\beta) & \frac{1}{\alpha(\beta-1)} \\ \frac{1}{\alpha^2} & \frac{1}{\alpha(\beta-1)} & \frac{1}{\alpha^2(\beta-2)} \end{bmatrix} \tag{8.64}$$

Finally, the variances and covariances of the parameters of the LPIII distribution are found as, Condie (1977, 1979):

$$\mathrm{var}(\alpha) = \frac{1}{ND\alpha^2}\left[\frac{\psi'(\beta)}{(\beta-2)} - \frac{1}{(\beta-1)^2}\right] \tag{8.65}$$

$$\mathrm{var}(\beta) = \frac{2}{ND\alpha^4(\beta-2)} \tag{8.66}$$

$$\mathrm{var}(x_0) = \frac{1}{ND\alpha^2}\left[\beta\psi'(\beta) - 1\right] \tag{8.67}$$

$$\mathrm{cov}(\alpha,\beta) = -\frac{1}{ND\alpha^3}\left[\frac{1}{(\beta-2)} - \frac{1}{(\beta-1)}\right] \tag{8.68}$$

$$\mathrm{cov}(\alpha,x_0) = \frac{1}{ND\alpha^2}\left[\frac{1}{(\beta-1)} - \psi'(\beta)\right] \tag{8.69}$$

$$\mathrm{cov}(\beta,x_0) = -\frac{1}{ND\alpha^3}\left[\frac{\beta}{(\beta-1)} - 1\right] \tag{8.70}$$

and D is the determinant of the Fisher's information matrix with the following form, Condie (1977, 1979):

$$D = \frac{1}{(\beta-2)\alpha^4}\left[2\psi'(\beta) - \frac{(2\beta-3)}{(\beta-1)^2}\right] \tag{8.71}$$

The first order partial derivatives of x_T with respect to the parameters are evaluated as, Condie (1977):

$$\frac{\partial x_T}{\partial \alpha} = \left[\beta^{\frac{1}{3}} - \frac{1}{9\beta^{\frac{2}{3}}} + \frac{z_T}{3\beta^{\frac{1}{6}}}\right]^3 \tag{8.72}$$

$$\frac{\partial x_T}{\partial \beta} = 3\alpha\left[\beta^{\frac{1}{3}} - \frac{1}{9\beta^{\frac{2}{3}}} + \frac{z_T}{3\beta^{\frac{1}{6}}}\right]^2\left[\frac{1}{3\beta^{\frac{2}{3}}} + \frac{2}{27\beta^{\frac{5}{3}}} - \frac{z_T}{18\beta^{\frac{7}{6}}}\right] \tag{8.73}$$

$$\frac{\partial x_T}{\partial x_0} = 1 \tag{8.74}$$

The substitution of the results of Eqs. (8.54) to (8.74) into Eq. (8.53) will provide the value of the maximum likelihood estimator of the standard error of the estimate of a LPIII distribution.

Table 8.4 Estimation of ML standard errors and the two-sided 95% confidence limits for the LPIII distribution for gauging station Huites, Mexico

Tr (Years)	S_T (m^3/s)	95% Lower Limit (m^3/s)	Q_T (m^3/s)	95% Upper Limit (m^3/s)
2	190.4025	1401	1774	2147
5	449.4272	2493	3374	4255
10	826.8005	3268	4889	6509
20	1458.8702	3903	6763	9622
50	2866.3238	4328	9946	15,564
100	4544.9312	4113	13,022	21,930
500	11,737.5378	172	23,178	46,183
1000	16,995.0181	−4053	29,257	62,567

Two-sided Confidence Limits: $z_\alpha = 1.96$.

8.8.2.1 Example of estimation of ML standard errors and the two-sided 95% confidence limits for the LPIII distribution

Find the maximum likelihood estimators of the standard errors and the two-sided 95% confidence limits for the LPIII distribution for the flood sample data of gauging station Huites, Mexico (1942–1992), contained in Appendix A.

The estimation of maximum likelihood estimators of the standard errors and the two-sided 95% confidence limits for the LPIII distribution is made by using the maximum likelihood estimators of the parameters and then inserted in the Eqs. (8.53) to (8.62) into Eq. (8.41), using selected values of the return intervals. Table 8.4 summarizes these results.

8.9 Examples of Application for the LPIII Distribution Using Excel® Spreadsheets

8.9.1 Flood Frequency Analysis

By using the flood data from gauging station Huites, Mexico the following descriptive statistics were obtained and are displayed in Fig. 8.1.

8.9.1.1 MOM Method

MOM1 Method

By using an Excel® spreadsheet, the results contained in Fig. 8.2, were obtained for the MOM1 method applied to the LPIII distribution for gauging station Huites, Mexico.

FLOOD FREQUENCY ANALYSIS
Gauging Station: Huites, Mexico (1942-1992)

Descriptive Statistics		*Descriptive Statistics of the Natural Logarithms*		*Descriptive Statistics of the Logarithms Base 10*	
Mean	2498.9627	Mean	7.5383	Mean	3.2738
Sample Standard Error	311.0530	Sample Standard Error	0.1024	Standard Error	0.0445
Median	1775.7000	Median	7.4819	Median	3.2494
Mode	#N/D	Mode	#N/D	Mode	#N/D
Standard Deviation	2221.3626	Standard Deviation	0.7312	Standard Deviation	0.3176
Sample Variance	4934451.7500	Sample Variance	0.5347	Sample Variance	0.1008
Kurtosis	3.6701	Kurtosis	0.0081	Kurtosis	0.0081
Skewness	2.0123	Skewness	0.4627	Skewness	0.4627
Range	9707.5000	Range	3.1787	Range	1.3805
Minimum	421.8000	Minimum	6.0445	Minimum	2.6251
Maximum	10129.3000	Maximum	9.2232	Maximum	4.0056
Sum	127447.1000	Sum	384.4530	Sum	166.9658
Count	51.0000	Count	51.0000	Count	51.0000
Largest(1)	10129.3000	Largest(1)	9.2232	Largest(1)	4.0056
Smallest(1)	421.8000	Smallest(1)	6.0445	Smallest(1)	2.6251
Mean Confidence Level(95.0%)	624.7683	Mean Confidence Level(95.0%)	0.2057	Confidence Level(95.0%)	0.0893

Fig. 8.1 Descriptive statistics for the flood sample for Huites, Mexico

FLOOD FREQUENCY ANALYSIS
Gauging Station: Huites, Mexico (1942-1992)
Log-Pearson Type III (LPIII) Distribution
Method: MOM1

Estimation of Parameters and Goodness of Fit Tests			
Skewness Coef. (Gamma Correction) =	2.4194	Γ(β) =	1.3252
Location Parameter =	6.9338		
Scale Parameter =	0.8846		
Shape Parameter =	0.6834		
SEF = 509		MARD =	21

Design Values and Confidence Limits

Return Period Tr (years)	F(x)	Standard Normal Variate z_T	$\partial K_T/\partial\gamma$	Frequency Factor K_T	Logartithmic Squared root Standard Error $Ln(S_T)$	Squared root Standard Error S_T (m³/s)	Two-sided 95% Lower Limit (m³/s)	Design Values Q_T (m³/s)	Two-sided 95% Upper Limit (m³/s)
2	0.5	0.0000	-0.0927	-0.3412	0.1487	218.54	1035	1464	1892
5	0.8	0.8416	-0.1950	0.5205	0.2476	687.59	1401	2749	4096
10	0.9	1.2816	-0.1467	1.2261	0.2794	1303.51	2050	4605	7160
20	0.95	1.6449	-0.0326	1.9670	0.3988	3241.08	1563	7916	14268
50	0.98	2.0537	0.1932	2.9927	0.7042	12800.40	-8331	16758	41847
100	0.99	2.3263	0.4097	3.7999	1.0114	36070.67	-40460	30238	100937
500	0.998	2.8782	1.0348	5.7665	1.8960	414512.44	-685074	127370	939815
1000	0.999	3.0902	1.3490	6.6498	2.3357	1244058.29	-2195371	242983	2681337
Two-sided Confidence Limits:		z_α =	1.96						

Fig. 8.2 MOM-1 estimators for the parameters and standard errors, quantiles and confidence limits of the LPIII distribution for the flood sample for Huites, Mexico

MOM2 Method

By using an Excel® spreadsheet, the results contained in Fig. 8.3, were obtained for the MOM2 method applied to the LPIII distribution for gauging station Huites, Mexico. This method is not suitable for this sample of flood data.

WRC Method

By using an Excel® spreadsheet, the results contained in Fig. 8.4, were obtained for the WRC method applied to the LPIII distribution for gauging station Huites, Mexico.

FLOOD FREQUENCY ANALYSIS
Gauging Station: Huites, Mexico (1942-1992)
Log-Pearson Type III (LPIII) Distribution
Method: MOM2

Estimation of Parameters and Goodness of Fit Tests							
Skewness Coef. (Gamma Correction) =				2.4194			
μ_1' =	2498.96		μ_2' =	11082513	μ_3' =	72649498673	
B =	2.6813		Test α =	-0.26158		F(α) =	4.0954E-07
Location Parameter =		10.8569					
Scale Parameter =		-0.26158	<	0	not suitable for flood frequency analysis		
Shape Parameter =		13.05379					
SEF =	1194		MARD =	21			

Design Values and Confidence Limits

Return Period Tr (years)	F(x)	Standard Normal Variate z_T	$\partial K_T/\partial \gamma$	Frequency Factor K_T	Logartithmic Squared root Stand. Error $Ln(S_T)$	Squared root Stand. Error S_T (m^3/s)	Two-sided 95% Lower Limit (m^3/s)	Design Values Q_T (m^3/s)	Two-sided 95% Upper Limit (m^3/s)
2	0.5	0.0000	-0.0927	-0.3341	0.0000	0.0000	1861	1861	1861
5	0.8	0.8416	-0.1950	0.5276	0.0000	0.0000	799	799	799
10	0.9	1.2816	-0.1467	1.2332	0.0000	0.0000	488	488	488
20	0.95	1.6449	-0.0326	1.9741	0.0000	0.0000	316	316	316
50	0.98	2.0537	0.1932	2.9998	0.0000	0.0000	187	187	187
100	0.99	2.3263	0.4097	3.8070	0.0000	0.0000	130	130	130
500	0.998	2.8782	1.0348	5.7736	0.0000	0.0000	59	59	59
1000	0.999	3.0902	1.3490	6.6569	0.0000	0.0000	43	43	43
Two-sided Confidence Limits:	z_α =		1.96						

Fig. 8.3 MOM2 estimators for the parameters, standard errors, quantiles, and confidence limits of the LPIII distribution for the flood sample for Huites, Mexico

FLOOD FREQUENCY ANALYSIS
Gauging Station: Huites, Mexico (1942-1992)
Log-Pearson Type III (LPIII) Distribution
Method: WRC (Bulletin 17C-5-B, USGS(2019))

Estimation of Parameters and Goodness of Fit Tests					
Skewness Coef. (Gamma Correction) =		0.5563	p =	0.9616	
Mean ($Log_{10}(x)$) =	3.2738	Stand. Dev ($Log_{10}(x)$) =	0.3176		
Location Parameter =	1.9011	Scale Parameter =	0.0735	Shape Parameter =	18.6854
SEF =	509		MARD =	7	

Design Values and Confidence Limits

Return Period Tr (years)	F(x)	Standard Normal Variate z_T	q	Frequency Factor K_T	K_L	K_U	Two-sided 95% Lower Limit (m^3/s)	Design Values Q_T (m^3/s)	Two-sided 95% Upper Limit (m^3/s)
2	0.5	0.0000	-0.0669	-0.0919	-0.3761	0.1849	1427	1776	2151
5	0.8	0.8416	0.5681	0.8021	0.5100	1.1583	2728	3398	4382
10	0.9	1.2816	1.6795	1.3247	0.9887	1.7665	3871	4931	6836
20	0.95	1.6449	3.1153	1.7862	1.3983	2.3168	5223	6828	10223
50	0.98	2.0537	5.3965	2.3392	1.8798	2.9855	7427	10048	16670
100	0.99	2.3263	7.3674	2.7281	2.2146	3.4597	9487	13157	23578
500	0.998	2.8782	12.6462	3.5667	2.9302	4.4882	16009	23406	50021
1000	0.999	3.0902	15.1951	3.9077	3.2196	4.9081	19782	29529	67997
Two-sided Confidence Limits:	z_α =	1.96							

Fig. 8.4 WRC estimators for the parameters, standard errors, quantiles, and confidence limits of the LPIII distribution for the flood sample for Huites, Mexico

8.9.1.2 ML Method

By using an Excel® spreadsheet, the results contained in Fig. 8.5, were obtained for the ML method applied to the LPIII distribution for gauging station Huites, Mexico.

In Fig. 8.6 a comparison is made between the histogram and LPIII theoretical density for flood sample of Huites, Mexico.

FLOOD FREQUENCY ANALYSIS									
Gauging Station: Huites, Mexico (1942-1992)									
Log-Pearson Type III (LPIII) Distribution									
Method: ML									
Estimation of Parameters and Goodness of Fit Tests									
Test Location Parameter =		4.5143		Skewness Coef. (Gamma Correction) =			2.4194		
$F(x_o)$ =	-4.9871E-07		$\psi(\beta)$ =	2.8301	SEF =	526	MARD =	7	
Location Parameter =		4.5143	Scale Parameter =		0.1734	Shape Parameter =		17.4445	
Variance-Covariance Matrix									
$var(x_0)$ =	3.5488E+00	$cov(\alpha, x_0)$ =	2.2009E-01	$cov(\beta, x_0)$ =	-4.2620E+01	$var(\alpha)$ =	1.4840E-02	$cov(\alpha, \beta)$ =	-2.7596E+00
$var(\beta)$ =	5.2355E+02	Determinant =	5.3705E-03	$\psi'(\beta)$ =	0.0590				

Design Values and Confidence Limits

Return Period Tr (years)	F(x)	Standard Normal Variate z_T	$\partial x_T/\partial\alpha$	$\partial x_T/\partial\beta$	Squared root Logarithmic Stand. Error S_T	Squared root Stand. Error S_T (m^3/s)	Two-sided 95% Lower Limit (m^3/s)	Design Values Q_T (m^3/s)	Two-sided 95% Upper Limit (m^3/s)
2	0.5	0.0000	17.1133	0.1733	0.1071	190.4025	1401	1774	2147
5	0.8	0.8416	20.8237	0.1910	0.1328	449.5199	2494	3375	4256
10	0.9	1.2816	22.9605	0.2002	0.1683	826.5804	3268	4888	6508
20	0.95	1.6449	24.8316	0.2078	0.2140	1458.0255	3903	6761	9618
50	0.98	2.0537	27.0560	0.2162	0.2843	2864.2038	4328	9941	15555
100	0.99	2.3263	28.6104	0.2218	0.3422	4541.5402	4114	13016	21917
500	0.998	2.8782	31.9368	0.2329	0.4868	11730.4841	177	23169	46161
1000	0.999	3.0902	33.2808	0.2372	0.5523	16986.3266	-4046	29247	62541
Two-sided Confidence Limits:		z_α =	1.96						

Fig. 8.5 ML estimators for the parameters, standard errors, quantiles, and confidence limits of the LPIII distribution for gauging station Huites, Mexico

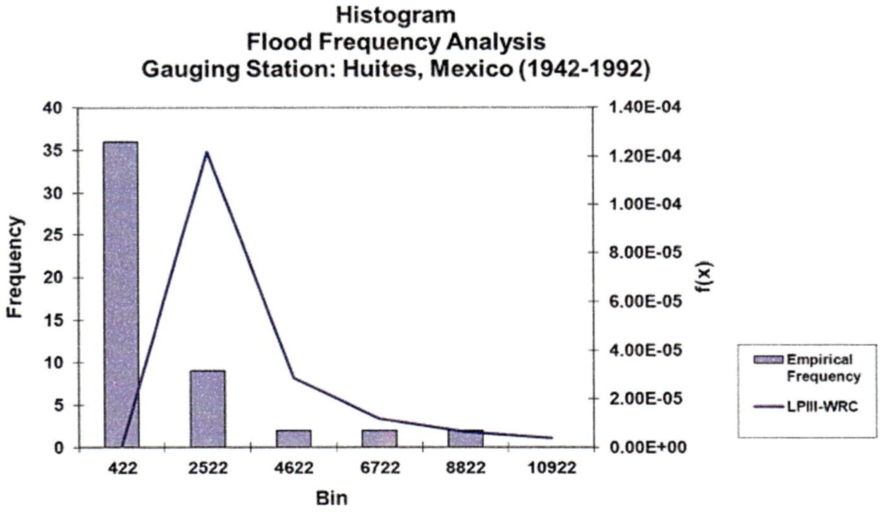

Fig. 8.6 Histogram and theoretical WRC LPIII density for flood data at Huites, Mexico

A graph containing the empirical and theoretical LPIII frequency curves is shown in Fig. 8.7.

A graphical depiction of the empirical and theoretical frequency curves and the confidence limits of the best fit provided, in this case that of WRC, is shown in Fig. 8.8.

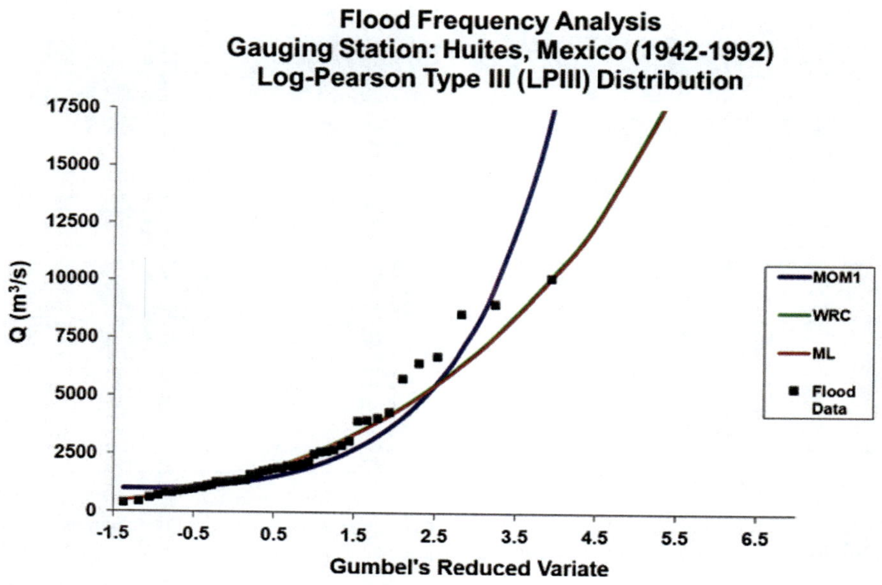

Fig. 8.7 Empirical and MOM1, WRC, and ML LPIII theoretical curves for the flood sample of Huites, Mexico

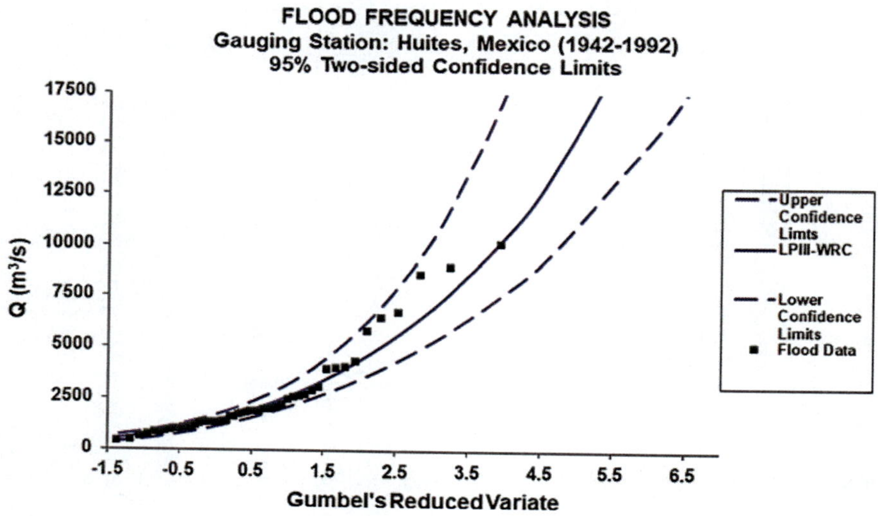

Fig. 8.8 Empirical and theoretical frequency curves and WRC confidence limits of the LPIII distribution for the flood data at Huites, Mexico

8.9.2 Rainfall Frequency Analysis

By using the 24 h, annual maximum rainfall data from rainfall at Chihuahua, Mexico the descriptive statistics results were obtained and are displayed in Fig. 8.9.

8.9.2.1 MOM Method

MOM1 Method

By using an Excel® spreadsheet, the results contained in Fig. 8.10, were obtained for the MOM1 method applied to the LPIII distribution for rainfall gauging station Chihuahua, Mexico.

MOM2 Method

By using an Excel® spreadsheet, the results contained in Fig. 8.11, were obtained for the MOM2 method applied to the LPIII distribution for rainfall at gauging station Chihuahua, Mexico.

8.9.2.2 ML Method

By using an Excel® spreadsheet, the results contained in Fig. 8.12, were obtained for the ML method applied to the LPIII distribution for rainfall at gauging station Chihuahua, Mexico.

In Fig. 8.13a comparison is made between the histogram and LPIII theoretical density for rainfall at Chihuahua, Mexico.

A graph containing the empirical and theoretical MOM1 LPIII frequency curves is shown in Fig. 8.14.

	A	B	C	D
1	24 Hr. Annual Maximum Rainfall Frequency Analysis			
2	Meteorological Station: Chihuahua, Mexico (1960-2005)			
3				
4	*Descriptive Statistics*		*Descriptive Statistics of the Natural Logarithms*	
5	Mean	42.0454	Mean	3.6871
6	Sample Standard Error	1.9784	Sample Standard Error	0.0487
7	Median	40.2400	Median	3.6948
8	Mode	#N/D	Mode	#N/D
9	Standard Deviation	13.4184	Standard Deviation	0.3302
10	Sample Variance	180.0535	Sample Variance	0.1090
11	Kurtosis	-0.7252	Kurtosis	-0.4776
12	Skewness	0.3591	Skewness	-0.2507
13	Range	54.3000	Range	1.4283
14	Minimum	17.1200	Minimum	2.8402
15	Maximum	71.4200	Maximum	4.2686
16	Sum	1934.0900	Sum	169.6058
17	Count	46.0000	Count	46.0000
18	Largest(1)	71.4200	Largest(1)	4.2686
19	Smallest(1)	17.1200	Smallest(1)	2.8402
20	Mean Confidence Level(95.0%)	3.9848	Mean Confidence Level(95.0%)	0.0981

Fig. 8.9 Descriptive statistics for 24 h, annual maximum rainfall data at Chihuahua, Mexico

	A	B	C	D	E	F	G	H	I	J
1	24 HR. ANNUAL MAXIMUM RAINFALL FREQUENCY ANALYSIS									
2	Meteorological Station: Chihuahua, Mexico (1960-2005)									
3	Log-Pearson Type III (LPIII) Distribution									
4	Method: MOM1									
5										
6	Estimation of Parameters and Goodness of Fit Tests									
7	Skewness Coef. (Gamma Correction) =			0.4400						
8	Location Parameter =		1.0526							
9	Scale Parameter =		0.0414							
10	Shape Parameter =		63.6578							
11	Γ(β) =	2.8103E+48								
12	SEF =	2.61		MARD =	4.24					
13										
14	Design Values and Confidence Limits									
15–19	Return Period Tr (years)	F(x)	Standard Normal Variate z_T	$\partial K_T/\partial\gamma$	Frequency Factor K_T	Logartithmic Squared root Stand. Error S_T	Squared root Stand. Error S_T (cm)	Two-sided 95% Lower Limit (cm)	Design Values P_T (cm)	Two-sided 95% Upper Limit (cm)
20	2	0.5	0.0000	-0.1640	-0.0729	0.0927	3.6566	32	39	47
21	5	0.8	0.8416	-0.0839	0.8124	0.1103	5.8003	41	52	64
22	10	0.9	1.2816	0.0602	1.3184	0.1396	8.6020	45	61	78
23	20	0.95	1.6449	0.2359	1.7596	0.1808	12.7813	45	70	95
24	50	0.98	2.0537	0.4984	2.2819	0.2470	20.5040	42	82	122
25	100	0.99	2.3263	0.7129	2.6457	0.3026	28.0904	36	91	147
26	Two-sided Confidence Limits:		z_α =	1.96						

Fig. 8.10 MOM-1 estimators of the parameters, standard errors, quantiles, and confidence limits for 24 h, annual maximum rainfall data at meteorological station Chihuahua, Mexico

	A	B	C	D	E	F	G	H	I	J
1	24 HR. ANNUAL MAXIMUM RAINFALL FREQUENCY ANALYSIS									
2	Meteorological Station: Chihuahua, Mexico (1960-2005)									
3	Log-Pearson Type III (LPIII) Distribution									
4	Method: MOM2									
5										
6	Estimation of Parameters and Goodness of Fit Tests									
7	Skewness Coef. (Gamma Correction) =			0.4400						
8	μ_1' =	42.05		μ_2' =	176.1393		μ_3' =	811.8712		
9	B =	1.9586		C =	-0.9602					
10	Location Parameter =		1.3391		Test α =	-53687091.2		F(α) =	########	
11	Scale Parameter =		-5.4E+07	<	0	not suitable for rainfall frequency analysis				
12	Shape Parameter =		-0.1348							
13	SEF =	#¡REF!		MARD =	#¡REF!					
14										
15	Design Values and Confidence Limits									
16–20	Return Period Tr (years)	F(x)	v	Standard Normal Variate z_T	$\partial K_T/\partial\gamma$	Frequency Factor K_T	Logartithmic Squared root Stand. Error S_T (m³/s)	Two-sided 95% Lower Limit (m³/s)	Design Values Q_T (m³/s)	Two-sided 95% Upper Limit (m³/s)
21	2	0.5	1.1774	0.0000	-0.1640	-0.0729	#¡NUM!	#¡NUM!	#¡NUM!	#¡NUM!
22	5	0.8	1.7941	0.8415	-0.0839	0.8122	#¡NUM!	#¡NUM!	#¡NUM!	#¡NUM!
23	10	0.9	2.1460	1.2817	0.0602	1.3186	#¡NUM!	#¡NUM!	#¡NUM!	#¡NUM!
24	20	0.95	2.4477	1.6452	0.2361	1.7600	#¡NUM!	#¡NUM!	#¡NUM!	#¡NUM!
25	50	0.98	2.7971	2.0542	0.4987	2.2825	#¡NUM!	#¡NUM!	#¡NUM!	#¡NUM!
26	100	0.99	3.0349	2.3268	0.7133	2.6463	#¡NUM!	#¡NUM!	#¡NUM!	#¡NUM!
27	Two-sided Confidence Limits:		z_α =	1.96						

Fig. 8.11 MOM-2 estimators of the parameters, standard errors, quantiles, and confidence limits for 24 h, annual maximum rainfall data at meteorological station Chihuahua, Mexico

A graphical depiction of the empirical and theoretical frequency curves and the confidence limits of the best fit provided, in this case that of MOM1, is shown in Fig. 8.15.

24 HR. ANNUAL MAXIMUM RAINFALL FREQUENCY ANALYSIS
Meteorological Station: Chihuahua, Mexico (1960-2005)
Log-Pearson Type III (LPIII) Distribution
Method: ML

Estimation of Parameters and Goodness of Fit Tests						
Test Location Parameter =	1.0526	$F(x_o)$ =	-6.0426E+01	$\Gamma(\beta)$ =	5.4453E+01	
Skewness Coef. (Gamma Correction) =		0.4400	$\psi(\beta)$ =	4.1141	$\psi'(\beta)$ =	0.0163
Location Parameter =	1.0526			Log-likelihood function =		5672.8083
Scale Parameter =	0.0578	No convergence was attained				
Shape Parameter =	61.6943					
SEF = 64	MARD =	100	AIC =	-11339.6167		

Variance-Covariance Matrix					
$var(x_0)$ =	2.3725E+01	$cov(\alpha, x_0)$ =	3.9302E-01	$cov(\beta, x_0)$ =	-8.2947E+02
$var(\alpha)$ =	6.6566E-03	$cov(\alpha, \beta)$ =	-1.3895E+01	$var(\beta)$ =	2.9165E+04
Determinant =	2.2322E-03				

Design Values and Confidence Limits

Return Period Tr (years)	F(x)	v	Standard Normal Variate z_T	$\partial x_T/\partial\alpha$	$\partial x_T/\partial\beta$	Squared root Logarithmic Stand. Error S_T	Squared root Stand. Error S_T (cm)	Two-sided 95% Lower Limit (cm)	Design Values P_T (cm)	Two-sided 95% Upper Limi (cm)
2	0.5	1.1774	0.0000	61.3615	0.0578	0.0719	7.1690	86	100	114
5	0.8	1.7941	0.8415	68.1854	0.0609	0.0835	12.3546	124	148	172
10	0.9	2.1460	1.2817	71.9493	0.0626	0.1025	18.8618	147	184	221
20	0.95	2.4477	1.6452	75.1591	0.0639	0.1285	28.5124	165	221	277
50	0.98	2.7971	2.0542	78.8831	0.0654	0.1698	46.8238	183	274	366
100	0.99	3.0349	2.3268	81.4324	0.0664	0.2043	65.4225	190	318	446
Two-sided Confidence Limits:		z_α =	1.96							

Fig. 8.12 ML estimation of the parameters and standard errors, quantiles, and confidence limits for the 24 h, annual maximum rainfall data at meteorological station Chihuahua, Mexico

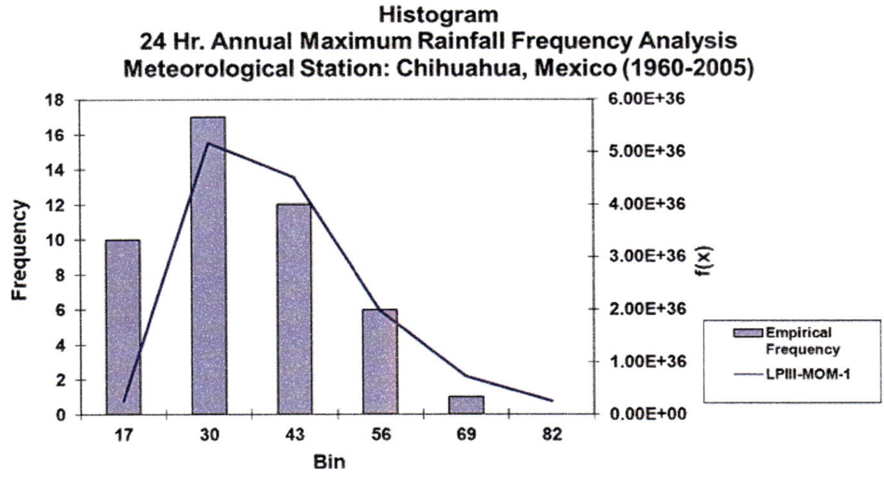

Fig. 8.13 Histogram and theoretical LPIII density for 24 h, annual maximum rainfall data at meteorological station Chihuahua, Mexico

8.9.3 Maximum Significant Wave Height Frequency Analysis

By using the maximum significant wave height data, Castillo (1988), the descriptive statistics results were obtained and are displayed in Fig. 8.16.

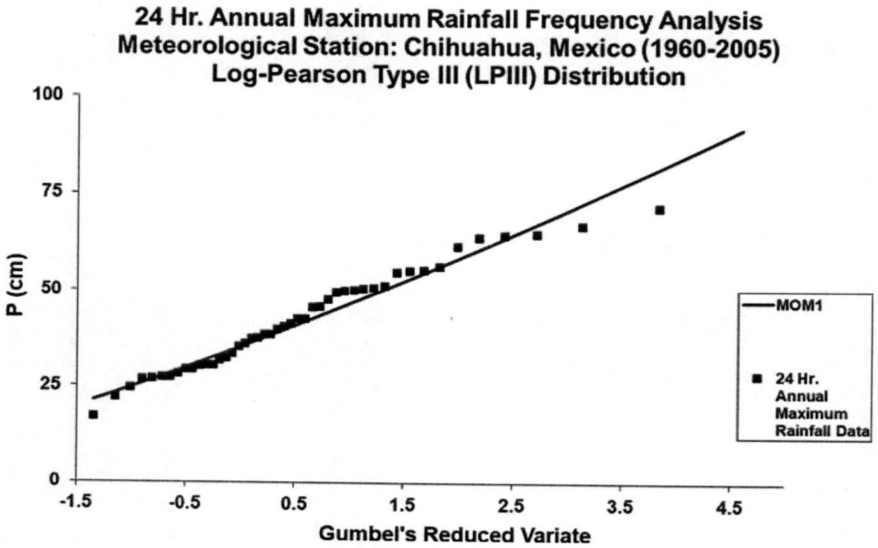

Fig. 8.14 Empirical and theoretical MOM1 LPIII distributions for 24 h, annual maximum rainfall data at meteorological station Chihuahua, Mexico

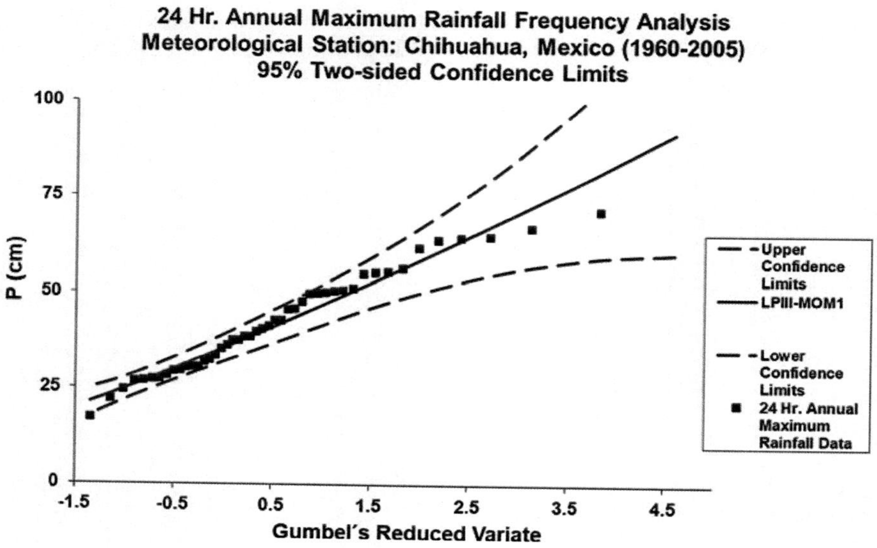

Fig. 8.15 Empirical and theoretical frequency curves and MOM1 confidence limits for 24 h, annual maximum rainfall data at meteorological station Chihuahua, Mexico

Maximum Significant Wave Height Frequency Analysis

Descriptive Statistics		*Descriptive Statistics of the Natural Logarithms*	
Mean	4.4247	Mean	1.3628
Sample Standard Error	0.3053	Sample Standard Error	0.0748
Median	4.1666	Median	1.4270
Mode	3.5509	Mode	1.2672
Standard Deviation	2.1585	Standard Deviation	0.5287
Sample Variance	4.6591	Sample Variance	0.2796
Kurtosis	1.0740	Kurtosis	0.6324
Skewness	0.9053	Skewness	-0.6102
Range	10.4485	Range	2.5478
Minimum	0.8870	Minimum	-0.1199
Maximum	11.3355	Maximum	2.4279
Sum	221.2326	Sum	68.1403
Count	50.0000	Count	50.0000
Largest(1)	11.3355	Largest(1)	2.4279
Smallest(1)	0.8870	Smallest(1)	-0.1199
Mean Confidence Level(95.0%)	0.6134	Mean Confidence Level(95.0%)	0.1503

Fig. 8.16 Descriptive statistics for maximum significant wave height

MAXIMUM SIGNIFICANT WAVE HEIGHT FREQUENCY ANALYSIS
Log-Pearson Type III (LPIII) Distribution
Method: MOM1

Estimation of Parameters and Goodness of Fit Tests

Skewness Coef. (Gamma Correction) =	1.0922	$\Gamma(\beta)$ =	3.5467E+05
Location Parameter =	-0.3702		
Scale Parameter =	0.1613		
Shape Parameter =	10.7429		
SEF = 0.6		MARD =	10.7

Design Values and Confidence Limits

Return Period Tr (years)	F(x)	Standard Normal Variate z_T	$\partial K_T/\partial\gamma$	Frequency Factor K_T	Logartithmic Squared root Standard Error $Ln(S_T)$	Squared root Standard Error S_T (m³/s)	Two-sided 95% Lower Limit (m³/s)	Design Values Q_T (m³/s)	Two-sided 95% Upper Limit (m³/s)
2	0.5	0.0000	-0.1504	-0.1761	0.0881	0.33	3	4	4
5	0.8	0.8416	-0.1308	0.7420	0.1122	0.67	5	6	7
10	0.9	1.2816	-0.0117	1.3343	0.1510	1.19	6	8	10
20	0.95	1.6449	0.1527	1.8869	0.2130	2.17	6	10	14
50	0.98	2.0537	0.4169	2.5819	0.3180	4.41	5	14	22
100	0.99	2.3263	0.6436	3.0902	0.4090	7.09	3	17	31
Two-sided Confidence Limits:		z_α =	1.96						

Fig. 8.17 MOM-1 estimators of the standard errors, quantiles and confidence limits for maximum significant wave height

8.9.3.1 MOM Method

MOM1 Method

By using an Excel® spreadsheet, the results contained in Fig. 8.17, were obtained for the MOM1 method applied to the LPIII distribution to the maximum significant wave height data.

	A	B	C	D	E	F	G	H	I	J
1	MAXIMUM SIGNIFICANT WAVE HEIGHT FREQUENCY ANALYSIS									
2	Log-Pearson Type III (LPIII) Distribution									
3	Method: MOM2									
4	Estimation of Parameters and Goodness of Fit Tests									
5	Skewness Coef. (Gamma Correction) =				2.4194					
6	μ_1' =	2498.96		μ_2' =	11082513	μ_3' =	72649498673			
7	B =	2.6813		Test α =	-0.26158		F(α) =	4.0954E-07		
8	Location Parameter =		10.8569			SEF =	#¡REF!		MARD =	#¡REF!
9	Scale Parameter =		-0.26158	<	0	not suitable for maximum significant wave				
10	Shape Parameter =		13.05379			height frequency analysis				
11	Design Values and Confidence Limits									
12–16	Return Period Tr (years)	F(x)	Standard Normal Variate z_T	$\partial K_T/\partial \gamma$	Frequency Factor K_T	Logartithmic Squared root Stand. Error $Ln(S_T)$	Squared root Stand. Error S_T (m^3/s)	Two-sided 95% Lower Limit (m^3/s)	Design Values Q_T (m^3/s)	Two-sided 95% Upper Limit (m^3/s)
17	2	0.5	0.0000	-0.0927	-0.3341	0.0000	0.0000	1861	1861	1861
18	5	0.8	0.8416	-0.1950	0.5276	0.0000	0.0000	799	799	799
19	10	0.9	1.2816	-0.1467	1.2332	0.0000	0.0000	488	488	488
20	20	0.95	1.6449	-0.0326	1.9741	0.0000	0.0000	316	316	316
21	50	0.98	2.0537	0.1932	2.9998	0.0000	0.0000	187	187	187
22	100	0.99	2.3263	0.4097	3.8070	0.0000	0.0000	130	130	130
23	Two-sided Confidence Limits:		z_α =	1.96						

Fig. 8.18 MOM-2 estimators of the standard errors, quantiles and confidence limits for maximum significant wave height

MOM2 Method

By using an Excel® spreadsheet, the results contained in Fig. 8.18, were obtained for the MOM2 method applied to the LPIII distribution to the maximum significant wave height data.

8.9.3.2 ML Method

By using an Excel® spreadsheet, the results contained in Fig. 8.19, were obtained for the ML method applied to the LPIII distribution for the maximum significant wave height data.

In Fig. 8.20 a comparison is made between the histogram and LPIII theoretical density for the maximum significant wave height data.

A graph containing the empirical and theoretical MOM1 LPIII frequency curves is shown in Fig. 8.21.

A graphical depiction of the empirical and theoretical frequency curves and the confidence limits of the best fit provided, in this case that of MOM1, is shown in Fig. 8.22.

8.9.4 Annual Maximum Wind Speed Frequency Analysis

By using the annual maximum wind speed data, Castillo (1988), the descriptive statistics results were obtained and are displayed in Fig. 8.23.

	A	B	C	D	E	F	G	H	I	J
1	MAXIMUM SIGNIFICANT WAVE HEIGHT FREQUENCY ANALYSIS									
2	Log-Pearson Type III (LPIII) Distribution									
3	Method: ML									
4										
5	Estimation of Parameters and Goodness of Fit Tests									
6	Test Location Parameter =		4.5143			Skewness Coef. (Gamma Correction) =		1.0922		
7	$F(x_o)$ =	-4.9871E-07		$\psi(\beta)$ =	2.8301	SEF =	3101	MARD =	48779	
8	Location Parameter =		4.5143	Scale Parameter =		0.1734	Shape Parameter =		17.4445	
9										
10	Variance-Covariance Matrix									
11	$var(x_0)$ =	3.5488E+00	$cov(\alpha,x_0)$ =	2.2009E-01	$cov(\beta,x_0)$ =	-4.2620E+01	$var(\alpha)$ =	1.4840E-02	$cov(\alpha,\beta)$ =	-2.7596E+00
12	$var(\beta)$ =	5.2355E+02	Determinant =	5.3705E-03	$\psi'(\beta)$ =	0.0590				
13										
14	Design Values and Confidence Limits									
15–18	Return Period Tr (years)	F(x)	Standard Normal Variate z_T	$\partial x_T/\partial\alpha$	$\partial x_T/\partial\beta$	Squared root Logarithmic Stand. Error S_T	Squared root Stand. Error S_T (m^3/s)	Two-sided 95% Lower Limit (m^3/s)	Design Values Q_T (m^3/s)	Two-sided 95% Upper Limit (m^3/s)
19	2	0.5	0.0000	17.1133	0.1733	0.1071	190.4025	1401	1774	2147
20	5	0.8	0.8416	20.8237	0.1910	0.1328	449.5199	2494	3375	4256
21	10	0.9	1.2816	22.9605	0.2002	0.1683	826.5804	3268	4888	6508
22	20	0.95	1.6449	24.8316	0.2078	0.2140	1458.0255	3903	6761	9618
23	50	0.98	2.0537	27.0560	0.2162	0.2843	2864.2038	4328	9941	15555
24	100	0.99	2.3263	28.6104	0.2218	0.3422	4541.5402	4114	13016	21917
25	Two-sided Confidence Limits:		z_α =	1.96						

Fig. 8.19 ML estimation of the parameters and standard errors, quantiles and confidence limits for the maximum significant wave height

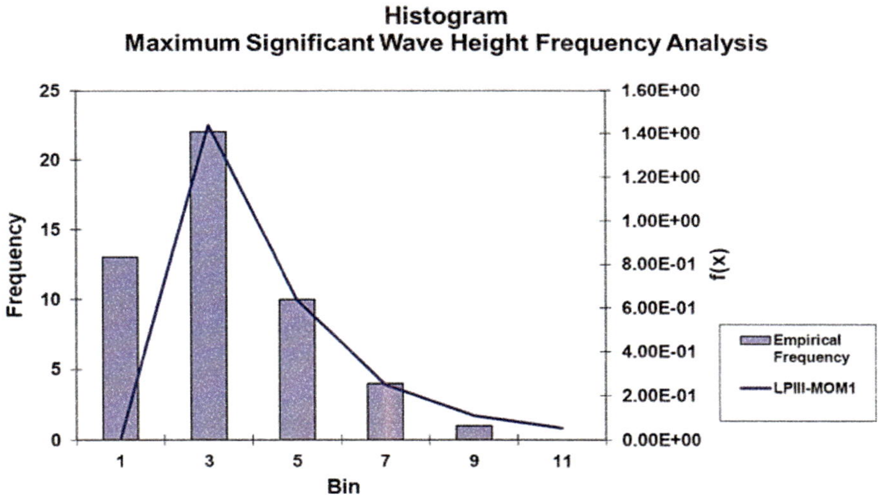

Fig. 8.20 Histogram and theoretical LPIII density for maximum significant wave height

8.9.4.1 MOM Method

MOM1 Method

By using an Excel® spreadsheet, the results contained in Fig. 8.24, were obtained for the MOM1 method applied to the LPIII distribution to the annual maximum wind speed data.

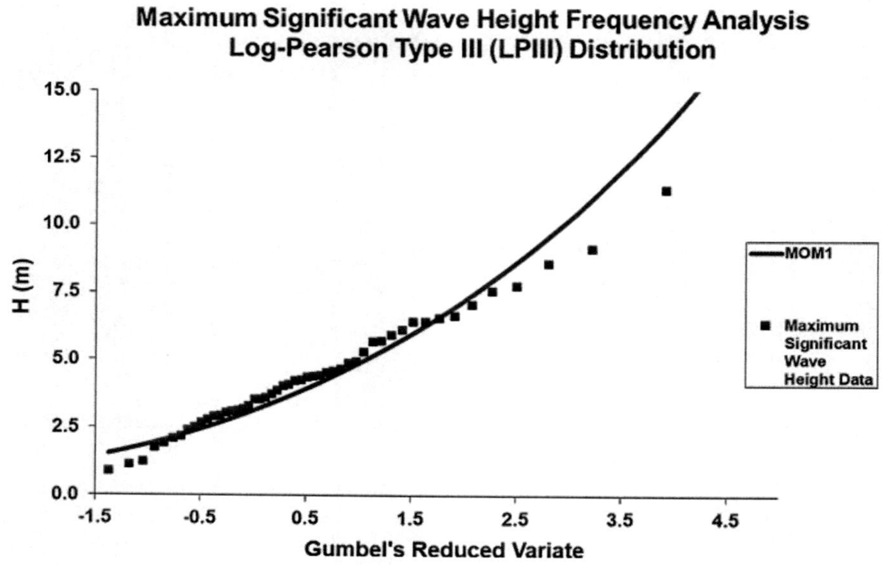

Fig. 8.21 Empirical and theoretical MOM1 LPIII distributions for maximum significant wave height

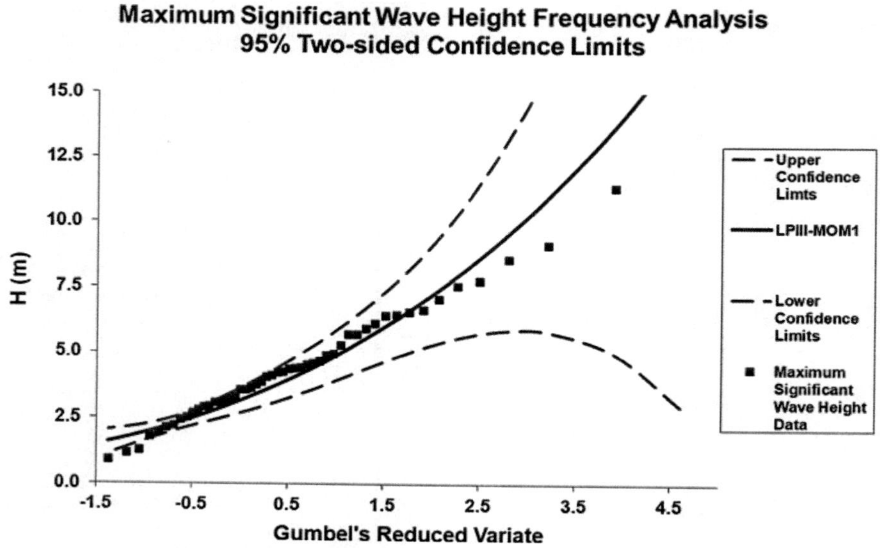

Fig. 8.22 Empirical and theoretical frequency curves and MOM1 confidence limits for maximum significant wave height

ANNUAL MAXIMUM WIND SPEED FREQUENCY ANALYSIS

Descriptive Statistics		*Descriptive Statistics of the Natural Logarithms*	
Mean	54.7964	Mean	3.9558
Sample Standard Error	2.9320	Sample Standard Error	0.0408
Median	47.9965	Median	3.8711
Mode	#N/D	Mode	#N/D
Standard Deviation	20.7327	Standard Deviation	0.2886
Sample Variance	429.8439	Sample Variance	0.0833
Kurtosis	12.5699	Kurtosis	3.6721
Skewness	3.0813	Skewness	1.6052
Range	121.5116	Range	1.4669
Minimum	36.4278	Minimum	3.5953
Maximum	157.9394	Maximum	5.0622
Sum	2739.8217	Sum	197.7906
Count	50.0000	Count	50.0000
Largest(1)	157.9394	Largest(1)	5.0622
Smallest(1)	36.4278	Smallest(1)	3.5953
Mean Confidence Level(95.0%)	5.8922	Mean Confidence Level(95.0%)	0.0820

Fig. 8.23 Descriptive statistics for 24 h, annual maximum wind speed

ANNUAL MAXIMUM WIND SPEED FREQUENCY ANALYSIS
Log-Pearson Type III (LPIII) Distribution
Method: MOM1

Estimation of Parameters and Goodness of Fit Tests

Skewness Coef. (Gamma Correction) =		3.7176	$\Gamma(\beta)$ =	8.9680E-01
Location Parameter =	3.5962			
Scale Parameter =	0.2316			
Shape Parameter =	1.5523			
SEF = 58		MARD =	2.25	

Design Values and Confidence Limits

Return Period Tr (years)	F(x)	Standard Normal Variate z_T	$\partial K_T/\partial\gamma$	Frequency Factor K_T	Logartithmic Squared root Standard Error $Ln(S_T)$	Squared root Standard Error S_T (m³/s)	Two-sided 95% Lower Limit (m³/s)	Design Values Q_T (m³/s)	Two-sided 95% Upper Limit (m³/s)
2	0.5	0.0000	-0.0157	-0.4122	0.0381	1.85	45	49	52
5	0.8	0.8416	-0.2067	0.2540	0.1851	11.79	40	63	86
10	0.9	1.2816	-0.2384	0.9706	0.2090	16.07	45	76	108
20	0.95	1.6449	-0.1963	1.8145	0.2143	19.79	53	92	130
50	0.98	2.0537	-0.0486	3.0860	0.3169	37.51	43	116	190
100	0.99	2.3263	0.1222	4.1479	0.4943	71.75	-1	139	280
Two-sided Confidence Limits:		z_α =	1.96						

Fig. 8.24 MOM-1 estimators of the standard errors, quantiles and confidence limits for 24 h, annual maximum wind speed

MOM2 Method

By using an Excel® spreadsheet, the results contained in Fig. 8.25, were obtained for the MO2 method applied to the LPIII distribution to the annual maximum wind speed data.

	A	B	C	D	E	F	G	H	I	J
1	ANNUAL MAXIMUM WIND SPEED FREQUENCY ANALYSIS									
2	Log-Pearson Type III (LPIII) Distribution									
3	Method: MOM2									
4										
5	Estimation of Parameters and Goodness of Fit Tests									
6	Skewness Coef. (Gamma Correction) =			3.7176	μ_1' =	54.80	μ_2' =	3423.8962		
7	μ_3' =	259617.2062		A =	3565.2227	B =	3.4741			
8	Test α =	0.2316	F(α) =	-3.5617E+03	not solution was found					
9	Location Parameter =		3.4542							
10	Scale Parameter =		0.1748							
11	Shape Parameter =		2.8599							
12	SEF =	0		MARD =	0					
13										
14	Design Values and Confidence Limits									
15–19	Return Period Tr (years)	F(x)	Standard Normal Variate z_T	$\partial K_T/\partial\gamma$	Frequency Factor K_T	Logartithmic Squared root Stand. Error Ln(S_T)	Squared root Stand. Error S_T (m³/s)	Two-sided 95% Lower Limit (m³/s)	Design Values Q_T (m³/s)	Two-sided 95% Upper Limit (m³/s)
20	2	0.5	0.0000	-0.0157	-0.3513	0.1591	7.8770	34	49	65
21	5	0.8	0.8416	-0.2067	0.3148	0.7731	55.1571	-43	65	173
22	10	0.9	1.2816	-0.2384	1.0314	1.3786	143.7247	-204	77	359
23	20	0.95	1.6449	-0.1963	1.8753	1.6983	241.5026	-382	91	565
24	50	0.98	2.0537	-0.0486	3.1469	15.7274	############	-749909989	113	749910215
25	100	0.99	2.3263	0.1222	4.2087	12.2541	############	-27193657	132	27193922
26	Two-sided Confidence Limits:		z_α =	1.96						

Fig. 8.25 MOM-2 estimators of the standard errors, quantiles and confidence limits for 24 h, annual maximum wind speed

8.9.4.2 ML Method

By using an Excel® spreadsheet, the results contained in Fig. 8.26, were obtained for the ML method applied to the LPIII distribution for the annual maximum wind speed data.

In Fig. 8.27 a comparison is made between the histogram and LPIII theoretical density for the annual maximum wind speed data.

	A	B	C	D	E	F	G	H	I	J
1	ANNUAL MAXIMUM WIND SPEED FREQUENCY ANALYSIS									
2	Log-Pearson Type III (LPIII) Distribution									
3	Method: ML									
4										
5	Estimation of Parameters and Goodness of Fit Tests									
6	Test Location Parameter =		1.9998	$F(x_o)$ =	5.3753E-07	ψ(β) =	4.2820			
7	Location Parameter =	1.9998	Scale Parameter =	0.0268		Skewness Coef. (Gamma Correction) =			3.7176	
8	Shape Parameter =		-1.2226	<	1	No feasible solution was found				
9	SEF =	#¡NUM!	MARD =	#¡NUM!						
10										
11	Variance-Covariance Matrix									
12	$var(x_o)$ =	2.8540E-05	$cov(\alpha, x_o)$ =	2.2983E-05	$cov(\beta, x_o)$ =	-1.6465E-05	var(α) =	7.1933E-06	cov(α,β) =	5.1091E-06
13	var(β) =	8.4860E-04	Determinant =	-2.8511E+07	ψ'(β) =	23.0170				
14										
15	Design Values and Confidence Limits									
16–19	Return Period Tr (years)	F(x)	Standard Normal Variate z_T	$\partial x_T/\partial\alpha$	$\partial x_T/\partial\beta$	Squared root Logarithmic Stand. Error S_T	Squared root Stand. Error S_T (m³/s)	Two-sided 95% Lower Limit (m³/s)	Design Values Q_T (m³/s)	Two-sided 95% Upper Limit (m³/s)
20	2	0.5	0.0000	#¡NUM!	#¡NUM!	#¡NUM!	#¡NUM!	#¡NUM!	#¡NUM!	#¡NUM!
21	5	0.8	0.8416	#¡NUM!	#¡NUM!	#¡NUM!	#¡NUM!	#¡NUM!	#¡NUM!	#¡NUM!
22	10	0.9	1.2816	#¡NUM!	#¡NUM!	#¡NUM!	#¡NUM!	#¡NUM!	#¡NUM!	#¡NUM!
23	20	0.95	1.6449	#¡NUM!	#¡NUM!	#¡NUM!	#¡NUM!	#¡NUM!	#¡NUM!	#¡NUM!
24	50	0.98	2.0537	#¡NUM!	#¡NUM!	#¡NUM!	#¡NUM!	#¡NUM!	#¡NUM!	#¡NUM!
25	100	0.99	2.3263	#¡NUM!	#¡NUM!	#¡NUM!	#¡NUM!	#¡NUM!	#¡NUM!	#¡NUM!
26	Two-sided Confidence Limits:		z_α =	1.96						

Fig. 8.26 ML estimation of the parameters and standard errors, quantiles and confidence limits for the annual maximum wind speed

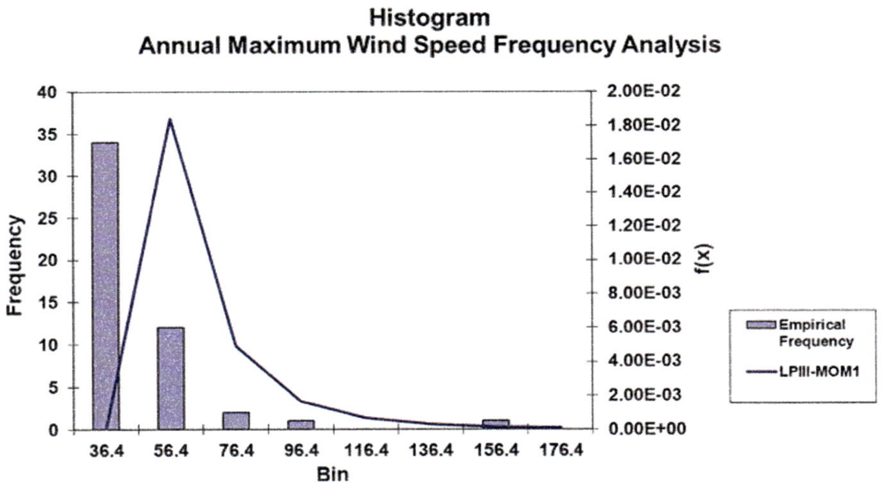

Fig. 8.27 Histogram and theoretical LPIII density for annual maximum wind speed

A graph containing the empirical and theoretical MOM1 LPIII frequency curves is shown in Fig. 8.28.

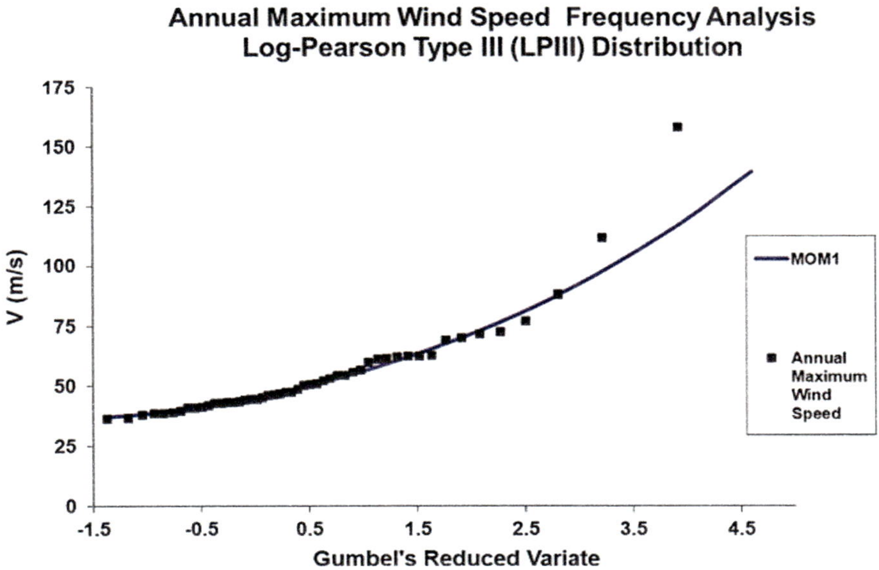

Fig. 8.28 Empirical and theoretical MOM1 LPIII distributions for annual maximum wind speed

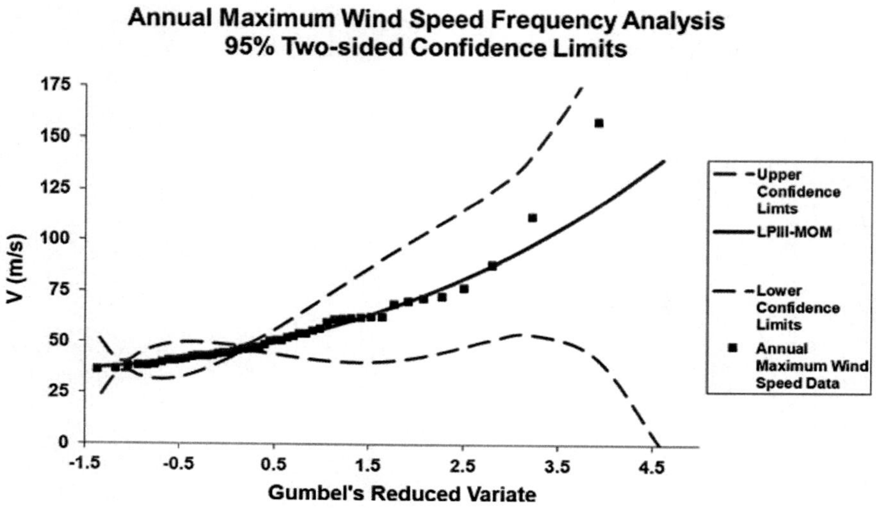

Fig. 8.29 Empirical and theoretical frequency curves and MOM1 confidence limits for annual maximum wind speed

A graphical depiction of the empirical and theoretical frequency curves and the confidence limits of the best fit provided, in this case that of MOM1, is shown in Fig. 8.29.

9 Extreme Value Type I Distribution

Enthusiasm is the electricity of life. How do you get it?
You act enthusiastic until you make it a habit. G. Parks

9.1 Introduction

The extreme value type I (EVI) distribution is one of the three particular solutions, independently found by Fisher-Tippett (1928) and Fréchet (1927), to the Stability Postulate that all the extremes must comply with. The EVI distribution, also known as Gumbel's distribution, or double exponential distribution, has been studied extensively by Gumbel (1958), he provided the means to use this distribution in statistics and in engineering practice and since then the EVI distribution is, by far, the most widely used member of the family of extreme value distributions in engineering practice.

9.2 Chapter Objectives

After reading this chapter, you will know how to:

(a) Recognize the distribution and density functions of the EVI distribution.
(b) Estimate the parameters of the EVI distribution.
(c) Compute the quantiles and confidence limits of the EVI distribution.
(d) Make a graphic display of your data and the EVI distribution.
(e) Develop an application of all the above using Excel® spreadsheets.

J. A. Raynal Villaseñor, *Frequency Analyses of Natural Extreme Events*,
Earth and Environmental Sciences Library,
https://doi.org/10.1007/978-3-030-86390-6_9

9.3 Probability Distribution and Density Functions

The probability distribution function of the extreme value type I (EVI) distribution is, NERC (1975):

$$F(x) = \exp\left\{-\exp\left[-\frac{(x - x_0)}{\alpha}\right]\right\} \tag{9.1}$$

where x_0 and α, are the location and scale parameters. *F (x)* is the probability distribution function of the random variable x and in the case of maxima natural extreme frequency analysis is equal to the non-exceedance probability, $Pr\,(X \leq x)$. The scale parameter must meet the condition that $\alpha > 0$. The domain of variable x in EVI distribution is $-\infty < x < \infty$.

The probability density function for the EVI distribution is, NERC (1975):

$$f(x) = \frac{1}{\alpha}\exp\left\{-\exp\left[-\frac{(x - x_0)}{\alpha}\right]\right\}\exp\left[-\frac{(x - x_0)}{\alpha}\right] \tag{9.2}$$

where $f\,(x)$ is the probability density function of the random variable x.

9.4 Estimation of the Parameters

9.4.1 MOM Method

The MOM method is based on the procedure devised to obtain the moments of inertia in statics. Fisher-Tippet (1928) adapted the method of moments to be used in statistics by considering the probability density function as the body to which the moments of inertia must be computed.

The population mean and variance of the EVI distribution are as follows:

$$\mu = x_0 + 0.5772\alpha \tag{9.3}$$

$$\sigma^2 = \frac{\pi^2}{6}\alpha^2 \tag{9.4}$$

The EVI distribution has a population value of 1.1396 on its skewness parameter.

For the EVI distribution, using the method of moments, the estimators of the parameters are obtained by first equating the population moments with the sample moments, and then simultaneously solving the resulting system of equations:

$$\mu = \overline{x} \tag{9.5}$$

$$\sigma^2 = \hat{\sigma}^2 = s^2 \tag{9.6}$$

then using Eqs. (9.3) and (9.4) in the left-hand side of Eqs. (9.5) and (9.6), respectively, and the expressions for the sample mean and sample variance, given in Chap. 2, in the right-hand side of Eqs. (9.5) and (9.6), respectively, we obtain the following expressions:

$$x_0 + 0.5772\alpha = \frac{1}{N}\sum_{i=1}^{N} x_i \tag{9.7}$$

$$\frac{\pi^2}{6}\alpha^2 = \frac{1}{N}\sum_{i=1}^{N} (x - \overline{x})^2 \tag{9.8}$$

The solution to the system of equations formed by Eqs. (9.7) and (9.8), provide the estimators of the MOM method for the EVI distribution:

$$\hat{x}_0 = \overline{x} - 0.45\hat{\sigma} = \overline{x} - 0.45s \tag{9.9}$$

$$\hat{\alpha} = \frac{\sqrt{6}}{\pi}\hat{\sigma} = \frac{\sqrt{6}}{\pi}s \tag{9.10}$$

where $\hat{\mu}$ or $\overline{x}$ is the sample mean, $\hat{\sigma}$ or s is the sample standard deviation and N is the sample size.

9.4.1.1 Example of Application of Estimation of the Parameters of the EVI Distribution Using the Method of Moments

Find the moments estimators for the parameters of the EVI distribution for the flood sample data of gauging station Huites, Mexico (1942–1992), contained in Appendix A.

The following statistics have already been obtained:

$$\hat{\mu} = \overline{x} = \frac{1}{N}\sum_{i=1}^{N} x_i = 2498.9627\ m^3/s$$

$$\hat{\sigma} = s = \left[\frac{1}{N}\sum_{i=1}^{N}(x_i - \hat{\mu})^2\right]^{1/2} = 2199.4767 m^3/s$$

So, the moment estimators of the parameters of the EVI distribution are as follows:

$$\begin{aligned}\hat{x}_0 &= \overline{x} - 0.45\hat{\sigma} = \overline{x} - 0.45s = 2498.9627 - 0.45(2199.4767)\\ &= 1499.35 m^3/s\end{aligned}$$

$$\hat{\alpha} = \frac{\sqrt{6}}{\pi}\hat{\sigma} = \frac{\sqrt{6}}{\pi}s = \frac{\sqrt{6}}{\pi}(2199.4767) = 1731.985 m^3/s$$

Finally, the moment estimators of the parameters of the EVI distribution for the flood sample data of gauging station Huites, Mexico, are:

location parameter:

$$\hat{x}_0 = 1499.35 m^3/s$$

and:

scale parameter:

$$\hat{\alpha} = 1731.985 m^3/s$$

9.4.2 ML Method

The likelihood function for the EVI distribution is as follows, NERC (1975):

$$L(x, x_0, \alpha) = \prod_{i=1}^{N} \left\{ \frac{1}{\alpha} \exp\left\{ -\exp\left[-\frac{(x_i - x_0)}{\alpha} \right] \right\} \exp\left[-\frac{(x_i - x_0)}{\alpha} \right] \right\} \tag{9.11}$$

By taking the natural logarithm of the previous equation, the log-likelihood function is obtained as, NERC (1975):

$$LL(x, x_0, \alpha) = \sum_{i=1}^{N} Ln \left\{ \frac{1}{\alpha} \exp\left\{ -\exp\left[-\frac{(x_i - x_0)}{\alpha} \right] \right\} \exp\left[-\frac{(x_i - x_0)}{\alpha} \right] \right\} \tag{9.12}$$

and the log-likelihood function for the EVI distribution is finally obtained as:

$$LL(x, x_0, \alpha) = -N Ln(\alpha) - \sum_{i=1}^{N} \left\{ \frac{(x_i - x_0)}{\alpha} + \exp\left[-\frac{(x_i - x_0)}{\alpha} \right] \right\} \tag{9.13}$$

Now, the classical approach to the method of maximum likelihood requires the computation of the first-order partial derivatives of the log-likelihood function with respect to each of its parameters, equating them equal to zero and then solving the resulting system of equations. So, the first-order partial derivatives are obtained as follows, NERC (1975):

$$\frac{\partial LL(x, x_0, \alpha)}{\partial x_0} = \frac{N}{\alpha} - \frac{1}{\alpha} \sum_{i=1}^{N} \left\{ \exp\left[-\frac{(x_i - x_0)}{\alpha} \right] \right\} = 0 \tag{9.14}$$

$$\frac{\partial LL(x, x_0, \alpha)}{\partial \alpha} = -\frac{N}{\alpha} + \frac{1}{\alpha^2}\sum_{i=1}^{N}(x_i - x_0) - \frac{1}{\alpha^2}\sum_{i=1}^{N}\left\{(x_i - x_0)\exp\left[-\frac{(x_i - x_0)}{\alpha}\right]\right\} = 0 \quad (9.15)$$

The exact solution to the system formed by Eqs. (9.14) and (9.15) is not known, so an iterative procedure is needed to evaluate the maximum likelihood estimators of the parameters of the EVI distribution. The iterative procedure is as follows, NERC (1975):

(1) Define a reduced variate as:

$$y = -\frac{(x - x_0)}{\alpha} \quad (9.16)$$

(2) Define parameters P and Q as follows:

$$P = N - \sum_{i=1}^{N}\exp(-y_i) \quad (9.17)$$

$$Q = N - \sum_{i=1}^{N} y_i + \sum_{i=1}^{N} y_i \exp(-y_i) \quad (9.18)$$

(3) Define the iterative procedure by:

$$(x_0)_{i+1} = (x_0)_i + \left(\delta_{x_0}\right)_i \quad (9.19)$$

$$(\alpha)_{i+1} = (\alpha)_i + (\delta_\alpha)_i \quad (9.20)$$

where the sub-index i refers to the iteration stage and δ are the differences between the estimator at iteration i and the true value for the maximum likelihood estimator for such parameter. The values of such differences are, NERC (1975):

$$\delta_i(x_0) = \frac{\alpha}{N}\left[1.11(P)_i - 0.26(Q)_i\right] \quad (9.21)$$

$$\delta_i(\alpha) = \frac{\alpha}{N}\left[0.26(P)_i - 0.61(Q)_i\right] \quad (9.22)$$

(4) Define a set of criteria of convergence in the following form:

$$\left|-\frac{\partial LL(x, x_0, \alpha)}{\partial x_0}\right| = \left|\frac{-P}{\alpha}\right| \approx 10^{-6} \quad (9.23)$$

$$\left|-\frac{\partial LL(x, x_0, \alpha)}{\partial \alpha}\right| = \left|\frac{Q}{\alpha}\right| \approx 10^{-6} \quad (9.24)$$

When conditions established by Eqs. (9.23) and (9.24) are met simultaneously, then the values of such parameters will correspond to the ML estimators of the parameters of the EVI distribution.

9.4.2.1 Example of Application of Estimation of the Parameters of the EVI Distribution Using the ML Method

Find the ML estimators for the parameters of the EVI distribution for the flood sample data of gauging station Huites, Mexico (1942–1992), contained in Appendix A.

The following statistics have already been obtained:

$$\hat{\mu} = \overline{x} = \frac{1}{N}\sum_{i=1}^{N} x_i = 2498.9627\ m^3/s$$

$$\hat{\sigma} = s = \left[\frac{1}{N}\sum_{i=1}^{N}(x_i - \hat{\mu})^2\right]^{1/2} = 2199.4767 m^3/s$$

Using the moment estimators, obtained previously, as initial values of the iterative scheme for the ML estimators for the parameters of the EVI distribution, then the initial values are:

$$\hat{x}_0 = 1509.198 m^3/s$$

$$\hat{\alpha} = 1714.921 m^3/s$$

Iteration 1

Defining an initial reduced variate as:

$$y = -\frac{(x - 1509.198)}{1714.921}$$

The parameters P and Q for the initial values are:

$$P = N - \sum_{i=1}^{N} \exp(y_i) = 51 - 43.6493 = 7.3507$$

$$Q = -N - \sum_{i=1}^{N} y_i + \sum_{i=1}^{N} y_i \exp(y_i) = 51 - 29.4346 - 2.3937 = 19.1717$$

Then, the initial deviations between the estimator at iteration 1 and the true value for the ML estimator for such parameters are:

$$\begin{aligned}
\delta_i(x_0) &= \frac{\alpha}{N}\left[1.11(P)_i - 0.26(Q)_i\right] \\
&= \left[\frac{1714.921}{51}\right][1.11(7.3507) - 0.26(19.1717)] \\
&= 106.7490
\end{aligned}$$

$$\begin{aligned}
\delta_i(\alpha) &= \frac{\alpha}{N}\left[0.26(P)_i - 0.61(Q)_i\right] \\
&= \left[\frac{1714.921}{51}\right][0.26(7.3507) - 0.61(19.1717)] \\
&= -328.9813
\end{aligned}$$

The new values of the ML parameters are:

$$(x_0)_{i+1} = (x_0)_i + \left(\delta_{x0}\right)_i = 1509.1982 + 106.7490 = 1615.9473$$

$$(\alpha)_{i+1} = (\alpha)_i + (\delta_\alpha)_i = 1714.9209 - 328.9813 = 1385.9396$$

The criteria of convergence are not met at this iteration:

$$\left| -\frac{\partial LL(x, x_0, \alpha)}{\partial x_0} \right| = \left| \frac{-P}{\alpha} \right| = \left| \frac{-7.3507}{1714.9209} \right| = 4.2863 \times 10^{-3} > 10^{-6}$$

$$\left| -\frac{\partial LL(x, x_0, \alpha)}{\partial \alpha} \right| = \left| \frac{Q}{\alpha} \right| = \left| \frac{19.1717}{1714.9209} \right| = 1.1179 \times 10^{-2} > 10^{-6}$$

After seven iterations, the final values of the parameters the ML procedure are as follows:

Iteration 7
The values of the parameters at this iteration are:

$$\hat{x}_0 = 1650.2591 m^3/s$$

$$\hat{\alpha} = 1212.0938 m^3/s$$

Defining a final reduced variate as:

$$y = -\frac{(x - 1650.2591)}{1212.0938}$$

The parameters P and Q for the final values are:

$$P = N - \sum_{i=1}^{N} \exp(y_i) = 51 - 50.9998 = 0.0002$$

$$Q = -N - \sum_{i=1}^{N} y_i + \sum_{i=1}^{N} y_i \exp(y_i) = 51 - 35.7100 - 15.2894 = 0.0006$$

Then the final deviations between the estimator at iteration 7 and the true value for the ML estimator for such parameters are:

$$\begin{aligned}\delta_i(x_0) &= \frac{\alpha}{N}\left[1.11(P)_i - 0.26(Q)_i\right]\\ &= \left[\frac{1212.0938}{51}\right][1.11(0.0002) - 0.26(0.0006)]\\ &= 0.0008\end{aligned}$$

$$\begin{aligned}\delta_i(\alpha) &= \frac{\alpha}{N}\left[0.26(P)_i - 0.61(Q)_i\right]\\ &= \left[\frac{1212.0938}{51}\right][0.26(0.0002) - 0.61(0.0006)]\\ &= -0.0073\end{aligned}$$

The final ML parameters are:

$$(x_0)_{i+1} = (x_0)_i + \left(\delta_{x_0}\right)_i = 1650.2591 + 0.0008 = 1650.2599$$

$$(\alpha)_{i+1} = (\alpha)_i + (\delta_\alpha)_i = 1212.0938 - 0.0073 = 1212.0865$$

The criteria of convergence are met for both parameters:

$$\left|-\frac{\partial LL(x, x_0, \alpha)}{\partial x_0}\right| = \left|\frac{-P}{\alpha}\right| = \left|\frac{0.0002}{1212.0938}\right| = 1.3468 \times 10^{-7} < 10^{-6}$$

$$\left|-\frac{\partial LL(x, x_0, \alpha)}{\partial \alpha}\right| = \left|\frac{Q}{\alpha}\right| = \left|\frac{0.0006}{1212.0938}\right| = 4.7344 \times 10^{-7} < 10^{-6}$$

Finally, the ML estimators of the parameters of the EVI distribution for the flood sample data of gauging station Huites, Mexico, are:

location parameter:

$$\hat{x}_0 = 1650.2599 m^3/s$$

and:

scale parameter:

$$\hat{\alpha} = 1212.0865 m^3/s$$

9.4.3 PWM Method

A distribution function may be characterized by its probability weighted moments (PWM) defined as, Greenwood et al. (1979):

$$M_{l,j,k} = E\left[X^l F^j (1-F)^k\right] = \int_0^1 [x(f)]^l F^j (1-F)^k dF \tag{9.25}$$

There are two different expressions to evaluate the population PWM defined in Eq. (9.25), they are, Greenwood et al. (1979):

$$M_{l,0,k} = \sum_{j=0}^{k} \binom{k}{j} (-1)^j M_{l,j,0} \tag{9.26}$$

and:

$$M_{l,j,0} = \sum_{k=0}^{j} \binom{j}{k} (-1)^k M_{l,0,k} \tag{9.27}$$

The conventional moments about the origin are represented by $M_{l,0,0.}$ The following convention is taken to simplify the notation, Greenwood et al. (1979):

$$M_{(k)} = M_{1,0,k} \tag{9.28}$$

An unbiased estimator of $M_{(k)}$ is, Landwher et al. (1979a):

$$M_{(k)} = \frac{1}{N} \sum_{i=1}^{N-k} x_i \frac{\binom{N-i}{k}}{\binom{N-1}{k}} \tag{9.29}$$

and k is a non-negative integer and the x_i´s, i = 1, 2, …, N, have been rank ordered from x_1 to x_N: $x_1 < x_2 < \ldots < x_{N.}$

So, the PMW for the EVI distribution are defined as, Greenwood et al. (1979):

$$M_{1,j,0} = \frac{x_0}{(1+j)} + \frac{\alpha[Ln(1+j) + 0.5772]}{(1+j)} \tag{9.30}$$

and the first two PWM are needed to construct the system of equations to obtain the PWM estimators of the EVI distribution:

$$M_{1,0,0} = M_{(0)} = x_0 + 0.5772\alpha = \overline{x} = \frac{1}{N}\sum_{i=1}^{N} x_i \tag{9.31}$$

$$\begin{aligned} M_{1,0,1} =& M_{(1)} = \frac{1}{2}[x_0 + \alpha(Ln(2) + 0.5772)] \\ =& \frac{1}{N}\sum_{i=1}^{N-1} x_i \frac{(N-i)}{(N-1)} \end{aligned} \tag{9.32}$$

$$\begin{aligned} M_{1,1,0} =& M_{(0)} - M_{(1)} = \frac{1}{2}[x_0 + \alpha[Ln(2) + 0.5772]] \\ =& \overline{x} - \frac{1}{N}\sum_{i=1}^{N-1} x_i \frac{(N-i)}{(N-1)} \end{aligned} \tag{9.33}$$

The solution to the system formed by Eqs. (9.31) and (9.33) are the PWM estimators of the parameters of the EVI distribution, Greenwood et al. (1979):

$$\hat{x}_0 = \overline{x} - 0.5772\hat{\alpha} \tag{9.34}$$

$$\hat{\alpha} = \frac{\left\{\overline{x} - \left[\frac{2}{N(N-1)}\right]\sum_{i=1}^{N-1} x_i(N-i)\right\}}{Ln(2)} \tag{9.35}$$

9.4.3.1 Example of Application of Estimation of the Parameters of the EVI Distribution Using the PWM Method

Find the PWM estimators for the EVI distribution for the flood sample data of gauging station Huites, Mexico (1942–1992), contained in Appendix A.

The following statistics have already been obtained:

$$\hat{\mu} = \overline{x} = \frac{1}{N}\sum_{i=1}^{N} x_i = 2498.9627\ m^3/s$$

$$\hat{\sigma} = s = \left[\frac{1}{N}\sum_{i=1}^{N}(x_i - \hat{\mu})^2\right]^{1/2} = 2199.4767 m^3/s$$

Now, the PWM are obtained as follows:

$$M_{(0)} = \hat{\mu} = \overline{x} = \frac{1}{N}\sum_{i=1}^{N} x_i = 2498.9627\ m^3/s$$

$$M_{(1)} = \frac{1}{N}\sum_{i=1}^{N-1} x_i \frac{(N-i)}{(N-1)} = 719.5557m^3/s$$

and the substitution of these values in Eqs. (9.34) and (9.35) give the PWM estimators of the EVI distribution as follows:

$$\hat{\alpha} = \frac{\left\{\overline{x} - \left[\frac{2}{N(N-1)}\right]\sum_{i=1}^{N-1} x_i(N-i)\right\}}{Ln(2)}$$

$$= \frac{[2498.9627 - 2(719.5557)]}{Ln(2)} = 1529.0423m^3/s$$

$$\hat{x}_0 = \overline{x} - 0.5772\alpha = 2498.9627 - 0.5772(1529.0423) = 1616.3995m^3/s$$

Finally, the PWM estimators of the parameters of the EVI distribution for the flood sample data of gauging station Huites, Mexico, are:

location parameter:

$$\hat{x}_0 = 1616.3995m^3/s$$

and:

scale parameter:

$$\hat{\alpha} = 1529.0423m^3/s$$

9.5 Estimation of Quantiles for the EVI Distribution

The quantiles for the EVI distribution are obtained by using the inverse form of the EVI distribution function:

$$x_T = \hat{x}_0 - \hat{\alpha}Ln(-Ln(F(x))) \qquad (9.36)$$

where x_T is the quantile value for a certain value of the distribution function $F(x)$.

The term Q_T is more frequently used in engineering instead of x_T and this also applies to T_r instead of $F(x)$; so, the following expression is widely used when it is needed to relate a maxima natural extreme event, Q_T, to a specific return period T_r:

$$Q_T = \hat{x}_0 - \hat{\alpha} Ln\left(-Ln\left(1 - \frac{1}{T_r}\right)\right) \tag{9.37}$$

where Q_T is a design value corresponding to a specific return period T_r.

9.5.1 Examples of Estimation of MOM, ML and PWM Quantiles for the EVI Distribution

Find the MOM, ML and PWM estimator of the quantiles 2, 5, 10, 20, 50, 100, 500, and 1,000 years of return period, for the EVI distribution for the flood sample data of gauging station Huites, Mexico (1942–1992), contained in Appendix A.

The estimation of MOM, ML and PWM quantiles for the EVI distribution is made by using the parameters estimated in the preceding sections and then inserted one pair of parameters at a time in the Eq. (9.37). Table 9.1 summarizes these results.

9.6 Goodness of Fit Test

The standard error of fit (SEF) for the EVI has the following form:

$$SEF = \left[\frac{\sum_{i=1}^{N}(xi - yi)^2}{(N-2)}\right]^{1/2} \tag{9.38}$$

Table 9.1 Estimation of MOM, ML and PWM quantiles for the EVI distribution for Gauging Station Huites, Mexico

Tr (Years)	MOM	ML	PWM
	Q (m³/s)	Q (m³/s)	Q (m³/s)
2	2138	2094	2177
5	4081	3468	3910
10	5368	4378	5057
20	6603	5250	6158
50	8201	6380	7583
100	9398	7226	8650
500	12,165	9182	11,117
1000	13,355	10,022	12,178

while the mean absolute relative deviation (MARD) remains the same as it was defined in Chap. 2:

$$MARD = \frac{100}{N}\sum_{i=1}^{N}\left|\frac{(x_i - y_i)}{x_i}\right| \tag{9.39}$$

where x_i are the sample historical values, y_i are the distribution function values corresponding to the same return periods of the historical values, N is the sample size.

9.6.1 Examples of Application of the SEF and MARD to the MOM, ML and PWM Estimators of the Parameters of the EVI Distribution

Find the values of the SEF and MARD of the MOM, ML and PWM estimators of the parameters of the EVI distribution for the flood sample data of gauging station Huites, Mexico (1942–1992), contained in Appendix A.

The values of SEF and MARD for the MOM, ML and PWM estimators of the parameters of the EVI distribution for the flood sample data of gauging station Huites, Mexico, have been obtained through the application of Eqs. (9.38) and (9.39) using the parameters obtained in previous examples. Table 9.2 contains a summary of both measures of goodness of fit tests.

For the flood sample data at gauging station Huites, Mexico, the best choice according to SEF measure is the MOM method. When using the MARD measure, the choice is the ML method. Such methods produce the least values on each of the mentioned measures of goodness of fit for this flood sample of data.

9.7 Estimation of Confidence Limits for the EVI Distribution

By assuming that the quantiles are normally distributed, the following form can be used to set the confidence limits on such quantiles:

$$x_l = x_T \pm z_\alpha S_T \tag{9.40}$$

Table 9.2 SEF and MARD measures for MOM, ML and PWM estimators of the parameters of the EVI distribution for the flood sample data of Gauging Station Huites, Mexico

Goodness of Fit Test	MOM	ML	PWM
SEF	**791**	1046	839
MARD	39	**21**	31

where x_l is the upper or lower confidence limit, depending on the sign in the preceding formula (+) for the upper confidence limit and (−) for the lower confidence limit, z_α is the standard normal variate corresponding to a confidence level α, and S_T is squared root of the standard error of the estimate. The evaluation procedures of the standard errors of the estimates will be described in the following subsections.

9.8 Estimation of Standard Errors for the EVI Distribution

9.8.1 MOM Method

The general form of the MOM estimator of the standard error of the estimate of a two-parameter distribution is, Kite (1988):

$$S_T^2 = \left(\frac{\partial x_T}{\partial m_1'}\right)^2 \text{var}(m_1') + \left(\frac{\partial x_T}{\partial m_2}\right)^2 \text{var}(m_2) + 2\left(\frac{\partial x_T}{\partial m_1'}\right)\left(\frac{\partial x_T}{\partial m_2}\right)\text{cov}(m_1', m_2) \quad (9.41)$$

and Eq. (9.39) can be simplified in terms of the frequency factor, K_T, as, Kite (1988):

$$S_T^2 = \frac{\mu_2}{N}\left[1 + K_T\hat{\gamma} + \frac{K_T^2}{4}(\hat{\kappa} - 1)\right] \quad (9.42)$$

So, the MOM estimator of the standard error of the estimate for the EVI distribution is:

$$S_T^2 = \frac{\sigma^2}{N}(1 + 1.1396K_T + 1.10K_T^2) \quad (9.43)$$

where K_T is the frequency factor given by, Kite (1988):

$$K_T = -\left[0.45 + 0.7797Ln\left(-Ln\left(1 - \frac{1}{T_r}\right)\right)\right] \quad (9.44)$$

And using this frequency factor, Eq. (9.43) can be simplified to:

$$S_T^2 = \frac{\alpha^2}{N}(1.1678 + 0.1919y_T + 1.0999y_T^2) \quad (9.45)$$

where y_T is, Kite (1988):

$$y_T = -Ln\left(-Ln\left(1 - \frac{1}{T_r}\right)\right) \quad (9.46)$$

Table 9.3 Estimation of MOM standard errors and the two-sided 95% confidence limits for the EVI distribution for Gauging Station Huites, Mexico

Tr (Years)	S_T(m³/s)	95% Lower Limit (m³/s)	Q_T(m³/s)	95% Upper Limit (m³/s)
2	262.70	1584	2138	2692
5	476.08	3148	4081	5015
10	643.02	4108	5368	6629
20	812.29	5011	6603	8195
50	1037.45	6167	8201	10,234
100	1208.54	7029	9398	11,767
500	1607.85	9014	12,165	15,316
1000	1780.50	9865	13,355	16,844

Two-sided Limits: $z_\alpha = 1.96$

9.8.1.1 Example of Estimation of MOM Standard Errors and the Two-Sided 95% Confidence Limits for the EVI Distribution

Find the MOM estimators of the standard errors and the two-sided 95% confidence limits for the EVI distribution for the flood sample data of gauging station Huites, Mexico (1942–1992), contained in Appendix A.

The estimation of MOM estimators of the standard errors and the two-sided 95% confidence limits for the EVI distribution is made by using the MOM estimators of the parameters and then inserted in the Eqs. (9.43) and (9.38), using selected values of the return intervals. Table 9.3 summarizes these results.

9.8.2 ML Method

The general form of the ML estimator of the standard error of the estimate of a two-parameter distribution is:

$$S_T^2 = \left(\frac{\partial x_T}{\partial \alpha}\right)^2 \text{var}(\alpha) + \left(\frac{\partial x_T}{\partial \beta}\right)^2 \text{var}(\beta) + 2\left(\frac{\partial x_T}{\partial \alpha}\right)\left(\frac{\partial x_T}{\partial \beta}\right)\text{cov}(\alpha, \beta) \quad (9.47)$$

and the variance–covariance matrix of the parameters for the EVI distribution is known to be, Kimball (1949):

$$[V] = \begin{bmatrix} \text{var}(x_0) & \text{cov}(x_0, \alpha) \\ \text{cov}(x_0, \alpha) & \text{var}(\alpha) \end{bmatrix} = \frac{\alpha^2}{N}\begin{bmatrix} 1.1086 \; 0.2570 \\ 0.2570 \; 0.6079 \end{bmatrix} \quad (9.48)$$

and from Eq. (9.36):

$$\frac{\partial x_T}{\partial \alpha} = -Ln\left[-Ln\left(1 - \frac{1}{Tr}\right)\right] \quad (9.49)$$

and:

$$\frac{\partial x_T}{\partial x_0} = 1 \tag{9.50}$$

The substitution of these results into Eq. (9.45) produces the ML estimator of the standard error of the estimate for the EVI distribution as, Kite (1988):

$$S_T^2 = \frac{\alpha^2}{N}\left(1.1086 + 0.5140 y_T + 0.6079 y_T^2\right) \tag{9.51}$$

9.8.2.1 Example of Estimation of ML Standard Errors and the Two-Sided 95% Confidence Limits for the EVI Distribution

Find the ML estimators of the standard errors and the two-sided 95% confidence limits for the EVI distribution for the flood sample data of gauging station Huites, Mexico (1942–1992), contained in Appendix A.

The estimation of ML estimators of the standard errors and the two-sided 95% confidence limits for the EVI distribution is made by using the ML estimators of the parameters and then inserted in the Eqs. (9.49) and (9.38), using selected values of the return intervals. Table 9.4 summarizes these results.

9.8.3 PWM Method

The general form of the PWM estimator of the standard error of the estimate of a two-parameter distribution is:

$$S_T^2 = \left(\frac{\partial x_T}{\partial \alpha}\right)^2 \operatorname{var}(\alpha) + \left(\frac{\partial x_T}{\partial \beta}\right)^2 \operatorname{var}(\beta)$$

Table 9.4 Estimation of ML standard errors and the two-sided 95% confidence limits for the EVI distribution for Gauging Station Huites, Mexico

Tr (Years)	y_T	S_T(m^3/s)	95% Lower Limit (m^3/s)	Q_T(m^3/s)	95% Upper Limit (m^3/s)
2	0.3665	199.2844	1704	2095	2485
5	1.4999	305.8464	2869	3468	4068
10	2.2504	392.3477	3609	4378	5147
20	2.9702	480.0019	4310	5250	6191
50	3.9019	596.9302	5210	6380	7550
100	4.6001	686.0152	5881	7226	8571
500	6.2136	894.4499	7429	9182	10,935
1000	6.9073	984.7277	8092	10,022	11,952

Two-sided Limits: $z_\alpha = 1.96$

$$+2\left(\frac{\partial x_T}{\partial \alpha}\right)\left(\frac{\partial x_T}{\partial \beta}\right)\text{cov}(\alpha, \beta) \tag{9.52}$$

The expressions for the variances and covariance shown in Eq. (9.52), have been obtained as, Hosking (1986a):

$$\text{var}(x_0) = 1.1128\frac{\alpha^2}{N} \tag{9.53}$$

$$\text{var}(\alpha) = 0.8046\frac{\alpha^2}{N} \tag{9.54}$$

$$\text{cov}(x_0, \alpha) = 0.2287\frac{\alpha^2}{N} \tag{9.55}$$

and:

$$\frac{\partial x_T}{\partial \alpha} = -Ln\left[-Ln\left(1 - \frac{1}{Tr}\right)\right] \tag{9.56}$$

and:

$$\frac{\partial x_T}{\partial x_0} = 1 \tag{9.57}$$

as before, then the PWM estimator of the standard error of the estimate for the EVI distribution is, Hosking (1986a):

$$S_T^2 = \frac{\alpha^2}{N}\left(1.1128 + 0.4574y_T + 0.8046y_T^2\right) \tag{9.58}$$

9.8.3.1 Example of Estimation of Standard Errors and the Two-Sided 95% Confidence PWM Limits for the EVI Distribution

Find the PWM estimators of the standard errors and the two-sided 95% confidence limits for the EVI distribution for the flood sample data of gauging station Huites, Mexico (1942–1992), contained in Appendix A.

The estimation of PWM estimators of the standard errors and the two-sided 95% confidence limits for the EVI distribution is made by using the PWM estimators of the parameters at selected values of the return intervals. Table 9.5 summarizes these results.

9.9 Examples of Application for the EVI Distribution Using Excel® Spreadsheets

9.9.1 Flood Frequency Analysis

By using the flood data from gauging station Huites, Mexico the following descriptive statistics were obtained and are shown in Fig. 9.1.

Table 9.5 Estimation of PWM standard errors and the two-sided 95% confidence limits for the EVI distribution for Gauging Station Huites, Mexico

Tr (Years)	y_T	$S_T(m^3/s)$	95% Lower Limit (m^3/s)	$Q_T(m^3/s)$	95% Upper Limit (m^3/s)
2	0.3665	252.2966	1682	2177	2671
5	1.4999	406.7547	3113	3910	4707
10	2.2504	533.8455	4011	5057	6104
20	2.9702	662.3404	4860	6158	7456
50	3.9019	833.3118	5949	7583	9216
100	4.6001	963.3311	6762	8650	10,538
500	6.2136	1267.0397	8634	11,117	13,601
1000	6.9073	1398.4397	9437	12,178	14,919

Two-sided Limits: $z_\alpha = 1.96$

Fig. 9.1. Descriptive statistics of the flood sample data at Huites, Mexico

	A	B
1	**Flood Frequency Analysis**	
2	**Gauging Station: Huites, Mexico (1942-1992)**	
3		
4	***Descriptive Statistics***	
5	Mean	2498.9627
6	Sample Standard Error	311.0530
7	Median	1775.7000
8	Mode	#N/D
9	Standard Deviation	2221.3626
10	Sample Variance	4934451.7500
11	Kurtosis	3.6701
12	Skewness	2.0123
13	Range	9707.5000
14	Minimum	421.8000
15	Maximum	10129.3000
16	Sum	127447.1000
17	Count	51.0000
18	Largest(1)	10129.3000
19	Smallest(1)	421.8000
20	Mean Confidence Level(95.0%)	624.7683

9.9.1.1 MOM Method

By using an Excel® spreadsheet, the results contained in Fig. 9.2, were obtained by the MOM method applied to the EVI distribution for gauging station Huites, Mexico.

9.9.1.2 ML Method

By using an Excel® spreadsheet, the results contained in Table 9.3, were obtained by the ML method applied to the EVI distribution for gauging station Huites, Mexico (Fig. 9.3).

FLOOD FREQUENCY ANALYSIS						
Gauging Station: Huites, Mexico (1942-1992)						
Extreme Value Type I (EVI) Distribution						
Method: MOM						
Estimation of Parameters and Goodness of Fit Tests						
Location Parameter =	1499.35	m^3/s				
Scale Parameter =	1731.985	m^3/s				
SEF =	790		MARD =	40		
Design Values and Confidence Limits						
Return Period Tr (years)	CDF F(x)	Freq. Factor K_T	Squared root Stand. Error S_T	Two-sided 95% Lower Limit	Design Values Q_T (m^3/s)	Two-sided 95% Upper Limit
2	0.5	-0.1642	285.5108	1575	2134	2694
5	0.8	0.7195	480.8158	3155	4097	5040
10	0.9	1.3046	649.4193	4124	5397	6670
20	0.95	1.8659	820.3734	5036	6644	8252
50	0.98	2.5923	1047.7680	6204	8257	10311
100	0.99	3.1367	1220.5669	7074	9467	11859
500	0.998	4.3947	1623.8474	9078	12261	15444
1000	0.999	4.9356	1798.2148	9938	13463	16987
Two-sided Confidence Limits:		z_α =	1.96			

Fig. 9.2 MOM estimators of the MOM estimators for the parameters, standard errors, quantiles, and confidence limits for the flood sample data at Huites, Mexico

FLOOD FREQUENCY ANALYSIS							
Gauging Station: Huites, Mexico (1942-1992)							
Extreme Value Type I (EVI) Distribution							
Method: ML							
Estimation of Parameters and Goodness of Fit Tests							
Initial Moment Estimators:	Location Parameter =	1499.34958	m^3/s	Scale Parameter=	1731.985255	m^3/s	
				Log-Likelihood function =	-448.8156		
Iterative Procedure:	i-th iteration:	Location Parameter =	1650.2615	(i+1)-th iteration:	Location Parameter=	1650.2604	
		Scale Parameter =	1212.0709		Scale Parameter=	1212.0810	
$\delta(X_0)$ =	-0.0011	$\delta(\alpha)$ =	0.0101	ABS($\partial LL/\partial x_0$) =	1.8696E-07		
ABS($\partial LL/\partial\alpha$) =	6.5680E-07	P =	-0.0002	Q =	-0.0008		
Final ML Estimators:	Location Parameter =	1650.2604	m^3/s	Scale Parameter =	1212.0810	m^3/s	
SEF =	1046	MARD =	21	AIC =	902		
Design Values and Confidence Limits							
Return Period Tr (Years)	F(x)	Frequency Factor K_T	y_T	Squared Root Stand. Error S_T	Two-sided 95% Lower Limit	Design Values Q_T (m^3/s)	Two-sided 95% Upper Limit
2	0.5	-0.1642	0.3665	199.2844	1704	2095	2485
5	0.8	0.7195	1.4999	305.8464	2869	3468	4068
10	0.9	1.3046	2.2504	392.3477	3609	4378	5147
20	0.95	1.8659	2.9702	480.0019	4310	5250	6191
50	0.98	2.5923	3.9019	596.9302	5210	6380	7550
100	0.99	3.1367	4.6001	686.0152	5881	7226	8571
500	0.998	4.3947	6.2136	894.4499	7429	9182	10935
1000	0.999	4.9356	6.9073	984.7277	8092	10022	11952
Two-sided Confidence Limits:		z_α =	1.96				

Fig. 9.3 ML estimators for the parameters and standard errors, quantiles and confidence limits of the EVI distribution for the flood sample at Huites. Mexico

9.9.1.3 PWM Method

By using an Excel® spreadsheet, the results contained in Figs. 9.4, were obtained by the PWM method applied to the EVI distribution for gauging station Huites, Mexico.

In Fig. 9.5 a comparison is made between the histogram and EVI theoretical density for flood sample of Huites, Mexico.

A graphical comparison between the empirical and theoretical frequency curves from the results provided by the three methods shown before is shown in Fig. 9.6.

A graphical depiction of the empirical and theoretical frequency curves and the confidence limits of the best fit provided, in this case that of MOM, is shown in Fig. 9.7.

9.9.2 Rainfall Frequency Analysis

By using the 24 h, annual maximum rainfall data from rainfall gauging station Chihuahua, Mexico the following statistical results were obtained and are shown in Fig. 9.8.

FLOOD FREQUENCY ANALYSIS
Gauging Station: Huites, Mexico (1942-1992)
Extreme Value Type I (EVI) Distribution
Method: PWM

Probability Weighted Moments

$M_{(0)}$ =	2498.9627	m^3/s
$M_{(1)}$ =	719.5557	m^3/s

Estimation of Parameters and Goodness of Fit Tests

Location Parameter =	1616.3995	m^3/s	
Scale Parameter =	1529.0423	m^3/s	
SEF =	839	MARD =	31

Design Values and Confidence Limits

Return Period Tr (years)	CDF F(x)	Freq. Factor K_T	Squared Root Stand. Error S_T	Two-sided 95% Lower Limit	Design Values Q_T (m3/s)	Two-sided 95% Upper Limit
2	0.5	-0.1642	252.2966	1682	2177	2671
5	0.8	0.7195	406.7547	3113	3910	4707
10	0.9	1.3046	533.8455	4011	5057	6104
20	0.95	1.8659	662.3404	4860	6158	7456
50	0.98	2.5923	833.3118	5949	7583	9216
100	0.99	3.1367	963.3311	6762	8650	10538
500	0.998	4.3947	1267.0397	8634	11117	13601
1000	0.999	4.9356	1398.4397	9437	12178	14919
Two-sided Confidence Limits:		z_α =	1.96			

Fig. 9.4 PWM estimators for the parameters, standard errors, quantiles, and confidence limits of the EVI distribution for the flood sample at Huites. Mexico

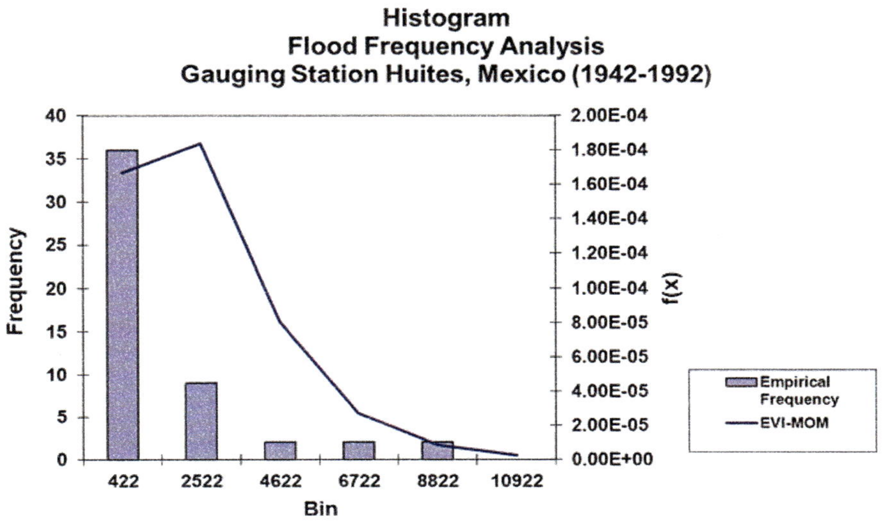

Fig. 9.5 Histogram and EVI theoretical density for the flood sample of Huites, Mexico

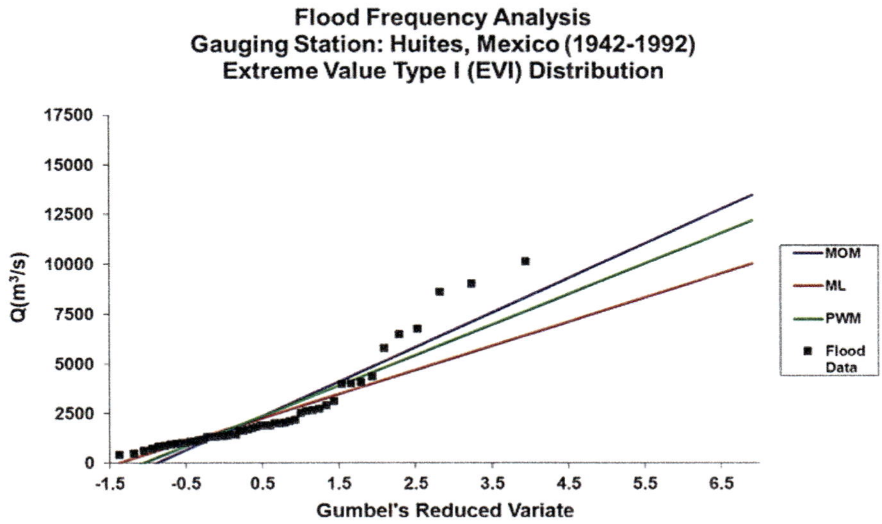

Fig. 9.6 Empirical and theoretical frequency curves for Gauging Station Huites, Mexico

9.9.2.1 MOM Method

By using an Excel® spreadsheet, the results contained in Fig. 9.9, were obtained for the MOM method applied to the EVI distribution for rainfall gauging station Chihuahua, Mexico.

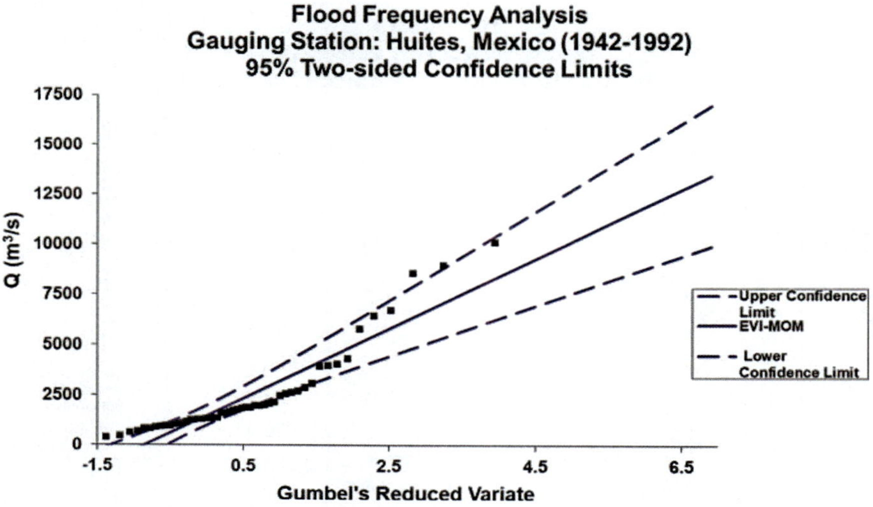

Fig. 9.7 Empirical and MOM theoretical frequency curves and confidence limits of flood sample of Huites, Mexico

	A	B	C
1	24 Hr. Annual Maximum Rainfall Frequency Analysis		
2	Meteorological Station: Chihuahua, Mexico (1960-2005)		
3			
4	*Descriptive Statistics*		
5	Mean	42.0454	
6	Sample Standard Error	1.9784	
7	Median	40.2400	
8	Mode	#N/D	
9	Standard Deviation	13.4184	
10	Sample Variance	180.0535	
11	Kurtosis	-0.7252	
12	Skewness	0.3591	
13	Range	54.3000	
14	Minimum	17.1200	
15	Maximum	71.4200	
16	Sum	1934.0900	
17	Count	46.0000	
18	Largest(1)	71.4200	
19	Smallest(1)	17.1200	
20	Mean Confidence Level(95.0%)	3.9848	

Fig. 9.8 Descriptive statistics for the 24 h, Annual Maximum Rainfall Data at Meteorological Station Chihuahua, Mexico

9.9.2.2 ML Method

By using an Excel® spreadsheet, the results contained in Fig. 9.10, were obtained for the ML method applied to the EVI distribution for rainfall gauging station Chihuahua, Mexico.

	A	B	C	D	E	F	G
1	24 HRS. ANNUAL MAXIMUM RAINFALL FREQUENCY ANALYSIS						
2	Meteorological Station: Chihuahua, Mexico (1960-2005)						
3	Extreme Value Type I (EVI) Distribution						
4	Method: MOM						
5							
6	Estimation of Parameters and Goodness of Fit Tests						
7	Location Parameter =	36.0072	cm				
8	Scale Parameter =	10.4623	cm				
9	SEF =	2.46		MARD =	4.95		
10							
11	Design Values and Confidence Limits						
12				Squared root	Two-sided	Design	Two-sided
13	Return Period	CDF	Freq. Factor	Stand. Error	95%	Values	95%
14	Tr (years)	F(x)	K_T	S_T	Lower Limit	P_T (cm)	Upper Limit
15	2	0.5	-0.1642	1.6657	37	40	43
16	5	0.8	0.7195	2.1124	48	52	56
17	10	0.9	1.3046	2.7981	54	60	65
18	20	0.95	1.8659	3.5697	60	67	74
19	50	0.98	2.5923	4.6423	68	77	86
20	100	0.99	3.1367	5.4738	73	84	95
21	Two-sided Confidence Limits:	z_α =	1.96				

Fig. 9.9 MOM estimators of the parameters, standard errors, quantiles, and confidence limits for 24 h, annual maximum rainfall data at Meteorological Station Chihuahua, Mexico

	A	B	C	D	E	F	G	H
1	24 HRS. ANNUAL MAXIMUM RAINFALL FREQUENCY ANALYSIS							
2	Meteorological Station: Chihuahua, Mexico (1960-2005)							
3	Extreme Value Type I (EVI) Distribution							
4	Method: ML							
5								
6	Estimation of Parameters and Goodness of Fit Tests							
7	Initial Moment Estimators:							
8	Location Parameter =	36.0072	cm	Scale Parameter=	10.4623	cm		
9	Iterative Procedure:							
10	i-th iteration:	Location Parameter =	35.6641	(i+1)-th iteration:	Location Parameter=	35.6641		
11		Scale Parameter =	11.3471		Scale Parameter=	11.3471		
12	$\delta(X_0)$ =	0.0000	$\delta(\alpha)$ =	0.0000				
13	P =	0.0000	Q =	-0.0001				
14	ABS($\partial LL/\partial x_0$) =	3.6237E-06	ABS($\partial LL/\partial\alpha$) =	4.5026E-06				
15	Final ML Estimators:	Location Parameter =	35.6641	cm	Scale Parameter =	11.3471	cm	
16	Log-Likelihood function =	-183.6016						
17	SEF =	2.21	MARD =	3.75	AIC =	371		
18								
19	Design Values and Confidence Limits							
20			Frequency		Squared Root	Two-sided	Design	Two-sided
21	Return Period		Factor		Stand. Error	95%	Values	95%
22	Tr (years)	F(x)	K_T	y_T	S_T	Lower Limit	P_T (cm)	Upper Limit
23	2	0.5	-0.1642	0.3665	1.9644	36	40	44
24	5	0.8	0.7195	1.4999	3.0148	47	53	59
25	10	0.9	1.3046	2.2504	3.8675	54	61	69
26	20	0.95	1.8659	2.9702	4.7315	60	69	79
27	50	0.98	2.5923	3.9019	5.8841	68	80	91
28	100	0.99	3.1367	4.6001	6.7623	75	88	101
29	Two-sided Confidence Limits:	z_α =	1.96					

Fig. 9.10 ML estimation of the parameters, standard errors, quantiles, and confidence limits for the 24 h, annual maximum rainfall data at Meteorological Station Chihuahua, Mexico

9.9.2.3 PWM Method

By using an Excel® spreadsheet, the results contained in Fig. 9.11, were obtained for the PWM method applied to the EVI distribution for 24 h annual maximum rainfall at meteorological station Chihuahua, Mexico.

	A	B	C	D	E	F	G
1	24 HRS. ANNUAL MAXIMUM RAINFALL FREQUENCY ANALYSIS						
2	Meteorological Station: Chihuahua, Mexico (1960-2005)						
3	Extreme Value Type I (EVI) Distribution						
4	Method: PWM						
5							
6	Probability Weighted Moments						
7	$M_{(0)}$ =	42.0454	cm				
8	$M_{(1)}$ =	17.1617	cm				
9							
10	Estimation of Parameters and Goodness of Fit Tests						
11	Location Parameter =	35.6151	cm				
12	Scale Parameter =	11.1406	cm				
13	SEF =	2.25		MARD =	3.98		
14							
15	Design Values and Confidence Limits						
16							
17	Return Period	CDF	Freq. Factor	Squared Root Stand. Error	Two-sided 95%	Design Values	Two-sided 95%
18	Tr (years)	F(x)	K_T	S_T	Lower Limit	P_T (cm)	Upper Limit
19	2	0.5	-0.1642	1.9356	36	40	43
20	5	0.8	0.7195	3.1205	46	52	58
21	10	0.9	1.3046	4.0955	53	61	69
22	20	0.95	1.8659	5.0813	59	69	79
23	50	0.98	2.5923	6.3930	67	79	92
24	100	0.99	3.1367	7.3904	72	87	101
25	Two-sided Confidence Limits:	z_{α} =	1.96				

Fig. 9.11 PWM estimators of the parameters, standard errors, quantiles, and confidence limits for 24 h, annual maximum rainfall data at Meteorological Station Chihuahua, Mexico

In Fig. 9.12 a comparison is made between the histogram and EVI theoretical density for 24 h annual maximum rainfall data at meteorological station Chihuahua, Mexico.

A graphical comparison between the empirical and theoretical frequency curves from the results provided by the three methods shown before is shown in Fig. 9.13.

A graphical depiction of the empirical and theoretical frequency curves and the confidence limits of the best fit provided, in this case that of, is shown in Fig. 9.14.

9.9.3 Maximum Significant Wave Height Frequency Analysis

By using the maximum significant wave height data, Castillo (1988), the following statistical results were obtained and are shown in Fig. 9.15.

9.9.3.1 MOM Method

By using an Excel® spreadsheet, the results contained in Fig. 9.16, were obtained for the MOM method applied to the EVI distribution for the maximum wind speed data, Castillo (1988).

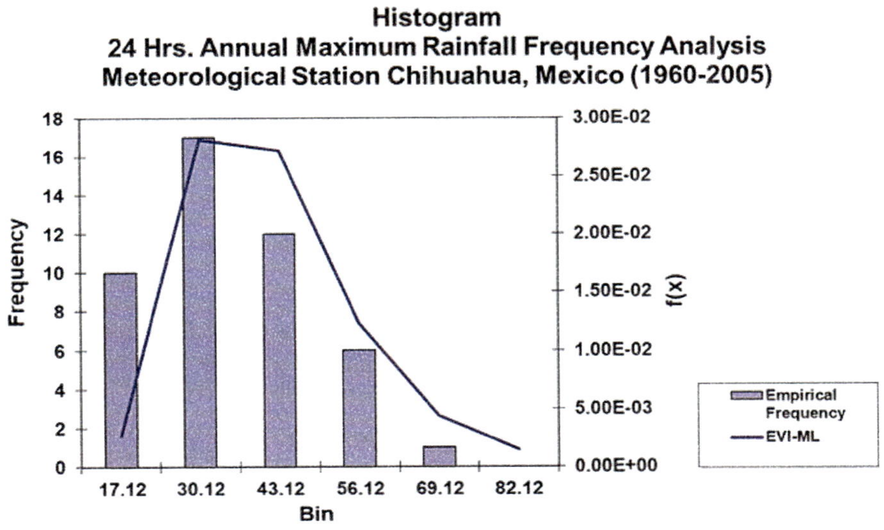

Fig. 9.12 Histogram and theoretical EVI density for 24 h, annual maximum rainfall data at Meteorological Station Chihuahua, Mexico

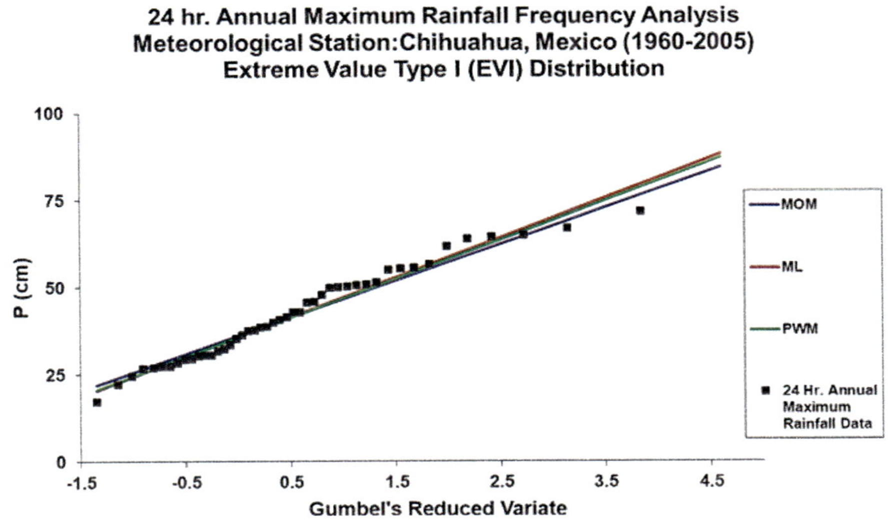

Fig. 9.13 Empirical and theoretical EVI distributions for 24 h, annual maximum rainfall data at Meteorological Station Chihuahua, Mexico

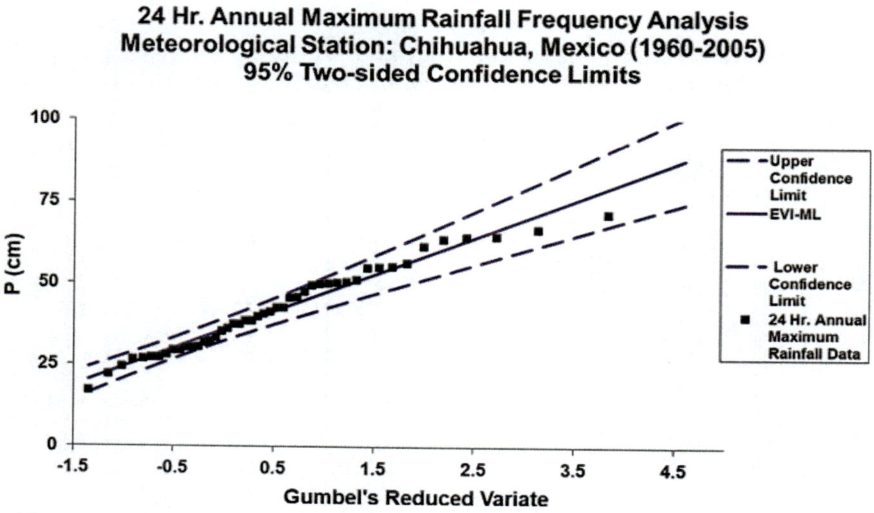

Fig. 9.14 Empirical and ML theoretical frequency curves and ML confidence limits for 24 h annual maximum rainfall data at Meteorological Station Chihuahua, Mexico

	A	B	C	D
1	MAXIMUM SIGNIFICANT WAVE HEIGHT FREQUENCY ANALYSIS			
2				
3	*Descriptive Statistics*			
4	Mean	4.4247		
5	Standard Error	0.3053		
6	Median	4.1666		
7	Mode	3.5509		
8	Standard Deviation	2.1585		
9	Sample Variance	4.6591		
10	Kurtosis	1.0740		
11	Skewness	0.9053		
12	Range	10.4485		
13	Minimum	0.8870		
14	Maximum	11.3355		
15	Sum	221.2326		
16	Count	50.0000		
17	Largest(1)	11.3355		
18	Smallest(1)	0.8870		
19	Confidence Level(95.0%)	0.6134		

Fig. 9.15 Descriptive statistics for the maximum significant wave height data

9.9.3.2 ML Method

By using an Excel® spreadsheet, the results contained in Fig. 9.17, were obtained for the ML method applied to the EVI distribution for rainfall gauging station Chihuahua, Mexico.

	A	B	C	D	E	F	G
1	MAXIMUM SIGNIFICANT WAVE HEIGHT FREQUENCY ANALYSIS						
2	Extreme Value Type I (EVI) Distribution						
3	Method: MOM						
4							
5	Estimation of Parameters and Goodness of Fit Tests						
6	Location Parameter =	3.4533	m				
7	Scale Parameter =	1.6830	m				
8	SEF =	0.25		MARD =	6.39		
9							
10	Design Values and Confidence Limits						
11–14	Return Period Tr (years)	F(x)	Frequency Factor K_T	Squared root Stand. Error S_T	Two-sided 95% Lower Limit (m)	Design Values H_T (m)	Two-sided 95% Upper Limit (m)
15	2	0.5	-0.1642	0.3296	3	4	5
16	5	0.8	0.7195	0.4180	5	6	7
17	10	0.9	1.3046	0.5537	6	7	8
18	20	0.95	1.8659	0.7064	7	8	10
19	50	0.98	2.5923	0.9187	8	10	12
20	100	0.99	3.1367	1.0832	9	11	13
21	Two-sided Confidence Limits:	z_α =	1.96				

Fig. 9.16 MOM estimators of the parameters, standard errors, quantiles, and confidence limits for maximum significant wave height data

	A	B	C	D	E	F	G	H
1	MAXIMUM SIGNIFICANT WAVE HEIGHT FREQUENCY ANALYSIS							
2	Extreme Value Type I (EVI) Distribution							
3	Method: ML							
4								
5	Estimation of Parameters and Goodness of Fit Tests							
6	Initial Moment Estimators:	Location Parameter =	3.4292	m				
7		Scale Parameter=	1.7247	m				
8	Iterative Procedure:							
9	i-th iteration:	Location Parameter =	5.5726	(i+1)-th iteration:	Location Parameter=	5.5727		
10		Scale Parameter =	2.5403		Scale Parameter=	2.5404		
11	$\delta(X_0)$ =	0.0001	$\delta(\alpha)$ =	0.0000	P =	0.0018		
12	Q =	0.0000	ABS($\partial LL/\partial x_0$) =	7.1358E-04	ABS($\partial LL/\partial \alpha$) =	2.9636E-07		
13	Final ML Estimators:	Location Parameter =	5.5727	m	Scale Parameter =	2.5404	m	
14	SEF =	3		MARD =	68			
15								
16–19	Return Period Tr (years)	F(x)	Frequency Factor KT	y_T	Squared Root Stand. Error S_T	Two-sided 95% Lower Limit (m)	Design Values H_T (m)	Two-sided 95% Upper Limit (m)
20	2	0.5	-0.1642	0.3665	0.4218	6	7	7
21	5	0.8	0.7195	1.4999	0.6474	8	9	11
22	10	0.9	1.3046	2.2504	0.8305	10	11	13
23	20	0.95	1.8659	2.9702	1.0160	11	13	15
24	50	0.98	2.5923	3.9019	1.2635	13	15	18
25	100	0.99	3.1367	4.6001	1.4521	14	17	20
26	Two-sided Confidence Limits:	z_α =	1.96					

Fig. 9.17 ML estimators of the parameters, standard errors, quantiles, and confidence limits for maximum significant wave height data

9.9.3.3 PWM Method

By using an Excel® spreadsheet, the results contained in Fig. 9.18, were obtained for the PWM method applied to the EVI distribution for meteorological station Chihuahua, Mexico.

In Fig. 9.19 a comparison is made between the histogram and EVI theoretical density for 24 h annual maximum rainfall data at meteorological station Chihuahua, Mexico.

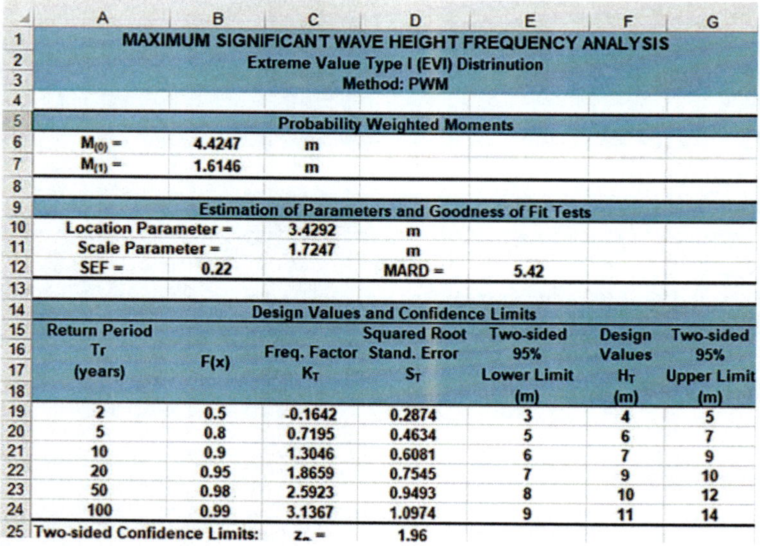

	A	B	C	D	E	F	G
1	MAXIMUM SIGNIFICANT WAVE HEIGHT FREQUENCY ANALYSIS						
2	Extreme Value Type I (EVI) Distrinution						
3	Method: PWM						
4							
5	Probability Weighted Moments						
6	$M_{(0)}$ =	4.4247	m				
7	$M_{(1)}$ =	1.6146	m				
8							
9	Estimation of Parameters and Goodness of Fit Tests						
10	Location Parameter =		3.4292	m			
11	Scale Parameter =		1.7247	m			
12	SEF =	0.22		MARD =	5.42		
13							
14	Design Values and Confidence Limits						
15–18	Return Period Tr (years)	F(x)	Freq. Factor K_T	Squared Root Stand. Error S_T	Two-sided 95% Lower Limit (m)	Design Values H_T (m)	Two-sided 95% Upper Limit (m)
19	2	0.5	-0.1642	0.2874	3	4	5
20	5	0.8	0.7195	0.4634	5	6	7
21	10	0.9	1.3046	0.6081	6	7	9
22	20	0.95	1.8659	0.7545	7	9	10
23	50	0.98	2.5923	0.9493	8	10	12
24	100	0.99	3.1367	1.0974	9	11	14
25	Two-sided Confidence Limits:		z_{α} =	1.96			

Fig. 9.18 PWM estimators of the parameters, standard errors, quantiles, and confidence limits for maximum significant wave height data

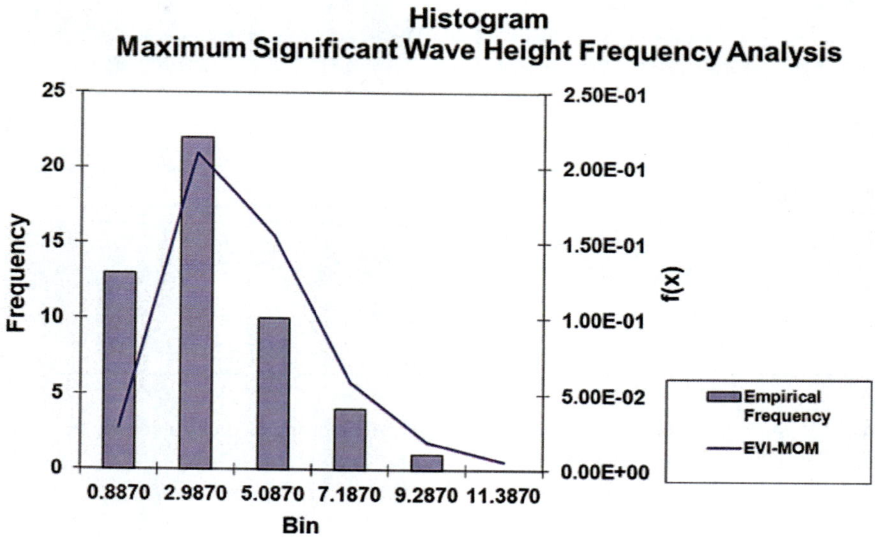

Fig. 9.19 Histogram and theoretical EVI density for maximum significant wave height data

A graphical comparison between the empirical and theoretical frequency curves from the results provided by the three methods shown before is shown in Fig. 9.20.

A graphical depiction of the empirical and theoretical frequency curves and the confidence limits of the best fit provided, in this case that of, is shown in Fig. 9.21.

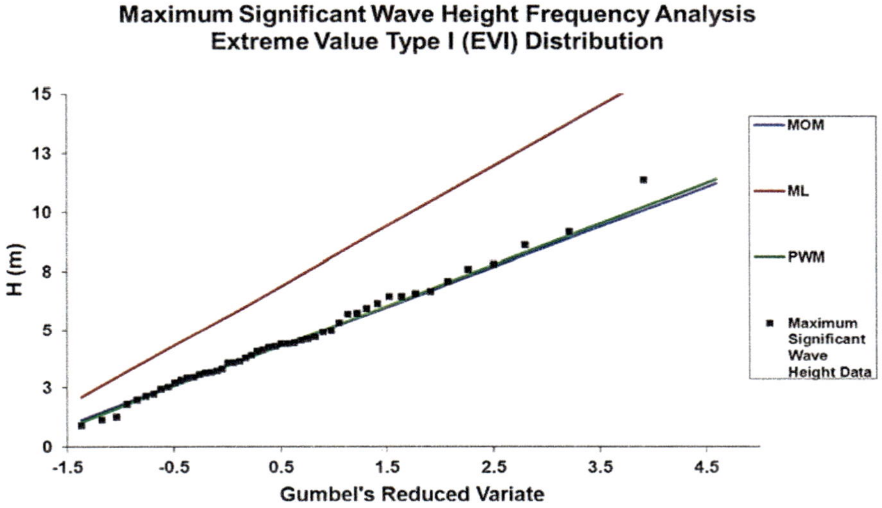

Fig. 9.20 Empirical and theoretical EVI distributions for maximum significant wave height data

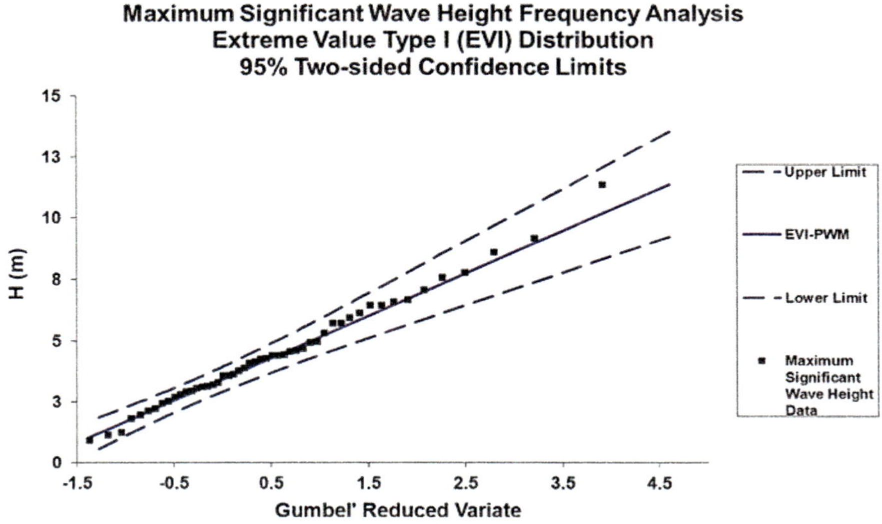

Fig. 9.21 Empirical and EVI-PWM theoretical frequency curves and PWM confidence limits for maximum significant wave height data

	A	B	C	D
1	ANNUAL MAXIMUM WIND SPEED FREQUENCY ANALYSIS			
2				
3	*Descriptive Statistics*			
4	Mean	54.7964		
5	Standard Error	311.0530		
6	Median	47.9965		
7	Mode	#N/D		
8	Standard Deviation	20.7327		
9	Sample Variance	429.8439		
10	Kurtosis	12.5699		
11	Skewness	3.0813		
12	Range	121.5116		
13	Minimum	36.4278		
14	Maximum	157.9394		
15	Sum	2739.8217		
16	Count	50.0000		
17	Largest(1)	157.9394		
18	Smallest(1)	36.4278		
19	Confidence Level(95.0%)	624.7686		

Fig. 9.22 Descriptive statistics for the maximum wind speed data

9.9.4 Annual Maximum Wind Speed Frequency Analysis

By using the annual maximum wind speed data, Castillo (1988), the following statistical results were obtained and are shown in Fig. 9.22.

9.9.4.1 MOM Methods

By using an Excel® spreadsheet, the results contained in Fig. 9.23, were obtained for the MOM method applied to the EVI distribution for the annual maximum wind speed data, Castillo (1988).

9.9.4.2 ML Method

By using an Excel® spreadsheet, the results contained in Fig. 9.24, were obtained for the ML method applied to the EVI distribution for annual maximum wind speed data.

9.9.4.3 PWM Method

By using an Excel® spreadsheet, the results contained in Fig. 9.25, were obtained for the PWM method applied to the EVI distribution for annual maximum wind speed data.

In Fig. 9.26 a comparison is made between the histogram and EVI theoretical density for annual maximum wind speed data.

A graphical comparison between the empirical and theoretical frequency curves from the results provided by the three methods shown before is shown in Fig. 9.27.

	A	B	C	D	E	F	G
1	ANNUAL MAXIMUM WIND SPEED FREQUENCY ANALYSIS						
2	Extreme Value Type I (EVI)						
3	Method: MOM						
4							
5	Estimation of Parameters and Goodness of Fit Tests						
6	Location Parameter =	45.4667	m/s				
7	Scale Parameter =	16.1652	m/s				
8	SEF =	9.05		MARD =	9.49		
9							
10	Design Values and Confidence Limits						
11				Squared root	Two-sided	Design	Two-sided
12	Return Period	CDF	Freq. Factor	Stand. Error	95%	Values	95%
13	Tr (years)	F(x)	K_T	S_T	Lower Limit	V_T (m/s)	Upper Limit
14	2	0.5	-0.1642	3.1660	45	51	58
15	5	0.8	0.7195	4.0152	62	70	78
16	10	0.9	1.3046	5.3184	71	82	92
17	20	0.95	1.8659	6.7851	80	93	107
18	50	0.98	2.5923	8.8239	91	109	126
19	100	0.99	3.1367	10.4043	99	120	140
20	Two-sided Confidence Limits:	z_α =	1.96				

Fig. 9.23 MOM estimators of the parameters, standard errors, quantiles, and confidence limits for maximum wind speed data

	A	B	C	D	E	F	G	H
1	ANNUAL MAXIMUM WIND SPEED ANALYSIS							
2	Extreme Value Type I (EVI)							
3	Method: ML							
4								
5	Estimation of Parameters and Goodness of Fit Tests							
6	Initial Moment Estimators:							
7	Location Parameter =		47.0424	m/s				
8	Scale Parameter=		13.4338	m/s				
9	Iterative Procedure:	i-th iteration		(i+1)-th iteration				
10		Location Parameter =	45.3374	$\delta(X_0)$ =	0.0000			
11		Scale Parameter =	6.1083	$\delta(\alpha)$ =	0.0000			
12	P =	-0.0003		Location Parameter=		45.3374		
13	Q =	-0.0006		Scale Parameter=		6.1083		
14	ABS($\partial LL/\partial x_0$) =	4.4979E-05	ABS($\partial LL/\partial \alpha$) =	1.0214E-04				
15	Final ML Estimators:							
16	Location Parameter =	45.3374	m/s					
17	Scale Parameter =	6.1083	m/s					
18	SEF =	16.02		MARD =	8.46			
19								
20	Design Values and Confidence Limits							
21					Squared Ro	Two-sided	Design	Two-sided
22	Return Period	CDF	Freq. Factor		Stand. Error	95%	Values	95%
23	Tr (years)	F(x)	K_T	y_T	S_T	Lower Limit	V (m/s)	Upper Limit
24	2	0.5	-0.1642	0.3665	1.0143	46	48	50
25	5	0.8	0.7195	1.4999	1.5567	51	54	58
26	10	0.9	1.3046	2.2504	1.9969	55	59	63
27	20	0.95	1.8659	2.9702	2.4431	59	63	68
28	50	0.98	2.5923	3.9019	3.0382	63	69	75
29	100	0.99	3.1367	4.6001	3.4916	67	73	80
30	Two-sided Confidence Limits:	z_α =	1.96					

Fig. 9.24 ML estimation of the parameters, standard errors, quantiles, and confidence limits for maximum wind speed data

	A	B	C	D	E	F	G
1	ANNUAL MAXIMUM WIND SPEED FREQUENCY ANALYSIS						
2	Extreme Value Type I (EVI)						
3	Method: PWM						
4							
5	Probability Weighted Moments						
6	$M_{(0)}$ =	54.7964	m/s				
7	$M_{(1)}$ =	22.7424	m/s				
8							
9	Estimation of Parameters and Goodness of Fit Tests						
10	Location Parameter =		47.0424	m/s			
11	Scale Parameter =		13.4338	m/s			
12	SEF =	9.53		MARD =	6.69		
13							
14	Design Values and Confidence Limits						
15				Squared Root	Two-sided	Design	Two-sided
16	Return Period	CDF	Freq. Factor	Stand. Error	95%	Values	95%
17	Tr (years)	F(x)	K_T	S_T	Lower Limit	V (m/s)	Upper Limit
18	2	0.5	-0.1642	2.2387	48	52	56
19	5	0.8	0.7195	3.6092	60	67	74
20	10	0.9	1.3046	4.7369	68	77	87
21	20	0.95	1.8659	5.8771	75	87	98
22	50	0.98	2.5923	7.3942	85	99	114
23	100	0.99	3.1367	8.5478	92	109	126
24	Two-sided Confidence Limits:		z_α =	1.96			

Fig. 9.25 PWM estimators of the parameters, standard errors, quantiles, and confidence limits for maximum wind speed data

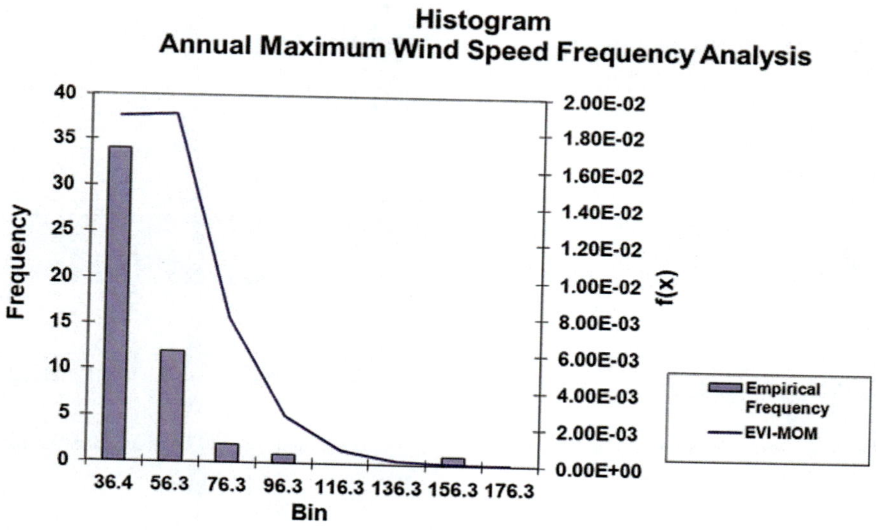

Fig. 9.26 Histogram and theoretical EVI density for maximum wind speed data

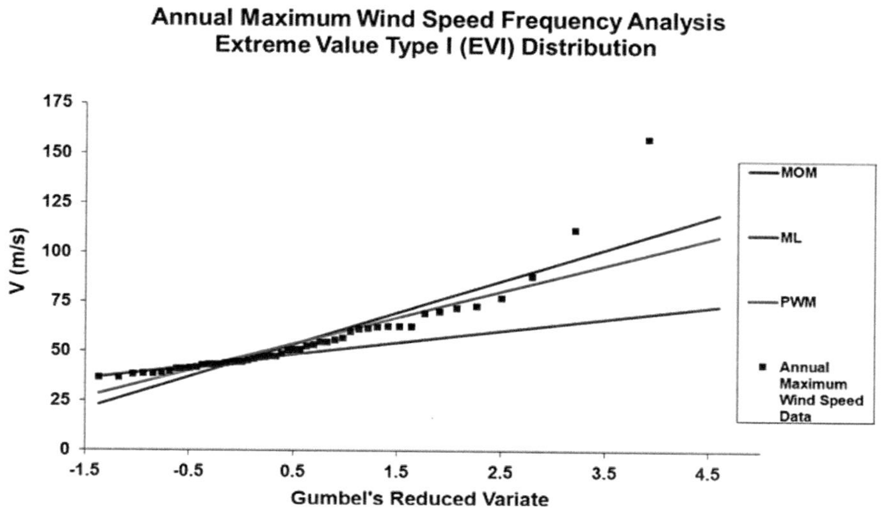

Fig. 9.27 Empirical and theoretical MOM, ML, and PWM EVI distributions for maximum wind speed data

A graphical depiction of the empirical and theoretical frequency curves and the confidence limits of the best fit provided, in this case that of MOM, is shown in Fig. 9.28.

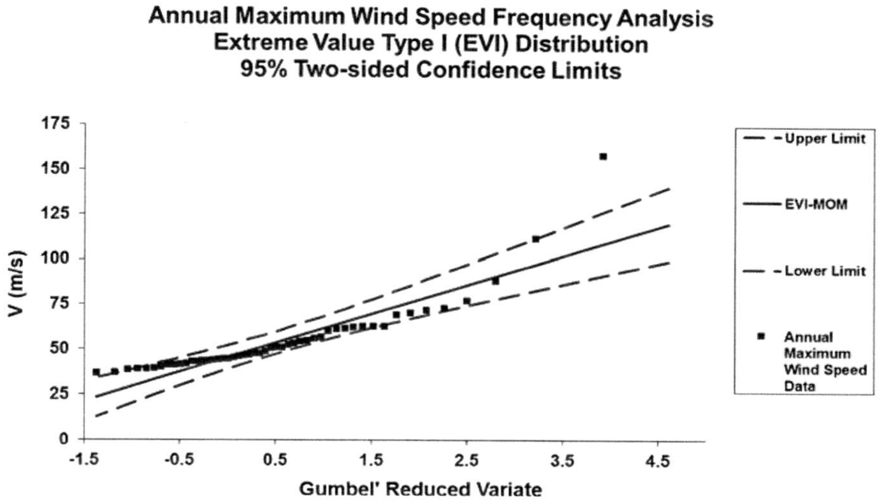

Fig. 9.28 Empirical and MOM theoretical frequency curves and ML confidence limits for maximum wind speed data

10 General Extreme Value Distribution

The trees that are slow to grow bear the best fruit. J. B. Poquelin (Moliere)

10.1 Introduction

The General Extreme Value (GEV) distribution is the general solution, found by Jenkinson (1955), to the Stability Postulate that all the extremes must comply with. The GEV distribution has been under study since 1955 and it has experienced a growing acceptance by the practicing engineers and scientists as computing devices have improved every single year since the 1980s. The GEV distribution, is a matter of fact, a family of distributions, see Fig. 10.1, and can directly represent the Extreme Value types II (EVII) and III (EVIII) distributions and when taking the limit when $\beta \rightarrow 0$, the GEV distribution can also represent the Extreme Value type I (EVI) distribution.

Jenkinson (1969) proposed a procedure to obtain the maximum likelihood estimators of the parameters of the GEV distribution. Hosking et al. (1985) obtained the probability weighted moments estimators of the parameters of the GEV distribution.

10.2 Chapter Objectives

After reading this chapter, you will know how to:

1. Recognize the distribution and density functions of the GEV distribution
2. Estimate the parameters of the GEV distribution

J. A. Raynal Villaseñor, *Frequency Analyses of Natural Extreme Events*,
Earth and Environmental Sciences Library,
https://doi.org/10.1007/978-3-030-86390-6_10

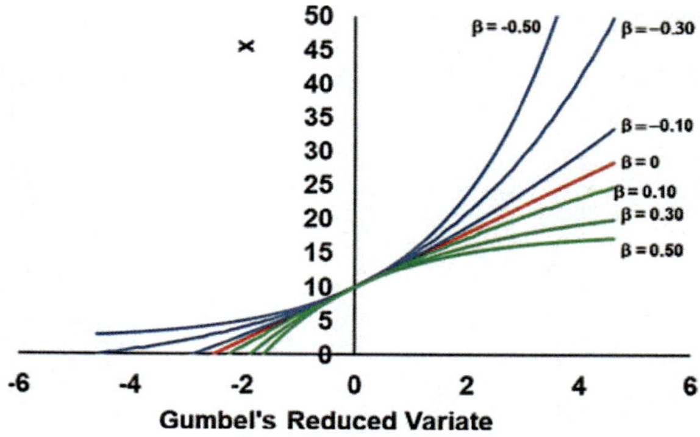

Fig. 10.1 The General Extreme Value Family of Distributions

3. Compute the quantiles and confidence limits of the GEV distribution
4. Make a graphic display of your data and the GEV distribution
5. Develop an application of all the above using Excel® spreadsheets

10.3 Probability Distribution and Density Functions

The probability distribution function of the general extreme value (GEV) distribution is, NERC (1975):

$$F(x) = \exp\left\{-\left[1 - \frac{\beta(x - x_0)}{\alpha}\right]^{\frac{1}{\beta}}\right\} \tag{10.1}$$

where x_0, α and β are the location, scale, and shape parameters, respectively. *F(x)* is the probability distribution function of the random variable x. The scale parameter must meet the condition that $\alpha > 0$. The domain of variable x in the GEV distribution is as follows:

1) For $\beta < 0$:

$$x_0 + \alpha/\beta \leq x < \infty \tag{10.2}$$

2) For $\beta > 0$:

$$-\infty \leq x < x_0 + \alpha/\beta \tag{10.3}$$

The probability density function for the GEV distribution is, NERC (1975):

$$f(x) = \frac{1}{\alpha}\exp\left\{-\left[1 - \frac{\beta(x - x_0)}{\alpha}\right]^{\frac{1}{\beta}}\right\}\left[1 - \frac{\beta(x - x_0)}{\alpha}\right]^{\frac{1}{\beta}-1} \tag{10.4}$$

10.4 Estimation of the Parameters

10.4.1 MOM Method

The population mean, variance and skewness of the GEV distribution may be expressed in terms of its reduced variates for $\beta < 0$ and $\beta > 0$:

1) For $\beta < 0$:

The reduced variate is:

$$y_2 = 1 - \frac{(x - x_0)\beta}{\alpha} \tag{10.5}$$

and its probability distribution and density functions are, NERC (1975):

$$G(y_2) = \exp\left[-y_2^{\frac{1}{\beta}}\right] \tag{10.6}$$

and:

$$g(y_2) = -\frac{y_2^{\frac{1}{\beta}-1}}{\beta}\exp\left[-y_2^{\frac{1}{\beta}}\right] \tag{10.7}$$

the domain is now $0 \leq y_2 < \infty$. The mean, variance and skewness may be expressed in terms of the reduced variate y_2 as follows, NERC (1975):

$$E(y_2) = \Gamma(1 + \beta) \tag{10.8}$$

and:

$$var(y_2) = \Gamma(1 + 2\beta) - \Gamma^2(1 + \beta) \tag{10.9}$$

and:

$$\gamma = \frac{\Gamma(1 + 3\beta) - 3\Gamma(1 + 2\beta)\Gamma(1 + \beta) + \Gamma^3(1 + \beta)}{\left[\Gamma(1 + 2\beta) - \Gamma^2(1 + \beta)\right]^{\frac{3}{2}}} \tag{10.10}$$

So, the mean and variance of the actual variable x, can be expressed as, NERC (1975):

$$\mu = x_0 + \frac{\alpha}{\beta}[1 - \Gamma(1 + \beta)] \tag{10.11}$$

and:

$$\sigma^2 = \left(\frac{\alpha}{\beta}\right)^2\left[\Gamma(1 + 2\beta) - \Gamma^2(1 + \beta)\right] \tag{10.12}$$

the skewness coefficient remains as shown in Eq. (10.10). The skewness coefficient, when $\beta < 0$, is always lesser than 1.1396.

1) For $\beta > 0$:

The reduced variate is:

$$-y_3 = 1 - \frac{(x - x_0)\beta}{\alpha} \tag{10.13}$$

and its probability distribution and density functions are, NERC (1975):

$$G(y_3) = \exp\left[-(-y_3)^{\frac{1}{\beta}}\right] \tag{10.14}$$

and:

$$g(y_3) = \frac{(-y_3)^{\frac{1}{\beta}-1}}{\beta} \exp\left[-(-y_3)^{\frac{1}{\beta}}\right] \tag{10.15}$$

the domain is now $-\infty \leq y_3 < 0$. Now, the mean, variance and skewness may be expressed in terms of the reduced variate y_3 as follows, NERC (1975):

$$E(y_3) = -\Gamma(1 + \beta) \tag{10.16}$$

and:

$$var(y_3) = \Gamma(1 + 2\beta) - \Gamma^2(1 + \beta) \tag{10.17}$$

and:

$$\gamma = -\left\{\frac{\Gamma(1 + 3\beta) - 3\Gamma(1 + 2\beta)\Gamma(1 + \beta) + \Gamma^3(1 + \beta)}{\left[\Gamma(1 + 2\beta) - \Gamma^2(1 + \beta)\right]^{\frac{3}{2}}}\right\} \tag{10.18}$$

So, the mean and variance of the actual variable x, can be expressed as in the same way when $\beta > 0$, NERC (1975):

$$\mu = x_0 + \frac{\alpha}{\beta}[1 - \Gamma(1 + \beta)] \tag{10.19}$$

and:

$$\sigma^2 = \left(\frac{\alpha}{\beta}\right)^2 \left[\Gamma(1 + 2\beta) - \Gamma^2(1 + \beta)\right] \tag{10.20}$$

the skewness coefficient remains as shown in Eq. (10.18). The skewness coefficient when $\beta > 0$ is always bigger than 1.1396.

For the GEV distribution, using the MOM method, the estimators of the parameters are obtained by first equating the population moments with the sample moments, and then simultaneously solving the resulting system of equations:

$$\mu = \bar{x} \tag{10.21}$$

$$\sigma^2 = \widehat{\sigma}^2 = s^2 \tag{10.22}$$

and:

$$\gamma = \widehat{\gamma} = g \tag{10.23}$$

then using Eqs. (10.11) and (10.12) and Eqs. (10.10) or (10.18) in the left-hand side of Eqs. (10.21)–(10.23), respectively, and the expressions for the sample mean, sample variance and sample skewness coefficient, given in Chap. 2, in the right-hand side of Eqs. (10.21), (10.22) and (10.23), respectively, we obtain the following expressions:

$$x_0 + \frac{\alpha}{\beta}[1 - \Gamma(1+\beta)] = \frac{1}{N}\sum_{i=1}^{N} x_i \tag{10.24}$$

$$\left(\frac{\alpha}{\beta}\right)^2 \left[\Gamma(1+2\beta) - \Gamma^2(1+\beta)\right] = \frac{1}{N}\sum_{i=1}^{N} (x - \bar{x})^2 \tag{10.25}$$

and for $\beta < 0$:

$$\left\{\frac{\Gamma(1+3\beta) - 3\Gamma(1+2\beta)\Gamma(1+\beta) + \Gamma^3(1+\beta)}{\left[\Gamma(1+2\beta) - \Gamma^2(1+\beta)\right]^{\frac{3}{2}}}\right\} = \frac{N}{(N-1)(N-2)\widehat{\sigma}^3}\sum_{i=1}^{N} (x_i - \widehat{\mu})^3 \tag{10.26}$$

and for $\beta > 0$:

$$-\left\{\frac{\Gamma(1+3\beta) - 3\Gamma(1+2\beta)\Gamma(1+\beta) + \Gamma^3(1+\beta)}{\left[\Gamma(1+2\beta) - \Gamma^2(1+\beta)\right]^{\frac{3}{2}}}\right\} = \frac{N}{(N-1)(N-2)\widehat{\sigma}^3}\sum_{i=1}^{N} (x_i - \widehat{\mu})^3 \tag{10.27}$$

The solution to the system of equations formed by Eqs. (10.24) and (10.25) and (10.26) or (10.27), provide the estimators of the MOM method for the GEV distribution:

$$\widehat{x}_0 = \widehat{A} - \frac{\widehat{\alpha}}{\widehat{\beta}} \tag{10.28}$$

$$\widehat{\alpha} = \widehat{B}\left|\widehat{\beta}\right| \tag{10.29}$$

where:

$$\widehat{A} = \widehat{\mu} + \frac{\widehat{\alpha}}{\widehat{\beta}}\Gamma(1+\widehat{\beta}) \tag{10.30}$$

$$\widehat{B} = \left(\frac{\widehat{\sigma}^2}{{\sigma_z}^2}\right)^{\frac{1}{2}} = \frac{\widehat{\sigma}}{\sigma_z} \tag{10.31}$$

The values of the mean and standard deviation of the z's are provided in Eqs. (10.8) or (10.16), depending on the sign of the shape parameter, and the squared root of Eq. (10.9), respectively. Equations (10.26) and (10.27) have been inverted to produce the following polynomial relationships:

1) For $\beta < 0$ and $1.1396 < \widehat{\gamma} \leq 19.0$:

$$\begin{aligned}\widehat{\beta} &= 0.25031 - 0.29219\widehat{\gamma} + 0.075357\widehat{\gamma}^2 - 0.0108833\widehat{\gamma}^3 \\ &\quad + 0.000904\widehat{\gamma}^4 - 0.000043\widehat{\gamma}^5\end{aligned} \tag{10.32}$$

2) For $\beta > 0$ $-11.4 \leq \widehat{\gamma} < 1.1396$:

$$\begin{aligned}\widehat{\beta} &= 0.279434 - 0.333535\widehat{\gamma} + 0.048306\widehat{\gamma}^2 + 0.023314\widehat{\gamma}^3 \\ &\quad + 0.003765\widehat{\gamma}^4 + 0.000263\widehat{\gamma}^5\end{aligned} \tag{10.33}$$

10.4.1.1 Example of Application of Estimation of the Parameters of the GEV Distribution Using the MOM Method

With the flood sample data of gauging station Huites, Mexico (1942–1992), contained in Appendix A, the following statistics have already been obtained:

$$\widehat{\mu} = \bar{x} = \frac{1}{N}\sum_{i=1}^{N} x_i = 2498.9627 \ \text{m}^3/\text{s}$$

$$\widehat{\sigma} = s = \left[\frac{1}{N}\sum_{i=1}^{N}(x_i - \widehat{\mu})^2\right]^{1/2} = 2199.4767 \ \text{m}^3/\text{s}$$

$$\widehat{\gamma} = \frac{N}{(N-1)(N-2)\widehat{\sigma}^3} \sum_{i=1}^{N} (x_i - \widehat{\mu})^3 = 2.0123$$

So, the MOM estimators of the parameters of the GEV distribution are as follows:

$$\widehat{\beta} = 0.25031 - 0.29219\widehat{\gamma} + 0.075357\widehat{\gamma}^2 - 0.0108833\widehat{\gamma}^3 + 0.000904\widehat{\gamma}^4 - 0.000043\widehat{\gamma}^5$$

$$\widehat{\beta} = 0.25031 - 0.29219(2.0123) + 0.075357(2.0123)^2 - 0.0108833(2.0123)^3$$
$$+0.000904(2.0123)^4 - 0.000043(2.0123)^5 = -0.1078$$

$$\widehat{x}_0 = A - \frac{\widehat{\alpha}}{\widehat{\beta}} = -12033.9 - \frac{1457.157}{(-0.1078)} = 1484.4493\text{m}^3/\text{s}$$
$$\widehat{\alpha} = B\left|\widehat{\beta}\right| = 13518.33(|-0.1078|) = 1457.1573\text{m}^3/\text{s}$$

where:

$$A = \widehat{\mu} - B\widehat{\mu}_z = 2498.9627 - 13518.33(1.075) = -12033.9$$

$$B = \left(\frac{\sigma^2}{\sigma_z{}^2}\right)^{\frac{1}{2}} = \frac{\sigma}{\sigma_z} = \frac{2199.4767}{\left[\Gamma(1+2(-0.1078)) - \Gamma^2(1+(-0.1078))\right]^{\frac{1}{2}}}$$
$$= \frac{2199.4767}{\left[1.18.22 - (1.075)^2\right]^{\frac{1}{2}}} = 13518.33$$

$$\mu_z = \Gamma(1+\beta) = \Gamma(0.8922) = 1.075$$

$$\sigma_z = \left[\Gamma(1+2\beta) - \Gamma^2(1+\beta)\right]^{\frac{1}{2}} = \left[\Gamma(0.7844) - \Gamma^2(0.8922)\right]^{\frac{1}{2}}$$
$$= \left[1.1822 - (1.075)^2\right]^{\frac{1}{2}} = 0.1627$$

So, the moment estimators of the parameters of the GEV distribution are: location parameter:

$$\widehat{x}_0 = 1484.449\text{m}^3/\text{s}$$

scale parameter:

$$\widehat{\alpha} = 1457.157\text{m}^3/\text{s}$$

and:
shape parameter:

$$\widehat{\beta} = -0.1078$$

10.4.2 ML Method

The likelihood function for the GEV distribution is as follows:

$$L(x, x_0, \alpha, \beta) = \prod_{i=1}^{N}\left\{\frac{1}{\alpha}\exp\left\{-\left[1-\frac{\beta(x_i-x_0)}{\alpha}\right]^{\frac{1}{\beta}}\right\}\left[1-\frac{\beta(x_i-x_0)}{\alpha}\right]^{\frac{1}{\beta}-1}\right\} \tag{10.34}$$

By taking the natural logarithm of the previous equation, the log-likelihood function for the GEV distribution is obtained as:

$$LL(x, x_0, \alpha, \beta) = -NLn(\alpha) - \sum_{i=1}^{N}\left[1-\frac{\beta(x_i-x_0)}{\alpha}\right]^{\frac{1}{\beta}} + \left(\frac{1}{\beta}-1\right)\sum_{i=1}^{N}Ln\left[1-\frac{\beta(x_i-x_0)}{\alpha}\right] \tag{10.35}$$

and the log-likelihood function for the GEV distribution is finally obtained:

$$LL(x, x_0, \alpha, \beta) = -NLn(\alpha) - \sum_{i=1}^{N}\left\{\left[1-\frac{\beta(x_i-x_0)}{\alpha}\right]^{\frac{1}{\beta}} + \left(\frac{1}{\beta}-1\right)Ln\left[1-\frac{\beta(x_i-x_0)}{\alpha}\right]\right\} \tag{10.36}$$

Now, the classical approach to the ML method requires the computation of the first-order partial derivatives of the log-likelihood function with respect to each of its parameters, equating them equal to zero and then solving the resulting system of equations. So, the first-order partial derivatives are obtained as follows:

$$\frac{\partial LnL}{\partial x_0} = \frac{1}{\alpha}\left\{-\sum_{i=1}^{N}\left[1-\frac{\beta(x_i-x_0)}{\alpha}\right]^{\frac{1}{\beta}-1} + (1-\beta)\sum_{i=1}^{N}\left[1-\frac{\beta(x_i-x_0)}{\alpha}\right]^{-1}\right\} = 0 \tag{10.37}$$

$$\frac{\partial LL}{\partial \alpha} = \frac{1}{\alpha}\left\{-N - \sum_{i=1}^{N}\left[1-\frac{\beta(x_i-x_0)}{\alpha}\right]^{\frac{1}{\beta}-1}\frac{(x_i-x_0)}{\alpha} + (1-\beta)\sum_{i=1}^{N}\left[1-\frac{\beta(x_i-x_0)}{\alpha}\right]^{-1}\frac{(x_i-x_0)}{\alpha}\right\} = 0 \tag{10.38}$$

$$\frac{\partial LL}{\partial \beta} = \frac{1}{\beta}\left\{\sum_{i=1}^{N}\left[1-\frac{\beta(x_i-x_0)}{\alpha}\right]^{\frac{1}{\beta}}\left[\left(\frac{1}{\beta}\right)Ln\left(1-\frac{\beta(x_i-x_0)}{\alpha}\right) + \frac{(x_i-x_0)}{\alpha}\left[1-\frac{\beta(x_i-x_0)}{\alpha}\right]^{-1}\right]\right.$$

$$-(1-\beta)\sum_{i=1}^{N}\frac{(x_i-x_0)}{\alpha}\left[1-\frac{\beta(x_i-x_0)}{\alpha}\right]^{-1}-(\frac{1}{\beta})\sum_{i=1}^{N}Ln\left[1-\frac{\beta(x_i-x_0)}{\alpha}\right]\Bigg\}=0 \tag{10.39}$$

The exact solution to the system formed by Eqs. (10.37), (10.38) and (10.39) is not known, so an iterative procedure is needed to evaluate the ML estimators of the parameters of the GEV distribution. The iterative procedure is as follows:

1) Define a reduced variate as:

$$y_i=-\frac{1}{\beta}Ln\left[1-\frac{(x_i-x_0)\beta}{\alpha}\right] \tag{10.40}$$

2) Define parameters P, Q and R as follows:

$$P=N-\sum_{i=1}^{N}\exp(-y_i) \tag{10.41}$$

$$Q=\sum_{i=1}^{N}\exp[(\beta-1)y_i]-(1-\beta)\sum_{i=1}^{N}\exp(\beta y_i) \tag{10.42}$$

$$R=N-\sum_{i=1}^{N}y_i+\sum_{i=1}^{N}y_i\exp(-y_i) \tag{10.43}$$

3) Define the iterative procedure by:

$$(x_0)_{i+1}=(x_0)_i+\left(\delta_{x_0}\right)_i \tag{10.44}$$

$$(\alpha)_{i+1}=(\alpha)_i+(\delta_\alpha)_i \tag{10.45}$$

$$(\beta)_{i+1}=(\beta)_i+\left(\delta_\beta\right)_i \tag{10.46}$$

where the sub-index i refers to the iteration stage and δ are the differences between the estimator at iteration i and the true value for the ML estimator for such parameter.

The relationship between the differences between the estimator at iteration i and the true value for the ML estimator for such parameter (δ's) and the first partial derivatives of the Log-likelihood function with respect to the parameters of the GEV distribution has the following form, NERC (1975):

$$\begin{bmatrix}-\delta x_0\\-\delta\alpha\\-\delta\beta\end{bmatrix}_i=\begin{bmatrix}E\left(-\frac{\partial^2 LL}{\partial x_0{}^2}\right) & E\left(-\frac{\partial^2 LL}{\partial x_0\partial\alpha}\right) & E\left(-\frac{\partial^2 LL}{\partial x_0\partial\beta}\right)\\ E\left(-\frac{\partial^2 LL}{\partial\alpha\partial x_0}\right) & E\left(-\frac{\partial^2 LL}{\partial\alpha^2}\right) & E\left(-\frac{\partial^2 LL}{\partial\alpha\partial\beta}\right)\\ E\left(-\frac{\partial^2 LL}{\partial\beta\partial x_0}\right) & E\left(-\frac{\partial^2 LL}{\partial\beta\partial\alpha}\right) & E\left(-\frac{\partial^2 LL}{\partial\beta^2}\right)\end{bmatrix}_i^{-1}\begin{bmatrix}-\frac{\partial LL}{\partial x_0}\\-\frac{\partial LL}{\partial\alpha}\\-\frac{\partial LL}{\partial\beta}\end{bmatrix}_i \tag{10.47}$$

The first matrix in the right-hand side of the previous equation is the Fisher's information matrix, it can be stated as:

$$[I] = \begin{bmatrix} E\left(-\frac{\partial^2 LL}{\partial {x_0}^2}\right) & E\left(-\frac{\partial^2 LL}{\partial x_0 \partial \alpha}\right) & E\left(-\frac{\partial^2 LL}{\partial x_0 \partial \beta}\right) \\ E\left(-\frac{\partial^2 LL}{\partial \alpha \partial x_0}\right) & E\left(-\frac{\partial^2 LL}{\partial \alpha^2}\right) & E\left(-\frac{\partial^2 LL}{\partial \alpha \partial \beta}\right) \\ E\left(-\frac{\partial^2 LL}{\partial \beta \partial x_0}\right) & E\left(-\frac{\partial^2 LL}{\partial \beta \partial \alpha}\right) & E\left(-\frac{\partial^2 LL}{\partial \beta^2}\right) \end{bmatrix} \tag{10.48}$$

The expected values inside the Fisher's information matrix have been obtained by Prescott and Walden (1980) for the interval $-0.50 < \beta < 0.50$, as:

$$E\left(-\frac{\partial^2 LL}{\partial {x_0}^2}\right) = \frac{N}{\alpha^2} p \tag{10.49}$$

$$E\left(-\frac{\partial^2 LL}{\partial \alpha^2}\right) = \frac{N}{\alpha^2 \beta^2}[1 - 2(1-\beta)\,\Gamma(1-\beta) + p] \tag{10.50}$$

$$E\left(-\frac{\partial^2 LL}{\partial \beta^2}\right) = \frac{N}{\beta^2}\left[\frac{\pi^2}{6} + \left(1 - \varepsilon - \frac{1}{\beta}\right)^2 + \frac{2q}{\beta} + \frac{p}{\beta^2}\right] \tag{10.51}$$

$$E\left(-\frac{\partial^2 LL}{\partial x_0 \partial \alpha}\right) = \frac{N}{\alpha^2 \beta}[p - (1-\beta)\Gamma(1-\beta)] \tag{10.52}$$

$$E\left(-\frac{\partial^2 LL}{\partial x_0 \partial \beta}\right) = \frac{N}{\alpha \beta}\left[-\frac{p}{\beta} - q\right] \tag{10.53}$$

$$E\left(-\frac{\partial^2 LL}{\partial \alpha \partial \beta}\right) = \frac{N}{\alpha \beta^2}\left[1 - \varepsilon - \frac{[1-(1-\beta)\Gamma(1-\beta)]}{\beta} - \frac{p}{\beta} - q\right] \tag{10.54}$$

where:

$$p = (1-\beta)^2 \Gamma(1-2\beta) \tag{10.55}$$

$$q = (1-\beta)\Gamma(1-\beta)\left[\psi(1-\beta) - \frac{(1-\beta)}{\beta}\right] \tag{10.56}$$

and $\Gamma(.)$ is the complete Gamma function, $\psi(.)$ is the Digamma function and ε is the Euler's constant (equal to 0.5772157).

The variance–covariance matrix of the parameters of the GEV distribution has the following form, NERC (1975):

$$[V] = \begin{bmatrix} Var(x_0) & Cov(x_0, \alpha) & Cov(x_0, \beta) \\ Cov(\alpha, x_0) & Var(\alpha) & Cov(\alpha, \beta) \\ Cov(\beta, x_0) & Cov(\beta, \alpha) & Var(\beta) \end{bmatrix}$$

$$= \begin{bmatrix} E\left(-\frac{\partial^2 LL}{\partial x_0^2}\right) & E\left(-\frac{\partial^2 LL}{\partial x_0 \partial \alpha}\right) & E\left(-\frac{\partial^2 LL}{\partial x_0 \partial \beta}\right) \\ E\left(-\frac{\partial^2 LL}{\partial \alpha \partial x_0}\right) & E\left(-\frac{\partial^2 LL}{\partial \alpha^2}\right) & E\left(-\frac{\partial^2 LL}{\partial \alpha \partial \beta}\right) \\ E\left(-\frac{\partial^2 LL}{\partial \beta \partial x_0}\right) & E\left(-\frac{\partial^2 LL}{\partial \beta \partial \alpha}\right) & E\left(-\frac{\partial^2 LL}{\partial \beta^2}\right) \end{bmatrix}^{-1} \tag{10.57}$$

An alternative way to express Eq. (10.57) is, NERC (1975):

$$[V] = \begin{bmatrix} Var(x_0) & Cov(x_0, \alpha) & Cov(x_0, \beta) \\ Cov(\alpha, x_0) & Var(\alpha) & Cov(\alpha, \beta) \\ Cov(\beta, x_0) & Cov(\beta, \alpha) & Var(\beta) \end{bmatrix} = \frac{1}{N}\begin{bmatrix} \alpha^2 b & \alpha^2 h & \alpha f \\ \alpha^2 h & \alpha^2 a & \alpha g \\ \alpha f & \alpha g & c \end{bmatrix} \tag{10.58}$$

where a, b, c, f, g, and h are the variance–covariance matrix coefficients for the GEV distribution. These coefficients have been evaluated as a function of the shape parameter; their values are shown in Table 10.1.

Now, Eq. (10.47) can be modified as:

$$\begin{bmatrix} \delta x_0 \\ \delta \alpha \\ \delta \beta \end{bmatrix}_i = -\frac{1}{N}\begin{bmatrix} \alpha^2 b & \alpha^2 h & \alpha f \\ \alpha^2 h & \alpha^2 a & \alpha g \\ \alpha f & \alpha g & c \end{bmatrix}_i \begin{bmatrix} \frac{Q}{\alpha} \\ \frac{(P+Q)}{\alpha\beta} \\ \frac{\left[R - \frac{(P+Q)}{\beta}\right]}{\beta} \end{bmatrix}_i \tag{10.59}$$

Then, the values of the differences between the estimator at iteration i and the true value for the maximum likelihood estimator for such parameter (δ's) are:

$$\delta_i(x_0) = -\frac{\alpha}{N}\left[b(Q)_i + \frac{h(P+Q)_i}{\beta} + \frac{f}{\beta}\left(R - \frac{(P+Q)}{\beta}\right)_i\right] \tag{10.60}$$

$$\delta_i(\alpha) = -\frac{\alpha}{N}\left[h(Q)_i + \frac{a(P+Q)_i}{\beta} + \frac{g}{\beta}\left(R - \frac{(P+Q)}{\beta}\right)_i\right] \tag{10.61}$$

Table 10.1 Exact coefficients of the variance–covariance matrix of the parameters of the GEV distribution

β	a	b	c	f	g	h
−0.4	1.04	1.29	0.83	0.26	−0.09	0.8
−0.3	0.91	1.29	0.74	0.27	−0.02	0.69
−0.2	0.81	1.28	0.64	0.27	0.05	0.57
−0.1	0.71	1.27	0.56	0.27	0.1	0.45
0	0.65[a]	1.25[a]	0.48[a]	0.26[a]	0.15[a]	0.34[a]
0.1	0.63	1.29	0.61	0.36	0.25	0.25
0.2	0.58	1.2	0.33	0.22	0.22	0.09
0.3	0.58	1.17	0.27	0.19	0.23	−0.03

[a]Values taken from Jenkinson (1969)

$$\delta_i(\beta) = -\frac{1}{N}\left[f(Q)_i + \frac{g(P+Q)_i}{\beta} + \frac{c}{\beta}\left(R - \frac{(P+Q)}{\beta}\right)_i\right] \tag{10.62}$$

4) Define a set of criteria of convergence in the following form:

$$\left|-\frac{\partial LL(x, x_0, \alpha, \beta)}{\partial x_0}\right| = \left|\frac{Q}{\alpha}\right| \approx 10^{-6} \tag{10.63}$$

$$\left|-\frac{\partial LL(x, x_0, \alpha, \beta)}{\partial \alpha}\right| = \left|\frac{(P+Q)}{\alpha\beta}\right| \approx 10^{-6} \tag{10.64}$$

$$\left|-\frac{\partial LL(x, x_0, \alpha, \beta)}{\partial \beta}\right| = \left|\frac{1}{\beta}\left[R - \frac{(P+Q)}{\beta}\right]\right| \approx 10^{-6} \tag{10.65}$$

When conditions established by Eqs. (10.63)–(10.65) are met simultaneously, then the values of such parameters will correspond to the ML estimators of the parameters of the GEV distribution.

10.4.2.1 Example of Application of Estimation of the Parameters of the GEV Distribution Using the ML Method

With the flood sample data of gauging station Huites, Mexico (1942–1992), contained in table Appendix A, the following statistics have already been obtained:

$$\hat{\mu} = \bar{x} = \frac{1}{N}\sum_{i=1}^{N} x_i = 2498.9627 \ \ \mathrm{m}^3/\mathrm{s}$$

$$\hat{\sigma} = s = \left[\frac{1}{N}\sum_{i=1}^{N}(x_i - \hat{\mu})^2\right]^{1/2} = 2199.4767\mathrm{m}^3/\mathrm{s}$$

Using the MOM estimators as initial values of the iterative scheme for the maximum likelihood estimators for the parameters of the GEV distribution, so the initial values are:

$$\hat{x}_0 = 1484.4493 \ \mathrm{m}^3/\mathrm{s}$$

$$\hat{\alpha} = 1457.1573\mathrm{m}^3/\mathrm{s}$$

$$\hat{\beta} = -0.1078$$

Iteration 1

Defining an initial reduced variate as:

$$y_i = -\frac{1}{(-0.1078)}Ln\left[1 - \frac{(x_i - 1484.4493)(-0.1078)}{1457.1573}\right]$$

The parameters P, Q and R for the initial values are:

$$P = N - \sum_{i=1}^{N} \exp(-y_i) = 51 - 44.0996 = 6.9004$$

$$Q = \sum_{i=1}^{N} \exp[(\beta - 1)y_i] - (1 - \beta)\sum_{i=1}^{N} \exp(\beta y_i) = 44.6029$$
$$- (1 - (-0.1078))(48.3124) = -8.9175$$

$$R = N - \sum_{i=1}^{N} y_i + \sum_{i=1}^{N} y_i \exp(-y_i) = 51 - 29.6015 + (-4.029) = 17.3695$$

Then, the initial deviations between the estimator at iteration 1 and the true value for the ML estimator for such parameters are:

$$\delta_i(x_0) = -\frac{\alpha}{N}\left[b(Q)_i + \frac{h(P+Q)_i}{\beta} + \frac{f}{\beta}\left(R - \frac{(P+Q)}{\beta}\right)_i\right]$$

$$\delta_i(x_0) = -\frac{1457.1573}{51}[(1.2664)(-8.9175)$$
$$+ \frac{(0.4639)(6.9004 + (-8.9175))}{(-0.1078)} + \frac{(0.2682)}{(-0.1078)}$$
$$\left(17.3695 - \frac{(6.9004 + (-8.9175))}{(-0.1078)}\right)\Bigg] = -20.7983$$

$$\delta_i(\alpha) = -\frac{\alpha}{N}\left[h(Q)_i + \frac{a(P+Q)_i}{\beta} + \frac{g}{\beta}\left(R - \frac{(P+Q)}{\beta}\right)_i\right]$$

$$\delta_i(\alpha) = -\frac{1457.1573}{51}[(0.4639)(-8.9175)$$
$$+ \frac{(0.7247)(6.9004 + (-8.9175))}{(-0.1078)} + \frac{0.0955}{(-0.1078)}$$
$$\left(17.3695 - \frac{(6.9004 + (-8.9175))}{(-0.1078)}\right) = -303.1894$$

$$\delta_i(\beta) = -\frac{1}{N}\left[f(Q)_i + \frac{g(P+Q)_i}{\beta} + \frac{c}{\beta}\left(R - \frac{(P+Q)}{\beta}\right)_i\right]$$

$$\delta_i(\beta) = -\frac{1}{51}\left[(0.2682)(-8.9175) + \frac{(0.0955)(6.9004 + (-8.9175))}{(-0.1078)} + \frac{0.5625}{(-0.1078)}\right.$$
$$\left(17.3695 - \frac{((6.9004 + (-8.9175))}{(-0.1078)}\right) = -0.1254$$

The new parameters are:

$$(x_0)_{i+1} = (x_0)_i + \left(\delta_{x0}\right)_i = 1484.4493 - 20.7983 = 1463.6510$$

$$(\alpha)_{i+1} = (\alpha)_i + (\delta_\alpha)_i = 1457.1573 - 303.1894 = 1153.9679$$

$$(\beta)_{i+1} = (\beta)_i + \left(\delta_\beta\right)_i = -0.1078 + (-0.1254) = -0.2332$$

The criteria of convergence are not met at this iteration:

$$\left|-\frac{\partial LL(x, x_0, \alpha, \beta)}{\partial x_0}\right| = \left|\frac{Q}{\alpha}\right| = \left|\frac{-8.9175}{1457.1573}\right| = 6.1198 \times 10^{-3} > 10^{-6}$$

$$\left|-\frac{\partial LL(x, x_0, \alpha, \beta)}{\partial \alpha}\right| = \left|\frac{(P+Q)}{\alpha\beta}\right| = \left|\frac{(6.9004 + (-8.9175)}{(1457.1573)(-0.1078)}\right|$$
$$= 1.2841 \times 10^{-2} > 0^{-6}$$

$$\left|-\frac{\partial LL(x, x_0, \alpha, \beta)}{\partial \beta}\right| = \left|\frac{1}{\beta}\left[R - \frac{(P+Q)}{\beta}\right]\right|$$
$$= \left|\frac{1}{(-0.1078)}\left[17.3695 - \frac{(6.9004 + (-8.9175)}{(-0.1078)}\right]\right|$$

$$= 1.245 \times 10 > 10^{-6}$$

After seven iterations, the final values of the procedure are as follows:
Iteration 7
Defining a final reduced variate as:

$$y_i = -\frac{1}{(-0.1078)} Ln\left[1 - \frac{(x_i - 1402.6182)(-0.4554)}{874.9148}\right]$$

The parameters P, Q and R for the final values are:

$$P = N - \sum_{i=1}^{N} \exp(y_i) = 51 - 51.00000009 = -9.4773 \times 10^{-8}$$

$$Q = \sum_{i=1}^{N} \exp[(\beta - 1)y_i] - (1 - \beta) \sum_{i=1}^{N} \exp(\beta y_i)$$
$$= 65.69700564 - (1 - (-0.4554))(45.14073528)$$

$$Q = 1.7932 \times 10^{-7}$$

$$R = N - \sum_{i=1}^{N} y_i + \sum_{i=1}^{N} y_i \exp(-y_i) = 51 - 29.46842940$$
$$+ (-21.53157058) = 1.9322 \times 10^{-8}$$

Then the final deviations between the estimator at iteration 7 and the true value for the ML estimator for such parameters are:

$$\delta_i(x_0) = -\frac{874.9148}{51}$$
$$\left[(1.2919)(1.7932 \times 10^{-7}) + \frac{(0.8597)(\text{-}9.4773 \times 10^{-8} + (1.7932 \times 10^{-7})}{(-0.4554)}\right.$$
$$\left. + \frac{(0.2574)}{(-0.4554)}\left(1.9322 \times 10^{-8} - \frac{(\text{-}9.4773 \times 10^{-8} + 1.7932 \times 10^{-7})}{(-0.4554)}\right)\right]$$
$$= 7.5167 \times 10^{-7}$$

$$\delta_i(\alpha) = -\frac{874.9148}{51}$$
$$\left[(0.8597)\left(1.7932 \times 10^{-7}\right) + \frac{(1.1242)(\text{-}9.4773 \times 10^{-8} + 1.7932 \times 10^{-7})}{(-0.4554)}\right.$$
$$+ \frac{(-0.1302)}{(-0.4554)}\left(1.9322 \times 10^{-8} - \frac{(\text{-}9.4773 \times 10^{-8} + 1.7932 \times 10^{-7})}{(-0.4554)}\right)$$
$$= \text{-}6.9340 \times 10^{-8}$$

$$\delta_i(\beta) = -\frac{1}{51}\left[(0.2574)\left(1.7932 \times 10^{-7}\right) + \frac{(-0.1302)(\text{-}9.4773 \times 10^{-8} + 1.7932 \times 10^{-7})}{(-0.4554)}\right.$$
$$+ \frac{0.8914}{(-0.4554)}\left(1.9322 \times 10^{-8} - \frac{(\text{-}9.4773 \times 10^{-8} + 1.7932 \times 10^{-7})}{(-0.4554)}\right)$$
$$= 6.4883 \times 10^{-9}$$

The final parameters are:

$$(x_0)_{i+1} = (x_0)_i + \left(\delta_{x0}\right)_i = 1402.6182 - 7.5167 \times 10^{-7} = 1402.6182$$

$$(\alpha)_{i+1} = (\alpha)_i + (\delta_\alpha)_i = 874.9148 - 6.9340 \times 10^{-8} = 874.9148$$

$$(\beta)_{i+1} = (\beta)_i + \left(\delta_\beta\right)_i = -0.4554 + 6.4883 \times 10^{-9} = -0.4554$$

The criteria of convergence are met for all parameters:

$$\left| -\frac{\partial LL(x, x_0, \alpha, \beta)}{\partial x_0} \right| = \left| \frac{1.7932 \times 10^{-7}}{874.9148} \right| = 2.0495 \times 10^{-10} < 10^{-6}$$

$$\left|-\frac{\partial LL(x, x_0, \alpha, \beta)}{\partial \alpha}\right| = \left|\frac{-9.4773 \times 10^{-8} + 1.7932 \times 10^{-10}}{(874.9148)(-0.4554)}\right|$$
$$= 2.1220 \times 10^{-10} < 10^{-6}$$

$$\left|-\frac{\partial LL(x, x_0, \alpha, \beta)}{\partial \beta}\right|$$
$$= \left|\frac{1}{(-0.4554)}\left[1.9233 \times 10^{-8} - \frac{(-9.4773 \times 10^{-8} + 1.7932 \times 10^{-10})}{(-0.4554)}\right]\right|$$
$$= 4.5013 \times 10^{-7} < 10^{-6}$$

So, the ML estimators of the parameters of the GEV distribution are:
location parameter:

$$\widehat{x}_0 = 1402.6182 \text{ m}^3/\text{s}$$

scale parameter:

$$\widehat{\alpha} = 874.9148 \text{ m}^3/\text{s}$$

shape parameter:

$$\widehat{\beta} = -0.4554$$

10.4.3 PWM Method

A distribution function may be characterized by its PWM's defined as, Greenwood et al. (1979):

$$M_{l,j,k} = E\left[X^l F^j (1-F)^k\right] = \int_0^1 [x(f)]^l F^j (1-F)^k dF \tag{10.66}$$

There are two different expressions to evaluate the population PWM defined in Eq. (10.66), they are, Greenwood et al. (1979):

$$M_{l,0,k} = \sum_{j=0}^{k} \binom{k}{j} (-1)^j M_{l,j,0} \tag{10.67}$$

and:

$$M_{l,j,0} = \sum_{k=0}^{j} \binom{j}{k} (-1)^k M_{l,0,k} \tag{10.68}$$

The conventional moments about the origin are represented by $M_{1,0,0.}$ The following convention is taken to simplify the notation, Greenwood et al. (1979):

$$M_{(k)} = M_{1,0,k} \tag{10.69}$$

An unbiased estimator of $M_{(k)}$ is, Maciunas Landwher et al. (1979a):

$$M_{(k)} = \frac{1}{N} \sum_{i=1}^{N-k} x_i \frac{\binom{N-i}{k}}{\binom{N-1}{k}} \tag{10.70}$$

and k is a non-negative integer and the x_i's, i = 1, 2, …, N, have been rank ordered from x_1 to x_N: $x_1 < x_2 < \ldots < x_N$.

Two different approaches have been proposed to obtain the parameters for the GEV distribution using the PWM method, they are the PWM1 (Hosking et al. (1985)) and the PMW2 (Raynal-Villasenor (1987)) methods. They will be described in the following two subsections.

10.4.3.1 PWM1 (Hosking et al 1985) Method

The PWM's for the GEV distribution are defined as, Hosking et al. (1985):

$$b_r = \frac{1}{(r+1)} \left\{ x_0 + \frac{\alpha\left[1 - (r+1)^{-\beta}\Gamma(1+\beta)\right]}{\beta} \right\} \tag{10.71}$$

for $\beta > -1$. Such PWM's do not exist for $\beta \leq -1$.

The first three PWM's are needed to construct the system of equations to obtain the PWM estimators of the GEV distribution, Hosking et al. (1985):

$$b_0 = x_0 + \frac{\alpha[1 - \Gamma(1+\beta)]}{\beta} = \bar{x} \tag{10.72}$$

$$2b_1 - b_0 = \frac{\alpha\left[(1 - 2^{-\beta})\Gamma(1+\beta)\right]}{\beta} = 2\left[\frac{1}{N(N-1)} \sum_{i=1}^{N} (i-1)x_i\right] - \bar{x} \tag{10.73}$$

$$(3b_2 - b_0)(2b_1 - b_0) = \frac{(1 - 3^{-\beta})}{(1 - 2^{-\beta})}$$

$$= \left[\frac{3}{N(N-1)(N-2)} \sum_{i=1}^{N} (i-1)(i-2)x_i - \bar{x} \right] \left[\frac{2}{N(N-1)} \sum_{i=1}^{N} (i-1)x_i - \bar{x} \right] \tag{10.74}$$

The solution to the system formed by Eqs. (10.72), (10.73) and (10.74) are the PWM estimators of the parameters of the GEV distribution are, Hosking et al. (1985):

$$\widehat{x}_0 = \bar{x} + \frac{\widehat{\alpha}\left[\Gamma(1+\widehat{\beta}) - 1\right]}{\widehat{\beta}} \tag{10.75}$$

$$\widehat{\alpha} = \frac{[2b_1 - b_0]\widehat{\beta}}{\Gamma(1+\widehat{\beta})\left(1 - 2^{-\widehat{\beta}}\right)} \tag{10.76}$$

$$\widehat{\beta} = 7.859c_H + 2.9554{c_H}^2 \tag{10.77}$$

where:

$$c_H = \frac{(2b_1 - b_0)}{(3b_2 - b_0)} - \frac{Ln(2)}{Ln(3)} \tag{10.78}$$

The Eq. (10.77) is an approximate solution to Eq. (10.74) in the range $-0.5 < \beta < 0.5$. The first three sample PWM's in this method are evaluated as, Hosking et al. (1985):

$$b_0 = \frac{1}{N} \sum_{i=1}^{N} x_i \tag{10.79}$$

$$b_1 = \frac{1}{N(N-1)} \sum_{i=1}^{N} x_i(i-1) \tag{10.80}$$

$$\boldsymbol{b_2} = \frac{1}{\boldsymbol{N(N-1)(N-2)}} \sum_{\boldsymbol{i=1}}^{\boldsymbol{N}} \boldsymbol{x_i(i-1)(i-2)} \tag{10.81}$$

10.4.3.2 PWM2 (Raynal-Villasenor 1987) Method

The PWM for the GEV distribution are defined as, Raynal-Villasenor (1987):

$$M_{(k)} = \left(x_0 + \frac{\alpha}{\beta}\right)\left[\frac{\Gamma(1+k)}{\Gamma(2+k)}\right] - \frac{\alpha\Gamma(1+\beta)}{\beta}$$
$$\left[1 - \frac{k}{2^{\beta+1}} + \frac{k(k-1)}{2!3^{\beta+1}} - \frac{k(k-1)(k-2)}{3!4^{\beta+1}} + \frac{k(k-1)(k-2)(k-3)}{4!5^{\beta+1}} \cdots\right] \tag{10.82}$$

for β > -1. Such PWM's do not exist for β ≤ −1.

The first three PWM's are needed to construct the system of equations to obtain the PWM estimators of the GEV distribution, Raynal-Villasenor (1987):

$$M_{1,0,0} = M_{(0)} = x_0 + \frac{\alpha}{\beta}[1 - \Gamma(1+\beta)] = \bar{x} \tag{10.83}$$

$$M_{1,1,0} = M_{(0)} - M_{(1)} = \frac{1}{2}\left\{x_0 + \frac{\alpha}{\beta}\left[1 - \frac{2\Gamma(1+\beta)}{2^{\beta+1}}\right]\right\} = \bar{x} - \frac{1}{N}\sum_{i=1}^{N-1} x_i \frac{(N-i)}{(N-1)} \tag{10.84}$$

$$M_{1,2,0} = M_{(0)} - 2M_{(1)} + M_{(2)} = \frac{1}{3}x_0$$
$$+ \frac{\alpha}{\beta}\left\{\frac{1}{3} - \Gamma(1+\beta)\left[2\left(1 - \frac{1}{2^{\beta+1}}\right) - \left(1 - \frac{2}{2^{\beta+1}} + \frac{1}{3^{\beta+1}}\right) - 1\right]\right\}$$
$$= \bar{x} - 2\left[\frac{1}{N(N-1)}\sum_{i=1}^{N-1}(N-i)x_i\right] + \frac{1}{N(N-1)(N-2)} \tag{10.85}$$

The solution to the system formed by Eqs. (10.83), (10.84) and (10.85) are the PWM estimators of the parameters of the GEV distribution, Raynal-Villasenor (1987):

$$\hat{x}_0 = \widehat{M}_{(0)} + \frac{\hat{\alpha}}{\hat{\beta}}\left[\Gamma\left(1+\hat{\beta}\right) - 1\right] \tag{10.86}$$

$$\hat{\alpha} = \frac{\left(\widehat{M}_{(0)} - 2\widehat{M}_{(1)}\right)\hat{\beta}}{\Gamma\left(1+\hat{\beta}\right)\left(1 - \frac{1}{2^{\hat{\beta}}}\right)} \tag{10.87}$$

$$F\left(\hat{\beta}\right) = 2^{\hat{\beta}}(C_{RV} + 2) + \left(\frac{2}{3}\right)^{\hat{\beta}} - (C_{RV} + 3) = 0 \tag{10.88}$$

where:

$$C_{RV} = \frac{3\widehat{M}_{(2)} - \widehat{M}_{(0)}}{\widehat{M}_{(0)} - 2\widehat{M}_{(1)}} \tag{10.89}$$

and:

$$\widehat{M}_{(0)} = \frac{1}{N}\sum_{i=1}^{N} x_i \tag{10.90}$$

$$\widehat{M}_{(1)} = \frac{1}{N(N-1)}\sum_{i=1}^{N-1} x_i(N-i) \tag{10.91}$$

$$\widehat{M}_{(2)} = \frac{1}{N(N-1)(N-2)}\sum_{i=1}^{N-2} x_i(N-i)(N-i-1) \tag{10.92}$$

10.4.3.3 Example of Application of Estimation of the Parameters of the GEV Distribution Using the PWM1 and PWM2 Methods

With the flood sample data of gauging station Huites, Mexico (1942–1992), contained in Appendix A, the following statistics have already been obtained:

$$\widehat{\mu} = \bar{x} = \frac{1}{N}\sum_{i=1}^{N} x_i = 2498.9627\ \text{m}^3/\text{s}$$

$$\widehat{\sigma} = s = \left[\frac{1}{N}\sum_{i=1}^{N}(x_i - \widehat{\mu})^2\right]^{1/2} = 2199.4767\ \text{m}^3/\text{s}$$

a) PWM1 Method

The PWM's are:

$$b_0 = \frac{1}{N}\sum_{i=1}^{N} x_i = 2498.967\ \text{m}^3/\text{s}$$

$$b_1 = \frac{1}{N(N-1)}\sum_{i=1}^{N} x_i(i-1) = 1779.4071\ \text{m}^3/\text{s}$$

$$b_2 = \frac{1}{N(N-1)(N-2)}\sum_{i=1}^{N} x_i(i-1)(i-2) = 1441.7070\ \text{m}^3/\text{s}$$

and:

$$c_H = \frac{(2b_1 - b_0)}{(3b_2 - b_0)} - \frac{Ln(2)}{Ln(3)} = \frac{(2(1779.4071) - 2498.9670)}{(3(1441.707) - 2498.9670)} - \frac{Ln(2)}{Ln(3)} = -0.0506$$

and the substitution of these values give the PWM estimators of the parameters of the GEV distribution as follows:

$$\widehat{\beta} = 7.859c_H + 2.9554{c_H}^2 = 7.859(-0.0506) + 2.9554(-0.0506)^2 = -0.3898$$

$$\widehat{\alpha} = \frac{[2b_1 - b_0]\widehat{\beta}}{\Gamma(1+\widehat{\beta})\left(1 - 2^{-\widehat{\beta}}\right)} = \frac{[2(1779.4071) - 249.967](-0.3898)}{(1.4662)\left(1 - 2^{-0.3898}\right)} = 908.3086 \text{ m}^3/\text{s}$$

$$\begin{aligned}\widehat{x}_0 = \bar{x} + \frac{\widehat{\alpha}\left[\Gamma(1+\widehat{\beta}) - 1\right]}{\widehat{\beta}} &= 2498.967 \\ &+ \frac{908.3086[1.4662 - 1]}{(-0.3898)} \\ &= 1412.5253 \text{ m}^3/\text{s}\end{aligned}$$

So, the PWM estimators of the parameters of the GEV distribution are:
location parameter:

$$\widehat{x}_0 = 1412.5253 \text{ m}^3/\text{s}$$

scale parameter:

$$\widehat{\alpha} = 908.3086 \text{ m}^3/\text{s}$$

and:
shape parameter:

$$\widehat{\beta} = -0.3898$$

b) PWM2 Method.
The PWM are obtained as follows:

$$M_{(0)} = \widehat{\mu} = \bar{x} = \frac{1}{N}\sum_{i=1}^{N} x_i = 2498.9627 \text{ m}^3/\text{s}$$

$$M_{(1)} = \frac{1}{N}\sum_{i=1}^{N-1} x_i \frac{(N-i)}{(N-1)} = 719.5557 \text{ m}^3/\text{s}$$

$$\widehat{M}_{(2)} = \frac{1}{N(N-1)(N-2)}\sum_{i=1}^{N-2} x_i(N-i)(N-i-1) = 381.8557 \text{ m}^3/\text{s}$$

and:

$$C_{RV} = \frac{3\widehat{M}_{(2)} - \widehat{M}_{(0)}}{\widehat{M}_{(0)} - 2\widehat{M}_{(1)}} = \frac{3(381.8557) - 2498.9627}{2498.9627 - 2(719.5557)} = -1.2770$$

now, the following function must be solved:

$$F\left(\widehat{\beta}\right) = 2^{\widehat{\beta}}(C_{RV} + 2) + \left(\frac{2}{3}\right)^{\widehat{\beta}} - (C_{RV} + 3) = 2^{\widehat{\beta}}(1.2770 + 2)$$
$$+ \left(\frac{2}{3}\right)^{\widehat{\beta}} - (-1.2770 + 3) = 0$$

applying the Newton–Raphson or the Bisection Methods for the finding of the roots of a single variable equation, one may find the following solution:

$$\widehat{\beta} = -0.3894$$

and the substitution of these values give the other two PWM estimators of the parameters of the GEV distribution as follows:

$$\widehat{\alpha} = \frac{\left(\widehat{M}_{(0)} - 2\widehat{M}_{(1)}\right)\widehat{\beta}}{\Gamma\left(1 + \widehat{\beta}\right)\left(1 - \frac{1}{2^{\widehat{\beta}}}\right)} = \frac{(2498.9627 - 2(719.5557))}{(1.4653)\left(1 - \frac{1}{2^{-0.3894}}\right)} = 908.9530 \text{ m}^3/\text{s}$$

$$\widehat{x}_0 = \widehat{M}_{(0)} + \frac{\widehat{\alpha}}{\widehat{\beta}}\left[\Gamma\left(1 + \widehat{\beta}\right) - 1\right] = 2498.9627 + \left(\frac{908.9530}{-0.3894}\right)(1.4653 - 1) = 1412.6485 \text{ m}^3/\text{s}$$

So, the PWM estimators of the parameters of the GEV distribution are:
location parameter:

$$\widehat{x}_0 = 1412.6485 \text{ m}^3/\text{s}$$

scale parameter:

$$\widehat{\alpha} = 908.9530 \text{ m}^3/\text{s}$$

and:
shape parameter:

$$\widehat{\beta} = -0.3894$$

The PWM1 and PWM2 provide results that are practically indistinguishable one to the another, so it is a matter of preference which method one would use. There is an advantage in using the PWM2 method when the shape parameter is lesser than −0.50, given that the PWM1 method is bounded to have a shape parameter bigger than –0.50. In the case of the PWM2 method, the shape parameter needs to be bigger than –1.0.

10.5 Estimation of Quantiles for the GEV Distribution

The quantiles for the GEV distribution are obtained by using the inverse form of the GEV distribution function:

$$x = x_0 + \frac{\alpha}{\beta}\{1 - [-Ln(F(x))]^{\beta}\} \tag{10.93}$$

where x_T is the quantile value for a certain value of the distribution function F(x).

The term Q_T is more frequently used in engineering instead of x_T and this also applies to T_r instead of *F(x)*; so, the following expression is widely used when it is needed to relate an event Q_T to a specific return period T_r:

$$Q_T = \widehat{x}_0 + \frac{\widehat{\alpha}}{\widehat{\beta}}\left\{1 - \left[-Ln(1 - \frac{1}{T_r})\right]^{\widehat{\beta}}\right\} \tag{10.94}$$

where Q_T is a design value corresponding to a specific return period T_r.

10.5.1 Examples of Estimation of MOM, ML and PWM Quantiles for the GEV Distribution

The estimation of MOM, ML and PWM quantiles for the GEV distribution is made by using the parameters estimated in the preceding sections and then inserted one pair of parameters at a time in the Eq. (10.94). Table 10.2 summarizes these results.

Table 10.2 Estimation of MOM, ML and PWM quantiles for the GEV distribution

Tr (years)	MOM	ML	PWM1	PWM2
	Q (m^3/s)	Q (m^3/s)	Q (m^3/s)	Q (m^3/s)
2	2025	1752	1770	1771
5	3870	3285	3264	3264
10	5222	4835	4684	4685
20	6626	6912	6499	6499
50	8613	10,839	9746	9745
100	10,238	15,090	13,082	13,077
500	14,496	32,024	25,339	25,318
1000	16,567	44,112	33,491	33,455

10.6 Goodness of Fit Test

The standard error of fit (SEF) for the GEV distribution has the following form:

$$SEF = \left[\frac{\sum_{i=1}^{N} (xi - yi)^2}{(N-3)} \right]^{1/2} \tag{10.95}$$

while the mean absolute relative deviation (MARD) remains the same as it was defined in Chap. 2:

$$MARD = \frac{100}{N} \sum_{i=1}^{N} \left| \frac{(x_i - y_i)}{x_i} \right| \tag{10.96}$$

where x_i are the sample historical values, y_i are the distribution function values, corresponding to the same return periods of the historical values, N is the sample size.

10.6.1 Examples of Application of the SEF and MARD to the MOM, ML and PWM Estimators of the Parameters of the GEV Distribution

Find the values of the SEF and MARD of the MOM, ML and PWM estimators of the parameters of the GEV distribution for the flood sample data of gauging station Huites, Mexico (1942–1992), Appendix A.

The values of SEF and MARD for the MOM, ML and PWM estimators of the parameters of the GEV distribution for the flood sample data of gauging station Huites, Mexico, have been obtained through the application of Eqs. (10.95) and (10.96) using the parameters obtained in previous examples. Table 10.3 contains a summary of both measures of goodness of fit tests.

For the flood sample data at gauging station Huites, Mexico the best choice according to SEF measure is the ML method. When using the MARD measure, the choice is either the ML method or the any of the PWM methods. Finally, the method with the best overall performance, for this sample of flood data, is the ML method.

Table 10.3 SEF and MARD Measures for MOM, ML, PMW1 and PWM2 estimators of the parameters of the GEV distribution for the flood sample data of gauging station Huites

Goodness of fit test	Method			
	MOM	ML	PWM1	PWM2
SEF	669	**527**	601	601
MARD	28	**7**	**7**	**7**

10.7 Estimation of Confidence Limits for the GEV Distribution

By assuming that the quantiles are normally distributed, the following form can be used to set the confidence limits on such quantiles:

$$x_l = x_T \pm z_\alpha S_T \tag{10.97}$$

where x_l is the upper or lower confidence limit, depending on the sign in the preceding formula (+) for the upper confidence limit and (−) for the lower confidence limit, z_α is the standard normal variate corresponding to a confidence level α, and S_T is squared root of the standard error of the estimate. The evaluation procedures of the standard errors of the estimates will be described in the following subsections.

10.8 Estimation of Standard Errors for the GEV Distribution

10.8.1 MOM Method

The general form of the MOM estimator of the standard error of the estimate of a three-parameters distribution is:

$$\begin{aligned} S_T{}^2 &= \left(\frac{\partial x_T}{\partial m_1{}'}\right)^2 var(m_1{}') + \left(\frac{\partial x_T}{\partial m_2}\right)^2 var(m_2) \\ &+ \left(\frac{\partial x_T}{\partial m_3}\right)^2 var(m_3) + 2\left(\frac{\partial x_T}{\partial m_1{}'}\right)\left(\frac{\partial x_T}{\partial m_2}\right) cov(m_1{}', m_2) \\ &+2\left(\frac{\partial x_T}{\partial m_1{}'}\right)\left(\frac{\partial x_T}{\partial m_3}\right) cov(m_1{}', m_3) + 2\left(\frac{\partial x_T}{\partial m_2}\right)\left(\frac{\partial x_T}{\partial m_3}\right) cov(m_2, m_3) \end{aligned} \tag{10.98}$$

and Eq. (10.98) can be simplified in terms of the frequency factor, K_T, as, Kite (1988):

$$\begin{aligned} S_T{}^2 &= \frac{\mu_2}{N}\left\{1 + K_T\hat{\gamma} + \frac{K_T{}^2}{4}(\hat{\kappa} - 1)\right. \\ &+ \frac{\partial K_T}{\partial \hat{\gamma}}\left[2\hat{\kappa} - 3\hat{\gamma}^2 - 6 + K_T\left(\hat{\lambda}_1 - \frac{6\hat{\gamma}\hat{\kappa}}{4} - \frac{10\hat{\gamma}}{4}\right)\right] \\ &\left.+\left(\frac{\partial K_T}{\partial \gamma}\right)^2\left[\hat{\lambda}_2 - 3\hat{\gamma}\hat{\lambda}_1 - 6\hat{\gamma} + \frac{9\hat{\gamma}^2\hat{\kappa}}{4} + \frac{35\hat{\gamma}^2}{4} + 9\right]\right\} \end{aligned} \tag{10.99}$$

where K_T is the frequency factor given by:

$$K_T = \frac{|\beta|B_K}{\beta}\left[A_K - \left(-Ln\left(1 - \frac{1}{T_r}\right)\right)^\beta\right] \tag{10.100}$$

$$\hat{\lambda}_1 = \frac{\left[\frac{\sum_{i=1}^{N}\left(x_i - \hat{\mu}\right)^5}{N}\right]}{\left(\sigma^2\right)^{2.5}} \tag{10.101}$$

$$\hat{\lambda}_2 = \frac{\left[\frac{\sum_{i=1}^{N}\left(x_i - \hat{\mu}\right)^6}{N}\right]}{\left(\sigma^2\right)^{3}} \tag{10.102}$$

and:

$$A_K = \Gamma(1+\beta) \tag{10.103}$$

$$B_K = \frac{1}{\left[\Gamma(1+2\beta) - \Gamma^2(1+\beta)\right]^{\frac{1}{2}}} \tag{10.104}$$

The first-order partial derivative of the frequency factor, K_T, with respect to the skewness coefficient, γ, can be evaluated by means of the chain rule for differentiation:

$$\frac{\partial K_T}{\partial \gamma} = \left(\frac{\partial K_T}{\partial \beta}\right)\left(\frac{\partial \beta}{\partial \gamma}\right) \tag{10.105}$$

and:

$$\frac{\partial K_T}{\partial \beta} = \frac{|\beta|}{\beta}\left\{\frac{\left[G_1P_1 - y^{\beta}Ln(y)\right] - \frac{1}{2}\left[G_1 - y^{\beta}\right]\left[G_2 - G_1{}^2\right]^{-1}\left[G_2P_2 - 2G_1{}^2P_1\right]}{\left[G_2 - G_1{}^2\right]^{\frac{1}{2}}}\right\} \tag{10.106}$$

$$\frac{\partial \gamma}{\partial \beta} = -\frac{|\beta|}{\beta}\left\{\frac{\left[G_2 - G_1{}^2\right]\left[G_3P_3 + 3G_1\left(P_1\left(G_2 - 2G_1{}^2\right) - G_2P_2\right)\right] - \frac{3}{2}\left[G_3 - 3G_1G_2 + 2G_1{}^3\right]\left[G_2P_2 - 2G_1{}^2P_1\right]}{\left[G_2 - G_1{}^2\right]^{2.5}}\right\} \tag{10.107}$$

where:

$$y = -Ln\left(1 - \frac{1}{T_r}\right) \tag{10.108}$$

$$G_r = \Gamma(1 + r\beta) \tag{10.109}$$

$$P_r = \psi(1 + r\beta) \tag{10.110}$$

and $\Gamma(.)$ and $\psi(.)$ are the complete Gamma and Digamma functions, previously defined.

Then, the first-order partial derivative of the frequency factor, K_T, with respect to the skewness coefficient, γ, is:

$$\frac{\partial K_T}{\partial \gamma} = \frac{\left[G_2 - G_1^2\right]^2 \left\{\left[G_1 P_1 - y^{\beta} Ln(y)\right] - \frac{1}{2}\left[G_1 - y^{\beta}\right]\left[G_2 - G_1^2\right]^{-1}\left[G_2 P_2 - 2G_1^2 P_1\right]\right\}}{\left\{\left[G_2 - G_1^2\right]\left[G_3 P_3 + 3G_1\left(P_1\left(G_2 - 2G_1^2\right) - G_2 P_2\right)\right] - \frac{3}{2}\left[G_3 - 3G_1 G_2 + 2G_1^3\right]\left[G_2 P_2 - 2G_1^2 P_1\right]\right\}} \tag{10.111}$$

The substitution of Eq. (10.111) into Eq. (10.99) will provide the moment estimator of the standard error of the variate for the GEV distribution.

10.8.1.1 Example of Estimation of MOM Standard Errors and the Two-Sided 95% Confidence Limits for the GEV Distribution

Find the MOM estimators of the standard errors and the two-sided 95% confidence limits for the GEV distribution for the flood sample data of gauging station Huites, Mexico (1942–1992), contained in Appendix A.

The estimation of MOM estimators of the standard errors and the two-sided 95% confidence limits for the GEV distribution is made by using the moment estimators of the parameters and then inserted in the Eqs. (10.100)–(10.109) and then insert these results into Eq. (10.99), using selected values of the return intervals. Table 10.4 summarizes these results.

Table 10.4 Estimation of MOM standard errors and the two-sided 95% confidence limits for the GEV distribution for gauging Station Huites, Mexico

Tr (years)	S_T (m^3/s)	95% lower limit (m^3/s)	Q_T (m^3/s)	95% Upper limit (m^3/s)
2	236.5943	1561	2025	2488
5	496.2512	2898	3870	4843
10	664.4137	3920	5222	6525
20	833.0345	4993	6626	8259
50	1066.7437	6522	8613	10,704
100	1255.6177	7777	10,238	12,699
500	1745.6571	11,075	14,496	17,918
1000	1982.5617	12,681	16,567	20,453

Two-sided limits: $z_{\alpha} = 1.96$.

10.8.2 ML Method

The general form of the ML estimator of the standard error of the estimate of a three-parameter distribution is, Kite (1988):

$$
\begin{aligned}
{S_T}^2 = & \left(\frac{\partial x_T}{\partial \alpha}\right)^2 var(\alpha) + \left(\frac{\partial x_T}{\partial \beta}\right)^2 var(\beta) + \left(\frac{\partial x_T}{\partial \gamma}\right)^2 var(\gamma) \\
& + 2\left(\frac{\partial x_T}{\partial \alpha}\right)\left(\frac{\partial x_T}{\partial \beta}\right) cov(\alpha, \beta) \\
& + 2\left(\frac{\partial x_T}{\partial \alpha}\right)\left(\frac{\partial x_T}{\partial \gamma}\right) cov(\alpha, \gamma) + 2\left(\frac{\partial x_T}{\partial \beta}\right)\left(\frac{\partial x_T}{\partial \gamma}\right) cov(\beta, \gamma)
\end{aligned} \tag{10.112}
$$

and the variance–covariance matrix of the parameters for the GEV distribution is known to be, NERC (1975):

$$
[V] = \begin{bmatrix} Var(x_0) & Cov(x_0, \alpha) & Cov(x_0, \beta) \\ Cov(\alpha, x_0) & Var(\alpha) & Cov(\alpha, \beta) \\ Cov(\beta, x_0) & Cov(\beta, \alpha) & Var(\beta) \end{bmatrix} = \frac{1}{N}\begin{bmatrix} \alpha^2 b & \alpha^2 h & \alpha f \\ \alpha^2 h & \alpha^2 a & \alpha g \\ \alpha f & \alpha g & c \end{bmatrix} \tag{10.113}
$$

and from Eq. (10.94):

$$
\frac{\partial x_T}{\partial x_0} = 1 \tag{10.114}
$$

$$
\frac{\partial x_T}{\partial \alpha} = \frac{1}{\beta}\left\{1 - \left[-Ln\left(1 - \frac{1}{Tr}\right)\right]^{\beta}\right\} \tag{10.115}
$$

$$
\begin{aligned}
\frac{\partial x_T}{\partial \beta} = & -\frac{\alpha}{\beta}\left\{\frac{1}{\beta}\left[1 - \left(-Ln\left(1 - \frac{1}{T_r}\right)\right)^{\beta}\right]\right. \\
& \left. - \left[-Ln\left(1 - \frac{1}{T_r}\right)\right]^{\beta}\left[Ln\left(-Ln\left(1 - \frac{1}{T_r}\right)\right)\right]\right\}
\end{aligned} \tag{10.116}
$$

So, Eq. (10.110) becomes to:

$$
\begin{aligned}
{S_T}^2 = & \frac{1}{N}\left[\alpha^2 b + \alpha^2 a\left(\frac{\partial x_T}{\partial \alpha}\right)^2 + c\left(\frac{\partial x_T}{\partial \beta}\right)^2\right. \\
& \left. + 2\alpha^2 h\left(\frac{\partial x_T}{\partial \alpha}\right) + 2\alpha f\left(\frac{\partial x_T}{\partial \beta}\right) + 2\alpha g\left(\frac{\partial x_T}{\partial \alpha}\right)\left(\frac{\partial x_T}{\partial \beta}\right)\right]
\end{aligned} \tag{10.117}
$$

10.8.2.1 Example of Estimation of ML Standard Errors and the two-Sided 95% Confidence Limits for the GEV Distribution

Find the ML estimators of the standard errors and the two-sided 95% confidence limits for the GEV distribution for the flood sample data of gauging station Huites, Mexico (1942–1992), contained in Appendix A.

The estimation of ML estimators of the standard errors and the two-sided 95% confidence limits for the GEV distribution is made by using the ML estimators of the parameters and then inserted in Eqs. (10.112)–(10.114) and into Eq. (10.115), using selected values of the return intervals. Table 10.5 summarizes these results.

10.8.3 PWM Method

The general form of the PWM estimator of the standard error of the estimate of a three-parameter distribution is:

$$
\begin{aligned}
{S_T}^2 = & \left(\frac{\partial x_T}{\partial \alpha}\right)^2 var(\alpha) + \left(\frac{\partial x_T}{\partial \beta}\right)^2 var(\beta) + \left(\frac{\partial x_T}{\partial \gamma}\right)^2 var(\gamma) \\
& + 2\left(\frac{\partial x_T}{\partial \alpha}\right)\left(\frac{\partial x_T}{\partial \beta}\right) cov(\alpha, \beta) \\
& + 2\left(\frac{\partial x_T}{\partial \alpha}\right)\left(\frac{\partial x_T}{\partial \gamma}\right) cov(\alpha, \gamma) + 2\left(\frac{\partial x_T}{\partial \beta}\right)\left(\frac{\partial x_T}{\partial \gamma}\right) cov(\beta, \gamma)
\end{aligned} \tag{10.118}
$$

The expressions for the variances and covariance shown in Eq. (10.118), have been obtained as, Hosking (1986a):

$$
var(x_0) = w_{11} \frac{\alpha^2}{N} \tag{10.119}
$$

Table 10.5 Estimation of ML standard errors and the two-sided 95% confidence limits for the GEV distribution for gauging Station huites, Mexico

Tr (years)	S_T(m^3/s)	95% Lower limit (m^3/s)	Q_T(m^3/s)	95% Upper limit (m^3/s)
2	41.6076	1670	1752	1833
5	189.8206	2913	3285	3657
10	368.1927	4113	4835	5557
20	635.4662	5666	6912	8157
50	1201.5362	8484	10,839	13,194
100	1876.9271	11,411	15,090	18,769
500	4952.3053	22,318	32,024	41,731
1000	7386.7345	29,634	44,112	58,590

Two-sided limits: $z_\alpha = 1.96$.

Table 10.6 Values of the elements of the variance–covariance matrix of the GEV distribution for the PWM1 method

β	w_{11}	w_{12}	w_{13}	w_{22}	w_{23}	w_{33}
−0.4	1.6637	1.3355	1.1405	1.8461	1.1628	2.9092
−0.3	1.4153	0.8912	0.5640	1.2574	0.4442	1.4090
−0.2	1.3322	0.6727	0.3926	1.0013	0.2697	0.9139
−0.1	1.2915	0.5104	0.3245	0.8440	0.2240	0.6815
0.0	1.2686	0.3704	0.2992	0.7390	0.2247	0.5633
0.1	1.2551	0.2411	0.2966	0.6708	0.2447	0.5103
0.2	1.2474	0.1177	0.3081	0.6330	0.2728	0.5021
0.3	1.2438	-0.0023	0.3297	0.6223	0.3033	0.5294
0.4	1.2433	-0.1205	0.3592	0.6368	0.3329	0.5880

$$var(\alpha) = w_{22}\frac{\alpha^2}{N} \tag{10.120}$$

$$var(\beta) = \frac{w_{33}}{N} \tag{10.121}$$

$$cov(x_0, \alpha) = w_{12}\frac{\alpha^2}{N} \tag{10.122}$$

$$cov(x_0, \beta) = w_{13}\frac{\alpha}{N} \tag{10.123}$$

$$cov(\alpha, \beta) = w_{23}\frac{\alpha}{N} \tag{10.124}$$

The values of the elements of the variance–covariance matrix of the parameters, the w's, of the GEV distribution for the PWM1 method had been obtained as shown in Table 10.6, Hosking et al. (1985):

The following equations had been derived, by a polynomial fitting procedure, to have an easy computational way to calculate the elements of the variance–covariance matrix of the GEV distribution for the PWM method:

$$\begin{aligned} w_{11} = {} & 1.2683 - 0.1948\beta + 0.5337\beta^2 + 0.3032\beta^3 \\ & - 1.2735\beta^4 - 14.804\beta^5 + 32.324\beta^6 \end{aligned} \tag{10.125}$$

$$\begin{aligned} w_{12} = {} & 0.37 - 1.3642\beta + 0.6255\beta^2 + 0.4311\beta^3 \\ & - 1.7568\beta^4 - 20.487\beta^5 + 44.537\beta^6 \end{aligned} \tag{10.126}$$

$$\begin{aligned} w_{13} = {} & 0.2985 - 0.1713\beta + 1.2877\beta^2 + 0.8003\beta^3 \\ & - 3.1341\beta^4 - 36.442\beta^5 + 79.481\beta^6 \end{aligned} \tag{10.127}$$

$$w_{22} = 0.7384 - 0.8909\beta + 1.9634\beta^2 + 0.6516 - 2.49\beta^4 - 28.304\beta^5 + 61.676\beta^6 \tag{10.128}$$

$$w_{23} = 0.2237 + 0.0588\beta + 1.1823\beta^2 + 1.2098\beta^3 - 4.4701\beta^4 - 50.353\beta^5 + 109.72\beta^6 \tag{10.129}$$

$$w_{33} = 0.5615 - 0.9349\beta + 3.6386\beta^2 + 2.0607\beta^3 - 7.6874\beta^4 - 89.654\beta^5 + 195.72\beta^6 \tag{10.130}$$

The first partial derivatives can be computed from Eq. (10.118) as:

$$\frac{\partial x_T}{\partial x_0} = 1 \tag{10.131}$$

$$\frac{\partial x_T}{\partial \alpha} = \frac{1}{\beta}\left\{1 - \left[-Ln\left(1 - \frac{1}{Tr}\right)\right]^{\beta}\right\} \tag{10.132}$$

$$\frac{\partial x_T}{\partial \beta} = -\frac{\alpha}{\beta}\left\{\frac{1}{\beta}\left[1 - \left(-Ln\left(1 - \frac{1}{T_r}\right)\right)^{\beta}\right] + \left[-Ln\left(1 - \frac{1}{T_r}\right)\right]^{\beta}\left[Ln\left(-Ln\left(1 - \frac{1}{T_r}\right)\right)\right]\right\} \tag{10.133}$$

By inserting Eqs. (10.119)–(10.133) into Eq. (10.118) will produce the PWM estimator of the standard error of the estimate for the GEV distribution.

10.8.3.1 Example of Estimation of Standard Errors and the Two-Sided 95% Confidence PWM1 Limits for the GEV Distribution

Find the PWM1 method for PWM1 estimators of the standard errors and the two-sided 95% confidence limits for the GEV distribution for the flood sample data of gauging station Huites, Mexico (1942–1992), contained in Appendix A.

The estimation of PWM1 estimators of the standard errors and the two-sided 95% confidence limits for the GEV distribution is made by using the PWM1 estimators of the parameters and then inserted in the Eqs. (10.117)–(10.125) and then into Eq. (10.116), using selected values of the return intervals. Table 10.7 summarizes these results.

10.8.3.2 Example of Estimation of Standard Errors and the Two-Sided 95% Confidence PWM2 Limits for the GEV Distribution

Find the PWM2 Method for PWM estimators of the standard errors and the two-sided 95% confidence limits for the GEV distribution for the flood sample data of gauging station Huites, Mexico (1942–1992), Appendix A.

Table 10.7 Estimation of PWM1 Method for Standard Errors and the Two-sided 95% Confidence Limits for the GEV Distribution for Gauging Station Huites, Mexico

Tr (years)	$S_T(m^3/s)$	95% Lower limit (m^3/s)	$Q_T(m^3/s)$	95% Upper limit (m^3/s)
2	229.7448	1320	1770	2221
5	201.3460	2869	3264	3658
10	158.2209	4374	4684	4994
20	149.7219	6205	6499	6792
50	376.8621	9008	9746	10,485
100	738.1738	11,635	13,082	14,529
500	2495.3537	20,448	25,339	30,230
1000	3897.7225	25,851	33,491	41,130

Two-sided Limits: $z_\alpha = 1.96$

Table 10.8 Estimation of PWM2 Method for standard errors and the two-sided 95% confidence limits for the GEV distribution for gauging station Huites, Mexico

Tr (years)	$S_T(m^3/s)$	95% Lower limit (m^3/s)	$Q_T(m^3/s)$	95% Upper limit (m^3/s)
2	229.7448	1320	1770	2221
5	201.3460	2869	3264	3658
10	158.2209	4374	4684	4994
20	149.7219	6205	6499	6792
50	376.8621	9008	9746	10,485
100	738.1738	11,635	13,082	14,529
500	2495.3537	20,448	25,339	30,230
1000	3897.7225	25,851	33,491	41,130

Two-sided limits: $z_\alpha = 1.96$

The estimation of PWM2 Method for PWM estimators of the standard errors and the two-sided 95% confidence limits for the GEV distribution is made by using the PWM estimators of the parameters and then inserted in the Eqs. (10.119)–(10.133) and then into Eq. (10.118), using selected values of the return intervals. Table 10.8 summarizes these results.

10.9 Examples of Application for the GEV Distribution Using Excel® Spreadsheets

10.9.1 Flood Frequency Analysis

By using the flood data from gauging station Huites, Mexico the following descriptive statistics were obtained and are shown in Fig. 10.2.

	A	B	C
1	FLOOD FREQUENCY ANALYSIS		
2	Gauging Station: Huites, Sinaloa, Mexico (1942-1992)		
3			
4	*Descriptive Statistics*		
5	Mean	2498.9627	
6	Sample Standard Error	311.0530	
7	Median	1775.7000	
8	Mode	#N/D	
9	Standard Deviation	2221.3626	
10	Sample Variance	4934451.7500	
11	Kurtosis	3.6701	
12	Skewness	2.0123	
13	Range	9707.5000	
14	Minimum	421.8000	
15	Maximum	10129.3000	
16	Sum	127447.1000	
17	Count	51.0000	
18	Largest(1)	10129.3000	
19	Smallest(1)	421.8000	
20	Mean Confidence Level(95.0%)	624.7683	

Fig. 10.2 Descriptive statistics for the flood sample for Huites, Mexico

10.9.1.1 MOM Method

By using an Excel® spreadsheet, the results contained in Fig. 10.3, were obtained by the MOM method applied to the GEV distribution for gauging station Huites, Mexico.

10.9.1.2 ML Method

By using an Excel® spreadsheet, the results contained in Figs. 10.4 and 10.5, were obtained by the ML method applied to the GEV distribution for gauging station Huites, Mexico.

10.9.1.3 PWM1 (Hosking et al (1985)) Method

By using an Excel® spreadsheet, the results contained in Figs. 10.6 and 10.7 were obtained by the PWM1 Method applied to the GEV distribution for gauging station Huites, Mexico.

10.9.1.4 PWM2 (Raynal-Villasenor (1987)) Method

By using an Excel® spreadsheet, the results contained in Figs. 10.8 and 10.9 were obtained by the PWM2 Method applied to the GEV distribution for gauging station Huites, Mexico.

In Fig. 10.10 a comparison is made between the histogram and GEV theoretical density for flood sample of Huites, Mexico.

A graphical comparison between the empirical and theoretical frequency curves from the results provided by the three methods shown before is shown in Fig. 10.11.

	A	B	C	D	E	F	G	H	I
1	FLOOD FREQUENCY ANALYSIS								
2	Gauging Station: Huites, Sinaloa, Mexico (1942-1992)								
3	General Extreme Value (GEV) Distribution								
4	Method: MOM								
5									
6	Estimation of Parameters and Goodness of Fit Tests								
7	1+β =	0.8923	1+2β =	0.7845	1+3β =	0.6768			
8	Γ(1+β) =	1.0749	Γ(1+2β) =	1.1821	Γ(1+3β) =	1.3363			
9	ψ(1+β) =	-0.7699	ψ(1+2β) =	-1.0011	ψ(1+3β) =	-1.2875			
10	μ_z =	1.0749	σ_z =	0.1630	λ_1 =	17.5785			
11	A =	-12153.1205	B =	13630.6038	λ_2 =	54.7624			
12	Coefficient of Skewness =		2.0123						
13	Location Parameter =		1477.4834	m^3/s					
14	Scale Parameter =		1468.4313	m^3/s					
15	Shape Parameter =		-0.1077						
16	SEF =	677		MARD =	28				
17									
18	Design Values and Confidence Limits								
19–22	Return Period Tr (years)	F(x)	y	Frequency Factor K_T	$\partial K_T/\partial\gamma$	Squared Root Stand. Error S_T (m^3/s)	Two-sided 95% Lower Limit (m^3/s)	Design Values Q_T (m^3/s)	Two-sided 95% Upper Limit (m^3/s)
23	2	0.5	0.6931	-0.2127	0.0020	236.8960	1562	2026	2491
24	5	0.8	0.2231	0.6163	-0.0049	495.9828	2896	3868	4840
25	10	0.9	0.1054	1.2236	-0.0102	663.7878	3916	5217	6518
26	20	0.95	0.0513	1.8541	-0.0159	832.0364	4987	6618	8248
27	50	0.98	0.0202	2.7463	-0.0242	1065.2080	6512	8600	10687
28	100	0.99	0.0101	3.4761	-0.0312	1253.6321	7764	10221	12678
29	500	0.998	0.0020	5.3882	-0.0505	1742.4535	11053	14468	17883
30	1000	0.999	0.0010	6.3181	-0.0603	1978.7476	12655	16534	20412
31	Two-sided Confidence Limits:		z =	1.96					

Fig. 10.3 MOM estimators for the parameters, standard errors, Quantiles, and confidence limits of the GEV distribution for the flood sample for Huites, Mexico

	A	B	C	D	E	F	G
1	FLOOD FREQUENCY ANALYSIS						
2	Gauging Station: Huites, Sinaloa, Mexico (1942-1992)						
3	General Extreme Value Distribution						
4	Method: ML						
5							
6	Estimation of Parameters and Goodness of Fit Tests						
7	Initial Moment Estimators:			(1-β)–1 =	0.4554	(1-2β)–1 =	0.9108
8	Location Parameter =	1477.4834	m^3/s	1-β =	1.4554	1-2β =	1.9108
9	Scale Parameter =	1468.4313	m^3/s	Γ(1–β) =	0.8856	Γ(1–2β) =	0.9655
10	Shape Parameter =	-0.1077		ψ(1–β) =	-0.0061		
11	Iterative Procedure:			p =	2.0450	q =	4.1115
12	i-th iteration				(i+1)-th iteration		
13	Location Parameter =	1402.6182		2.0450	-1.6604	-0.8330	
14	Scale Parameter =	874.9148		-1.6604	2.2530	0.8084	
15	Shape Parameter =	-0.4554		-0.8330	0.8084	1.4805	
16	P =	-9.4773E-08				Determinant =	2.0760
17	Q =	1.7932E-07		a =	1.1242	f =	0.2574
18	R =	1.9322E-08		b =	1.2919	g =	-0.1302
19	ABS(∂LL/ ∂x_0) =	2.0495E-10		c =	0.8914	h =	0.8597
20	ABS(∂LL/ ∂α) =	2.1220E-10		δ(x_0) =	7.5167E-07		
21	ABS(∂LL/ ∂β) =	4.5013E-07		δ(α) =	-6.9340E-08		
22	Final MV Estimators:			δ(β) =	6.4883E-09		
23	Location Parameter =	1402.6182		Location Parameter=		1402.6182	
24	Scale Parameter =	874.9148		Scale Parameter =		874.9148	
25	Shape Parameter =	-0.4554		ShapeParameter =		-0.4554	
26	Log-Likelihood Function=	-439.3683					
27	SEF =	527	MARD =	6.4	AIC =	884.7	

Fig. 10.4 ML estimators for the parameters of the GEV distribution for the flood sample for Huites, Mexico

	A	B	C	D	E	F	G	H
29	Variance-Covariance Matrix							
30	var(x_0) =	19389.8422		cov(x_0,α) =	12903.9136			
31	var(α) =	16873.0650		cov(x_0,β)	4.4154			
32	var(β) =	0.0175		cov(α,β) =	-2.2329			
33								
34	Design Values and Confidence Limits							
35–38	Return Period Tr (years)	F(x)	$\partial x_T/\partial\alpha$	$\partial x_T/\partial\beta$	Squared Root Stand. Error S_T (m^3/s)	Two-sided 95% Lower Limit (m^3/s)	Design Values Q_T (m^3/s)	Two-sided 95% Upper Limit (m^3/s)
39	2	0.5	0.3989	1598.4268	41.6076	1670	1752	1833
40	5	0.8	2.1518	9839.9087	189.8206	2913	3285	3657
41	10	0.9	3.9230	19584.7462	368.1927	4113	4835	5557
42	20	0.95	6.2966	34166.9151	635.4662	5666	6912	8157
43	50	0.98	10.7851	65036.7138	1201.5362	8484	10839	13194
44	100	0.99	15.6440	101857.9038	1876.9271	11411	15090	18769
45	500	0.998	34.9994	269451.2231	4952.3053	22318	32024	41731
46	1000	0.999	48.8159	402066.4988	7386.7345	29634	44112	58590
47	Two-sided Confidence Limits:	z_α =	1.96					

Fig. 10.5 ML estimators of the standard errors, Quantiles and confidence limits for the flood sample for Huites, Mexico

	A	B	C	D	E	F	G	H
1	FLOOD FREQUENCY ANALYSIS							
2	Gauging Station: Huites, Sinaloa, Mexico (1942-1992)							
3	General Extreme Value (GEV) Distribution							
4	Method: PWM1							
5								
6	Probability Weighted Moments							
7	Hosking et al (1985) :							
8	b_0 =	2498.9627	m^3/s					
9	b_1 =	1779.4071	m^3/s					
10	b_2 =	1441.7070	m^3/s					
11	c_H =	-0.0506						
12								
13	Estimation of Parameters and Goodness of Fit Tests							
14	Hosking et al (1985) :							
15	1+β =	0.6102						
16	Γ(1+β) =	1.4662						
17	Location Parameter =		1412.5253	m^3/s				
18	Scale Parameter =		908.3086	m^3/s				
19	Shape Parameter =		-0.3898					
20	SEF =	601		MARD =	7			
21								
22	Variance-Covariance Matrix							
23	w_{11} =	1.6280	w_{12} =	1.2760	w_{13} =	1.0563		
24	w_{22} =	1.7655	w_{23} =	1.0553	w_{33} =	1.8393		
25	var(x_0) =	26336.4443		cov(x_0,α) =	20641.5553			
26	var(α) =	28560.9059		cov(x_0,β)	18.8134			
27	var(β) =	0.0361		cov(α,β) =	0.0207			

Fig. 10.6 PWM1 Method, PWM's and estimators for the parameters of the GEV distribution for the flood sample for Huites, Mexico

Design Values and Confidence Limits							
				Squared Root Stand. Error	Two-sided 95%	Design Values	Two-sided 95%
Return Period				S_T	Lower Limit	Q_T	Upper Limit
Tr (years)	F(x)	$\partial x_T/\partial \alpha$	$\partial x_T/\partial \beta$	(m^3/s)	(m^3/s)	(m^3/s)	(m^3/s)
2	0.5	0.3940	-67.1411	29.5955	1712	1770	1828
5	0.8	2.0380	-1522.7754	70.7603	3125	3264	3402
10	0.9	3.6021	-4212.8190	141.8965	4406	4684	4962
20	0.95	5.5998	-8979.6010	270.1558	5969	6499	7028
50	0.98	9.1752	-20230.0493	574.0137	8621	9746	10872
100	0.99	12.8475	-34462.4525	957.9138	11205	13082	14960
500	0.998	26.3421	-101765.6106	2767.2498	19915	25339	30763
1000	0.999	35.3164	-155369.8791	4204.8968	25249	33491	41732
Two-sided Confidence Limits:		z_α =	1.96				

Fig. 10.7 PWM1 estimators of the standard errors, Quantiles and confidence limits for the flood sample for Huites, Mexico

FLOOD FREQUENCY ANALYSIS					
Gauging Station: Huites, Sinaloa, Mexico (1942-1992)					
General Extreme Value (GEV) Distribution					
Method: PWM2					
Probability Weighted Moments					
$M_{(0)}$ =	2498.9627	m^3/s			
$M_{(1)}$ =	719.5557	m^3/s		C_{RV} =	-1.2770
$M_{(2)}$ =	381.8557	m^3/s			
Estimation of Parameters and Goodness of Fit Tests					
Test β =	-0.3894		$1+\beta$ =	0.6106	
$F(\beta)$ =	-4.8411E-07		$\Gamma(1+\beta)$ =	1.4654	
Location Parameter =		1412.6485	m^3/s		
Scale Parameter =		908.9530	m^3/s		
ShapeParameter =		-0.3894			
SEF =	601		MARD =	7	
Variance-Covariance Matrix					
w_{11} =	1.6269	w_{12} =	1.2741	w_{13} =	1.0538
w_{22} =	1.7630	w_{23} =	1.0521	w_{33} =	1.8359
$var(x_0)$ =	26355.9450		$cov(x_0,\alpha)$ =	20639.9854	
$var(\alpha)$ =	28559.8526		$cov(x_0,\beta)$	18.7808	
$var(\beta)$ =	0.0360		$cov(\alpha,\beta)$ =	0.0206	

Fig. 10.8 PWM2 Method, PWM's and estimators for the parameters of the GEV distribution for the flood sample for Huites, Mexico

A graphical depiction of the empirical and theoretical frequency curves and the confidence limits of the best fit provided, in this case that of ML, is shown in Fig. 10.12.

Return Period Tr (years)	F(x)	$\partial x_T/\partial\alpha$	$\partial x_T/\partial\beta$	Squared Root Stand. Error S_T (m³/s)	Two-sided 95% Lower Limit (m³/s)	Design Values Q_T (m³/s)	Two-sided 95% Upper Limit (m³/s)
2	0.5	0.3940	-67.1824	29.6018	1713	1771	1829
5	0.8	2.0373	-1523.2495	70.7450	3126	3264	3403
10	0.9	3.6003	-4213.2392	141.8124	4407	4685	4963
20	0.95	5.5960	-8978.6111	269.9128	5970	6499	7028
50	0.98	9.1668	-20222.1335	573.3021	8621	9745	10868
100	0.99	12.8331	-34441.4956	956.4981	11203	13077	14952
500	0.998	26.2995	-101650.9497	2761.6653	19905	25318	30731
1000	0.999	35.2514	-155159.1684	4195.4286	25232	33455	41678
Two-sided Confidence Limits:		z_α =	1.96				

Design Values and Confidence Limits

Fig. 10.9 PWM2 method estimators of the standard errors, Quantiles and confidence limits for the flood sample of Huites, Mexico

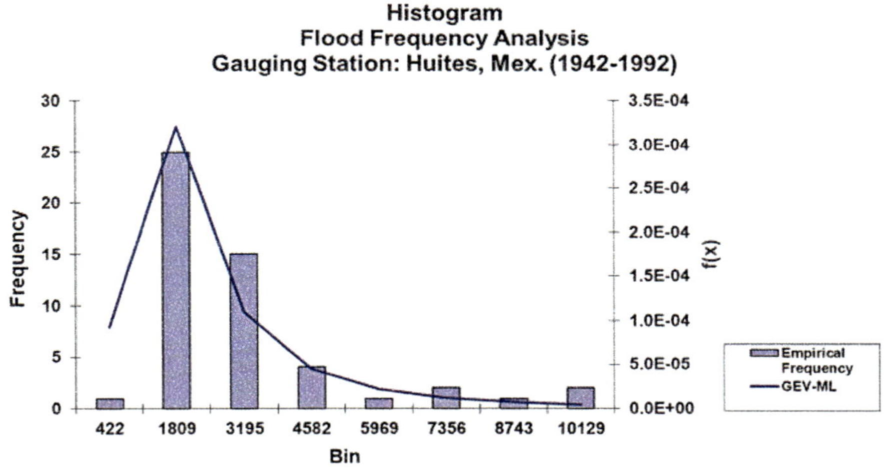

Fig. 10.10 Histogram and GEV theoretical density for the flood sample at Huites, Mexico

10.9.2 Rainfall Frequency Analysis

By using the 24 h, annual maximum rainfall data from rainfall at meteorological station Chihuahua, Mexico the following descriptive statistics were obtained and are shown in Fig. 10.13.

10.9.2.1 MOM Method

By using an Excel® spreadsheet, the results contained in Fig. 10.14, were obtained for the MOM method applied to the GEV distribution for 24 h annual maximum rainfall at meteorological station Chihuahua, Mexico.

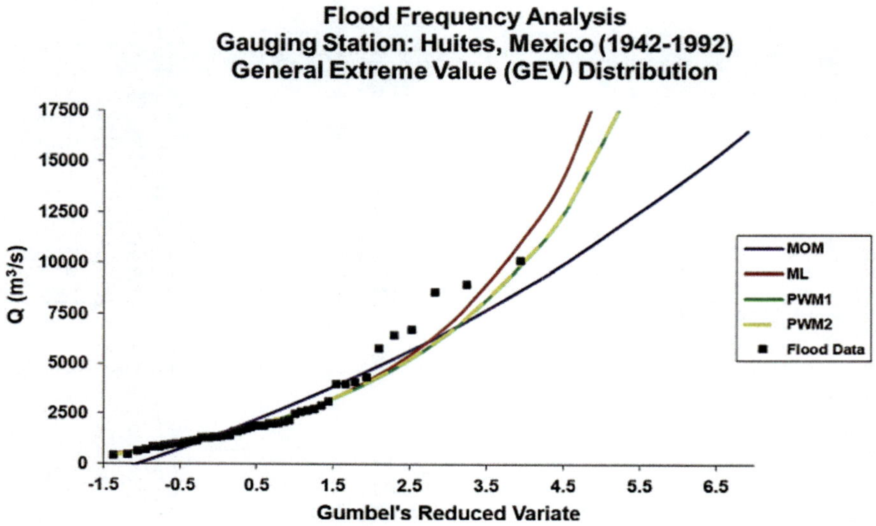

Fig. 10.11 Empirical and MOM, ML and PWM theoretical curves of GEV distribution of flood sample Huites, Mexico

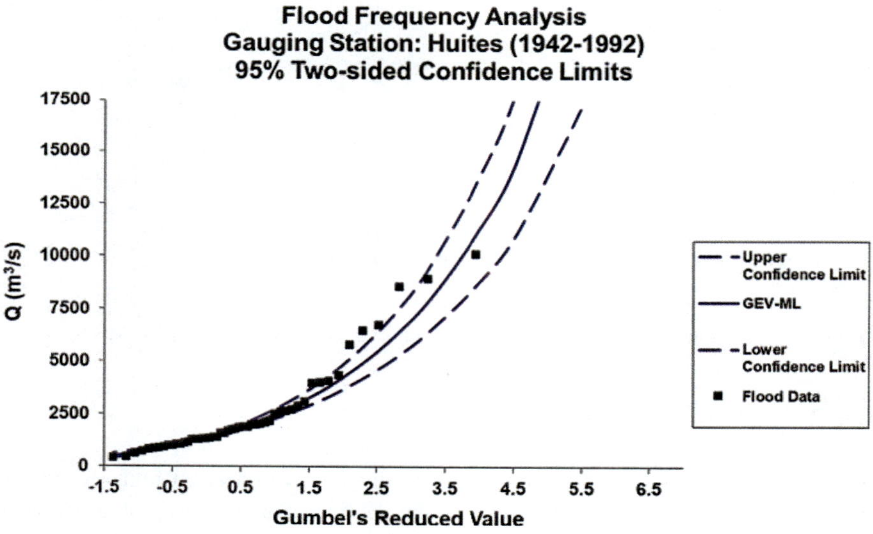

Fig. 10.12 Empirical and ML theoretical frequency curves and confidence limits of flood sample of Huites, Mexico

24 Hr. Annual Maximum Rainfall Frequency Analysis	
Meteorological Station: Chihuahua, Mexico (1960-2005)	
Descriptive Statistics	
Mean	42.0454
Sample Standard Error	1.9784
Median	40.2400
Mode	#N/D
Standard Deviation	13.4184
Sample Variance	180.0535
Kurtosis	-0.7252
Skewness	0.3591
Range	54.3000
Minimum	17.1200
Maximum	71.4200
Sum	1934.0900
Count	46.0000
Largest(1)	71.4200
Smallest(1)	17.1200
Mean Confidence Level(95.0%)	3.9848

Fig. 10.13 Descriptive statistics for the 24 h annual maximum rainfall data at Chihuahua, Mexico

24 Hr. Annual Maximum Rainfall Frequency Analysis
Meteorological Station: Chihuahua, Mexico (1960-2005)
General Extreme Value (GVE) Distribution
Method: MOM

Estimation of Parameters and Goodness of Fit Tests								
$1+\beta =$	1.4065	$1+2\beta =$	1.8130	$1+3\beta =$	2.2195	$\psi(1+\beta) =$	-0.0547	
$\Gamma(1+\beta) =$	0.8890	$\Gamma(1+2\beta) =$	0.9952	$\Gamma(1+3\beta) =$	1.6756	$\psi(1+2\beta) =$	0.2945	
$\psi(1+3\beta) =$	0.5554	$\mu_z =$	-0.8890	$\sigma_z =$	0.4526	$\lambda_1 =$	-3.5083	
$\lambda_2 =$	6.4043	A =	68.3997	B =	29.6449	Coefficient of Skewness =		0.3591
Location Parameter =		38.7548	cm	Scale Parameter =		12.0506	cm	
Shape Parameter =		0.4065	SEF =	2.94	MARD =	4.46		

Return Period Tr (years)	F(x)	y	Frequency Factor K_T	$\partial K_T/\partial\gamma$	Squared Root Stand. Error S_T (cm)	Two-sided 95% Lower Limit (cm)	Design Values P_T (cm)	Two-sided 95% Upper Limit (cm)
2	0.5	0.6931	0.0606	-0.3449	0.8141	41	43	44
5	0.8	0.2231	0.7633	-0.6370	1.3392	50	52	55
10	0.9	0.1054	1.0790	-0.5713	1.2552	54	57	59
20	0.95	0.0513	1.3035	-0.4177	1.0143	58	60	62
50	0.98	0.0202	1.5118	-0.1657	0.5810	61	62	63
100	0.99	0.0101	1.6235	0.0292	0.0684	64	64	64
Two-sided Confidence Limits:		$z_\alpha =$	1.96					

Fig. 10.14 MOM estimation of the parameters, standard errors, quantiles, and confidence limits for the 24 h annual maximum rainfall data at Chihuahua, Mexico

10.9.2.2 ML Method

By using an Excel® spreadsheet, the results contained in Figs. 10.15 and 10.16, were obtained for the ML method applied to the GEV distribution for 24 h annual maximum rainfall at meteorological station Chihuahua, Mexico.

	A	B	C	D	E	F	G
1	24 Hr. Annual Maximum Rainfall Frequency Analysis						
2	Meteorological Station: Chihuahua, Mexico (1960-2005)						
3	General Extreme Value (GVE) Distribution						
4	Method: ML						
5							
6	Estimation of Parameters and Goodness of Fit Tests						
7	Initial Moment Estimators:			(1-β)−1 =	-0.0684	(1-2β)−1 =	-0.1368
8	Location Parameter =	38.7548	cm	1-β =	0.9316	1-2β =	0.8632
9	Scale Parameter =	12.0506	cm	Γ(1−β) =	1.0444	Γ(1−2β) =	1.1001
10	Shape Parameter =	0.4065		ψ(1−β) =	-0.6957		
11	Iterative Procedure:			p =	0.9548	q =	-13.9287
12	i-th iteration				(i+1)-th iteration		
13	Location Parameter =	36.2886		0.9548	-0.2662	-0.4358	
14	Scale Parameter =	11.7754		-0.2662	1.8841	-0.4579	
15	Shape Parameter =	0.0684		-0.4358	-0.4579	0.4696	
16	P =	-0.4774			Determinant =	0.1472	
17	Q =	5.8304E-01		a =	1.7551	f =	6.4036
18	R =	1.2356E+00		b =	4.5846	g =	3.7566
19	ABS(∂LL/ ∂x₀) =	4.9513E-02		c =	11.7352	h =	2.2040
20	ABS(∂LL/ ∂α) =	1.3121E-01		δ(x₀) =	5.8615E+00		
21	ABS(∂LL/ ∂β) =	4.5250E+00		δ(α) =	3.3283E+00		
22	Final MV Estimators:			δ(β) =	9.4704E-01		
23	Location Parameter =	36.2886		Location Parameter=		42.1501	
24	Scale Parameter =	11.7754		Scale Parameter =		15.1037	
25	Shape Parameter =	0.0684		ShapeParameter =		1.0154	
26	Log-Likelihood Function=	-245.9534					
27	SEF =	1.47	MARD =	2.28	AIC =	498	

Fig. 10.15 ML estimation of the parameters for the 24 h annual maximum rainfall Data at Chihuahua, Mexico

	A	B	C	D	E	F	G	H
29	Variance-Covariance Matrix							
30	var(x₀) =	13.8196		cov(x₀,α) =	6.6435			
31	var(α) =	5.2904		cov(x₀,β) =	1.6392			
32	var(β) =	0.2551		cov(α,β) =	0.9616			
33								
34	Design Values and Confidence Limits							
35–38	Return Period Tr (years)	F(x)	∂x_T/∂α	∂x_T/∂β	Squared Root Stand. Error S_T (cm)	Two-sided 95% Lower Limit (cm)	Design Values P_T (cm)	Two-sided 95% Upper Limit (cm)
39	2	0.5	0.3620	-123.8476	58.6520	-74	41	156
40	5	0.8	1.4256	-478.4608	235.7219	-409	53	515
41	10	0.9	2.0857	-691.2107	50.4139	-38	61	160
42	20	0.95	2.6879	-880.0615	64.3091	-58	68	194
43	50	0.98	3.4246	-1103.9541	80.7763	-82	77	235
44	100	0.99	3.9467	-1257.5980	92.0722	-98	83	263
45	500	0.998	5.0619	-1570.7717	115.0824	-130	96	321
46	1000	0.999	5.5048	-1689.0638	123.7678	-141	101	344
47	Two-sided Confidence Limits:	z_α =	1.96					

Fig. 10.16 ML estimation of standard errors, Quantiles and confidence limits for the 24 h annual maximum rainfall data at Chihuahua, Mexico

10.9.2.3 PWM1 Method

By using an Excel® spreadsheet, the results contained in Fig. 10.17, were obtained by the PWM1 method applied to the GEV distribution for 24 h annual maximum rainfall data at meteorological station Chihuahua, Mexico.

	A	B	C	D	E	F	G	H	I	J	K	L
1	24 Hr. Annual Maximum Rainfall Frequency Analysis											
2	Meteorological Station: Chihuahua, Mexico (1960-2005)											
3	General Extreme Value (GVE) Distribution											
4	Method: PWM1											
5												
6	Probability Weighted Moments											
7	Hosking et al (1985) :											
8	b_0 =	42.0454	cm	b_1 =	24.8837	cm	b_2 =	17.9994	cm	c_H =	0.0151	
9												
10	Estimation of Parameters and Goodness of Fit Tests											
11	1+β =	1.1195	Γ(1+β) =	0.9439								
12	Location Parameter =		36.2716	cm	Scale Parameter =		12.2982	cm	Shape Parameter =		0.1195	
13	SEF =	1.33	MARD =	2.14								
14												
15	Variance-Covariance Matrix											
16	w_{11} =	1.2584	w_{12} =	0.2239	w_{13} =	0.3107	w_{22} =	0.6712	w_{23} =	0.2675	w_{33} =	0.5371
17	var(x_0) =	2.6650	cov(x_0,α) =	0.4742	cov(x_0,β)	0.0535	var(α) =	1.4214	cov(α,β) =	0.0037	var(β) =	0.0075
18												
19	Design Values and Confidence Limits											
20–23	Return Period Tr (years)	F(x)	$\partial x_T/\partial\alpha$	$\partial x_T/\partial\beta$	Squared Root Stand. Error S_T (cm)	Two-sided 95% Lower Limit (cm)	Design Values P_T (cm)	Two-sided 95% Upper Limit (cm)				
24	2	0.5	0.3586	-0.8023	0.2085	40	41	41				
25	5	0.8	1.3732	-12.2873	0.2995	53	53	54				
26	10	0.9	1.9732	-26.0824	0.4134	60	61	61				
27	20	0.95	2.5003	-42.9702	0.5651	66	67	68				
28	50	0.98	3.1186	-69.0328	0.8120	73	75	76				
29	100	0.99	3.5389	-90.9785	1.0257	78	80	82				
30	Two-sided Confidence Limits:		z_α =	1.96								

Fig. 10.17 PWM1 estimation of the parameters, standard errors, Quantiles, and confidence limits for the 24 h annual maximum rainfall data at Chihuahua, Mexico

	A	B	C	D	E	F	G	H	I	J
1	MAXIMUM SIGNIFICANT WAVE HEIGHT FREQUENCY ANALYSIS									
2	Log-Pearson Type III (LPIII) Distribution									
3	Method: MOM2									
4	Estimation of Parameters and Goodness of Fit Tests									
5	Skewness Coef. (Gamma Correction) =				2.4194					
6	μ_1' =	2498.96		μ_2' =	11082513	μ_3' =	72649498673			
7	B =	2.6813		Test α =	-0.26158		F(α) =	4.0954E-07		
8	Location Parameter =		10.8569							
9	Scale Parameter =		-0.26158	<	0	not suitable for maximum significant wave				
10	Shape Parameter =		13.05379			height frequency analysis				
11	SEF =	#¡REF!		MARD =	#¡REF!					
12	Design Values and Confidence Limits									
13–17	Return Period Tr (years)	F(x)	Standard Normal Variate z_T	$\partial K_T/\partial\gamma$	Frequency Factor K_T	Logartithmic Squared root Stand. Error Ln(S_T)	Squared root Stand. Error S_T (m^3/s)	Two-sided 95% Lower Limit (m^3/s)	Design Values Q_T (m^3/s)	Two-sided 95% Upper Limit (m^3/s)
18	2	0.5	0.0000	-0.0927	-0.3341	0.0000	0.0000	1861	1861	1861
19	5	0.8	0.8416	-0.1950	0.5276	0.0000	0.0000	799	799	799
20	10	0.9	1.2816	-0.1467	1.2332	0.0000	0.0000	488	488	488
21	20	0.95	1.6449	-0.0326	1.9741	0.0000	0.0000	316	316	316
22	50	0.98	2.0537	0.1932	2.9998	0.0000	0.0000	187	187	187
23	100	0.99	2.3263	0.4097	3.8070	0.0000	0.0000	130	130	130
24	Two-sided Confidence Limits:		z_α =	1.96						

Fig. 10.18 PWM2 estimation of the parameters, standard errors, Quantiles, and confidence limits for the 24 h annual maximum rainfall data at Chihuahua, Mexico

10.9.2.4 PWM2 Method

By using an Excel® spreadsheet, the results contained in Fig. 10.18, were obtained by the PWM2 method applied to the GEV distribution for 24 h annual maximum rainfall data at meteorological station Chihuahua, Mexico.

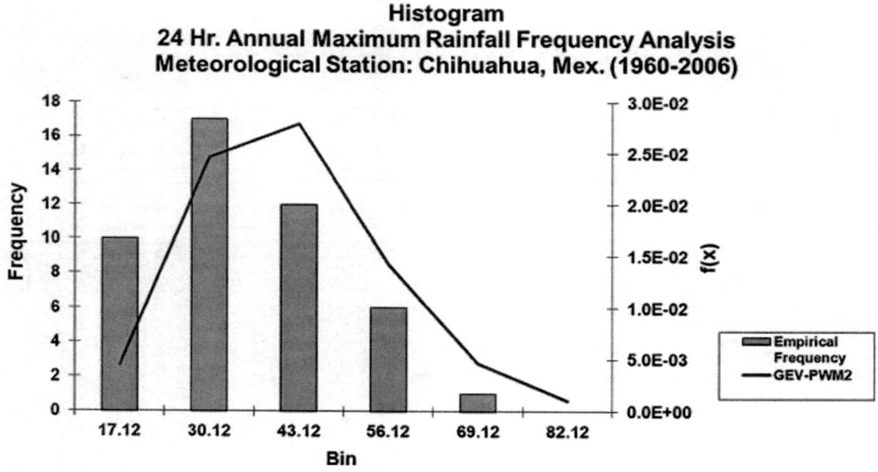

Fig. 10.19 Histogram and theoretical GEV density for 24 h annual maximum rainfall data at Chihuahua, Mexico

In Fig. 10.19 a comparison is made between the histogram and GEV theoretical density for 24 h annual maximum rainfall data at meteorological station Chihuahua, Mexico.

A graphical comparison between the empirical and theoretical frequency curves from the results provided by the three methods shown before is shown in Fig. 10.20.

A graphical depiction of the empirical and theoretical frequency curves and the confidence limits of the best fit provided, in this case that of ML, is shown in Fig. 10.21.

10.9.3 Maximum Significant Wave Height Frequency Analysis

By using the 24 h annual rainfall data from maximum significant wave height data, Castillo (1988), the following descriptive statistics were obtained and are shown in Fig. 10.22.

10.9.3.1 MOM Method

By using an Excel® spreadsheet, the results contained in Fig. 10.23, were obtained for the MOM method applied to the GEV distribution for maximum significant wave height data.

10.9.3.2 ML Method

By using an Excel® spreadsheet, the results contained in Figs. 10.24 and 10.25, were obtained for the ML method applied to the GEV distribution for maximum significant wave height data. But no convergence was attained when applying this method to this set of data.

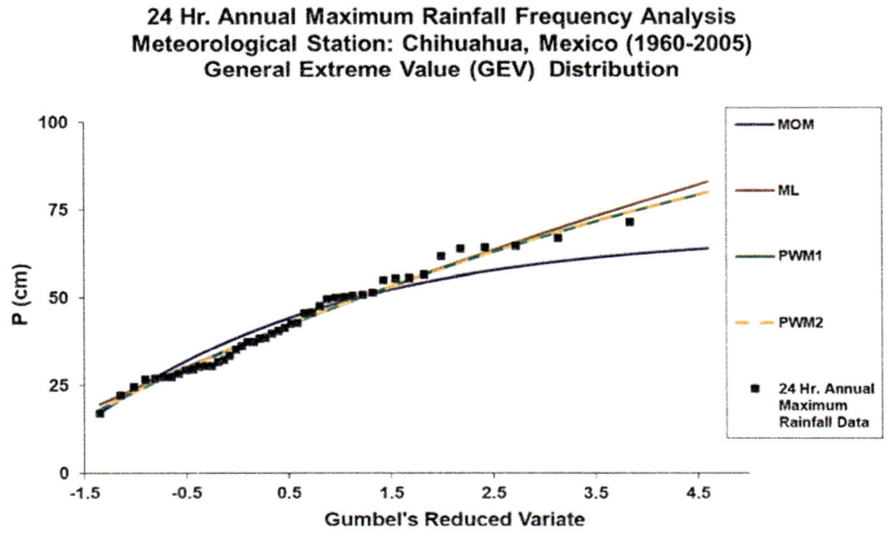

Fig. 10.20 Empirical and theoretical GEV distributions for 24 h annual maximum rainfall data at Chihuahua, Mexico

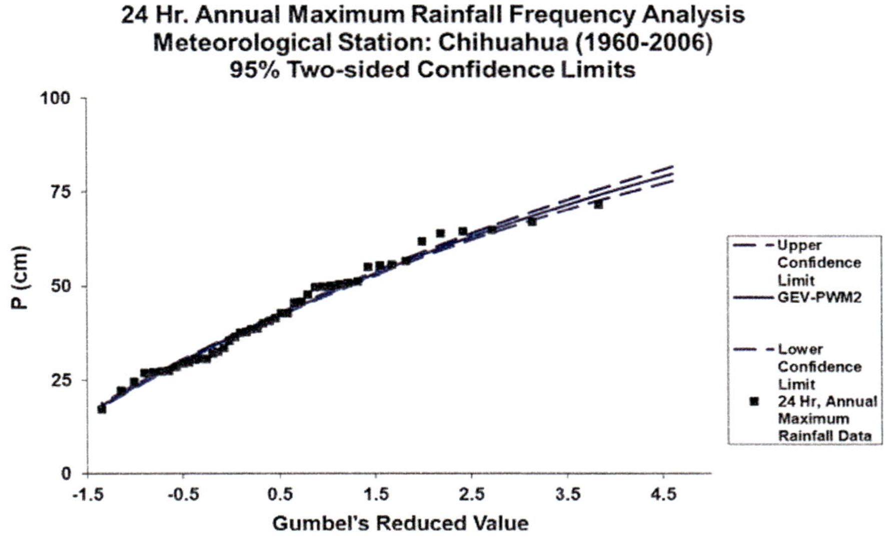

Fig. 10.21 Empirical and PWM2 GEV theoretical frequency curves and confidence limits for 24 h annual maximum rainfall data at Chihuahua, Mexico

	A	B	C
1	Maximum Significant Wave Height Frequency Analysis		
2			
3	*Descriptive Statistics*		
4	Mean	4.4247	
5	Sample Standard Error	0.3053	
6	Median	4.1666	
7	Mode	3.5509	
8	Standard Deviation	2.1585	
9	Sample Variance	4.6591	
10	Kurtosis	1.0740	
11	Skewness	0.9053	
12	Range	10.4485	
13	Minimum	0.8870	
14	Maximum	11.3355	
15	Sum	221.2326	
16	Count	50.0000	
17	Largest(1)	11.3355	
18	Smallest(1)	0.8870	
19	Mean Confidence Level(95.0%)	0.6134	

Fig. 10.22 Descriptive statistics for maximum significant wave height data

	A	B	C	D	E	F	G	H	I
1	MAXIMUM SIGNIFICANT WAVE HEIGHT FREQUENCY ANALYSIS								
2	General Extreme Value (GEV) Distribution								
3	Method: MOM								
4									
5	Estimation of Parameters and Goodness of Fit Tests								
6	$1+\beta=$	1.0371	$1+2\beta=$	1.0741	$1+3\beta=$	1.1112	$\psi(1+\beta)=$	-0.5178	
7	$\Gamma(1+\beta)=$	0.9799	$\Gamma(1+2\beta)=$	0.9623	$\Gamma(1+3\beta)=$	0.9469	$\psi(1+2\beta)=$	-0.4615	
8	$\psi(1+3\beta)=$	-0.4078	$\mu_z=$	-0.9799	$\sigma_z=$	0.0456	$\lambda_1=$	5.76E+01	
9	$\lambda_2=$	2.36E+02	A =	50.8586	B =	47.3861	Coefficient of Skewness =		0.9053
10	Location Parameter =		3.4725	m	Scale Parameter =		1.7566	m	
11	Shape Parameter =		0.037	SEF =	0.26	MARD =	2.28		
12									
13	Design Values and Confidence Limits								
14–17	Return Period Tr (years)	F(x)	y	Frequency Factor K_T	$\partial K_T/\partial\gamma$	Squared Root Stand. Error S_T (m)	Two-sided 95% Lower Limit (m)	Design Values H_T (m)	Two-sided 95% Upper Limit (m)
18	2	0.5	0.6931	-0.1449	-0.0003	0.2854	4	4	5
19	5	0.8	0.2231	0.7462	0.0017	0.4029	5	6	7
20	10	0.9	0.1054	1.3159	0.0030	0.4744	6	7	8
21	20	0.95	0.0513	1.8477	0.0042	0.5397	7	8	9
22	50	0.98	0.0202	2.5153	0.0057	0.6206	9	10	11
23	100	0.99	0.0101	3.0007	0.0068	0.6790	10	11	12
24	Two-sided Confidence Limits:		$z_\alpha=$	1.96					

Fig. 10.23 MOM estimation of the parameters, standard errors, quantiles, and confidence limits for maximum significant wave height data

10.9.3.3 PWM1 Method

By using an Excel® spreadsheet, the results contained in Fig. 10.26, were obtained by the PWM1 Method applied to the GEV distribution for maximum significant wave height data. This set of data can not comply with the restrictions set for this method on the shape parameter of the GEV distribution.

	A	B	C	D	E	F	G
1	MAXIMUM SIGNIFICANT WAVE HEIGHT FREQUENCY ANALYSIS						
2	General Extreme Value (GEV) Distribution						
3	Method: ML						
4							
5	Estimation of Parameters and Goodness of Fit Tests						
6	Initial Moment Estimators:			$(1-\beta)-1 =$	-0.1190	$(1-2\beta)-1 =$	-0.2380
7	Location Parameter =	3.5001	m	$1-\beta =$	0.8810	$1-2\beta =$	0.7620
8	Scale Parameter =	1.9676	m	$\Gamma(1-\beta) =$	1.0844	$\Gamma(1-2\beta) =$	1.2091
9	Shape Parameter =	0.1190		$\psi(1-\beta) =$	-0.7920		
10	Iterative Procedure:			p =	0.9384	q =	-7.8295
11	i-th iteration				(i+1)-th iteration		
12	Location Parameter =	3.5001		0.9384	-0.1422	-0.4742	
13	Scale Parameter =	1.9676		-0.1422	1.9576	-0.6245	
14	Shape Parameter =	0.1190		-0.4742	-0.6245	1.0371	
15	P =	3.6569			Determinant =	0.9938	
16	Q =	1.8621E+00		a =	0.7530	f =	1.0236
17	R =	1.6430E+01		b =	1.6505	g =	0.6576
18	$ABS(\partial LL/\partial x_0) =$	9.4638E-01		c =	1.8282	h =	0.4464
19	$ABS(\partial LL/\partial \alpha) =$	2.3571E+01		$\delta(x_0) =$	9.2008E+00		
20	$ABS(\partial LL/\partial \beta) =$	2.5166E+02		$\delta(\alpha) =$	5.1050E+00		
21	Final MV Estimators:			$\delta(\beta) =$	8.5538E+00		
22	Location Parameter =	3.5001	m	Location Parameter=		12.7009	
23	Scale Parameter =	1.9676	m	Scale Parameter =		7.0726	
24	Shape Parameter =	0.1190		ShapeParameter =		8.6728	
25	Log-Likelihood Function=	-96.8977					
26	SEF =	0.30	MARD =	3.46	AIC =	200	

Fig. 10.24 ML estimation of the parameters for maximum significant wave height data

	A	B	C	D	E	F	G	H
33	Design Values and Confidence Limits							
34–37	Return Period Tr (years)	F(x)	$\partial x_T/\partial \alpha$	$\partial x_T/\partial \beta$	Squared Root Stand. Error S_T (m)	Two-sided 95% Lower Limit (m)	Design Values H_T (m)	Two-sided 95% Upper Limit (m)
38	2	0.5	0.3586	-11.7313	2.0074	0	4	8
39	5	0.8	1.3737	-43.4599	7.9249	-9	6	22
40	10	0.9	1.9742	-61.1100	11.2193	-15	7	29
41	20	0.95	2.5020	-75.8580	13.9693	-19	8	36
42	50	0.98	3.1214	-92.1625	17.0052	-24	10	43
43	100	0.99	3.5425	-102.5703	18.9400	-27	10	48
44	Two-sided Confidence Limits:	$z_\alpha =$	1.96					

Fig. 10.25 ML estimation of standard errors, quantiles and confidence limits for maximum significant wave height data

10.9.3.4 PWM2 Method

By using an Excel® spreadsheet, the results contained in Fig. 10.27, were obtained by the PWM2 Method applied to the GEV distribution for maximum significant wave height data.

In Fig. 10.28 a comparison is made between the histogram and GEV theoretical density for maximum significant wave height data.

A graphical comparison between the empirical and theoretical frequency curves from the results provided by the three methods shown before is shown in Fig. 10.29.

A graphical depiction of the empirical and theoretical frequency curves and the confidence limits of the best fit provided, in this case that of ML, is shown in Fig. 10.30.

	A	B	C	D	E	F	G	H	I	J	K	L
1	MAXIMUM SIGNIFICANT WAVE HEIGHT FREQUENCY ANALYSIS											
2	General Extreme Value (GEV) Distribution											
3	Method: PWM1											
4												
5	Probability Weighted Moments											
6	b_0 =	4.4247	m	b_1 =	2.0944	m	b_2 =	1.4089	m	c_H =	0.5605	
7												
8	Estimation of Parameters and Goodness of Fit Tests											
9	$1+\beta$ =	6.3334	$\Gamma(1+\beta)$ =	6067.5681								
10	Location Parameter =		4.1828	m	Scale Parameter =		-0.0002	m	Shape Parameter =		5.3334	
11	SEF =	2.05	MARD =	30.65			Shape parameter is outside of the range $-0.5 < \beta < 0.5$					
12												
13	Variance-Covariance Matrix											
14	w_{11} =	4532.4937	w_{12} =	6233.6713	w_{13} =	11145.5541	w_{22} =	8632.2965	w_{23} =	1.53E+04	w_{33} =	2.75E+04
15	var(x_0) =	0.0000	cov(x_0,α) =	0.0000	cov(x_0,β)	-0.0474	var(α) =	0.0000	cov(α,β) =	305.4459	var(β) =	549.9716
16												
17	Design Values and Confidence Limits											
18–21	Return Period Tr (years)	F(x)	$\partial x_T/\partial \alpha$	$\partial x_T/\partial \beta$	Squared Root Stand. Error S_T (m)	Two-sided 95% Lower Limit (m)	Design Values H_T (m)	Two-sided 95% Upper Limit (m)				
22	2	0.5	0.1609	0.0000	0.0021	4	4	4				
23	5	0.8	0.1874	0.0000	0.0029	4	4	4				
24	10	0.9	0.1875	0.0000	0.0029	4	4	4				
25	20	0.95	0.1875	0.0000	0.0029	4	4	4				
26	50	0.98	0.1875	0.0000	0.0029	4	4	4				
27	100	0.99	0.1875	0.0000	0.0029	4	4	4				
28	Two-sided Confidence Limits:		z_α =	1.96								

Fig. 10.26 PWM1 estimation of the parameters, standard errors, quantiles, and confidence limits for maximum significant wave height data

	A	B	C	D	E	F	G	H	I	J	K	L
1	MAXIMUM SIGNIFICANT WAVE HEIGHT FREQUENCY ANALYSIS											
2	General Extreme Value (GEV) Distribution											
3	Method: PWM2											
4	Probability Weighted Moments											
5	$M_{(0)}$ =	4.4247	m	$M_{(1)}$ =	1.5944	m	$M_{(2)}$ =	0.9089	m	C_{RV} =	-1.3739	
6	Estimation of Parameters and Goodness of Fit Tests											
7	Test β =	0.1190	$1+\beta$ =	1.1190	$\Gamma(1+\beta)$ =	0.9441	F(β) =	6.7297E-03				
8	Location Parameter =		3.5001	m	Scale Parameter =		1.9676	m	Shape Parameter =		0.1190	
9	SEF =	0.30	MARD =	3.46								
10	Variance-Covariance Matrix											
11	w_{11} =	1.2585	w_{12} =	0.2246	w_{13} =	0.3107	w_{22} =	0.6714	w_{23} =	0.2674	w_{33} =	0.5373
12	var(x_0) =	0.0974	cov(x_0,α) =	0.0174	cov(x_0,β)	0.0122	var(α) =	0.0520	cov(α,β) =	0.0053	var(β) =	0.0107
13												
14	Design Values and Confidence Limits											
15–18	Return Period Tr (years)	F(x)	$\partial x_T/\partial \alpha$	$\partial x_T/\partial \beta$	Squared Root Stand. Error S_T (m)	Two-sided 95% Lower Limit (m)	Design Values H_T (m)	Two-sided 95% Upper Limit (m)				
19	2	0.5	0.3586	-0.1284	0.0476	4	4	4				
20	5	0.8	1.3737	-1.9668	0.0667	6	6	6				
21	10	0.9	1.9742	-4.1759	0.0905	7	7	8				
22	20	0.95	2.5020	-6.8813	0.1232	8	8	9				
23	50	0.98	3.1214	-11.0581	0.1773	9	10	10				
24	100	0.99	3.5425	-14.5766	0.2247	10	10	11				
25	Two-sided Confidence Limits:		z_α =	1.96								

Fig. 10.27 PWM2 estimation of the parameters, standard errors, quantiles, and confidence limits for maximum significant wave height data

10.9.4 Annual Maximum Wind Speed Frequency Analysis

By using the annual maximum wind speed data, Castillo (1988), the following descriptive statistics were obtained and are shown in Fig. 10.31.

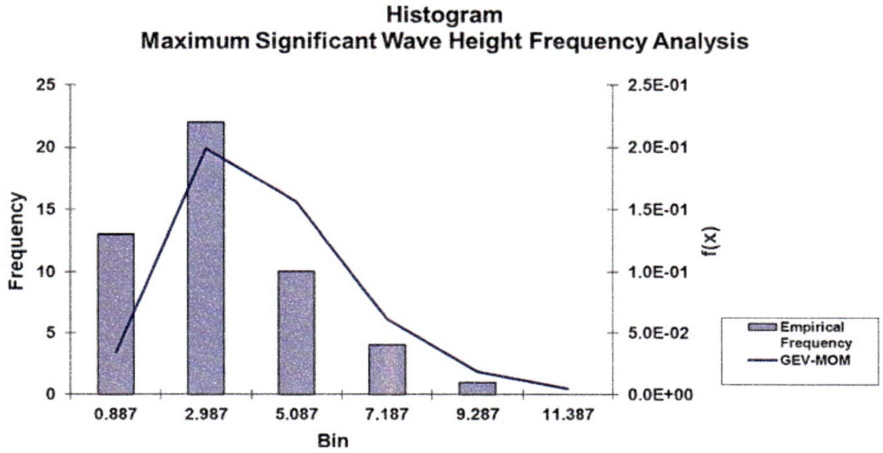

Fig. 10.28 Histogram and theoretical GEV density for maximum significant wave height data

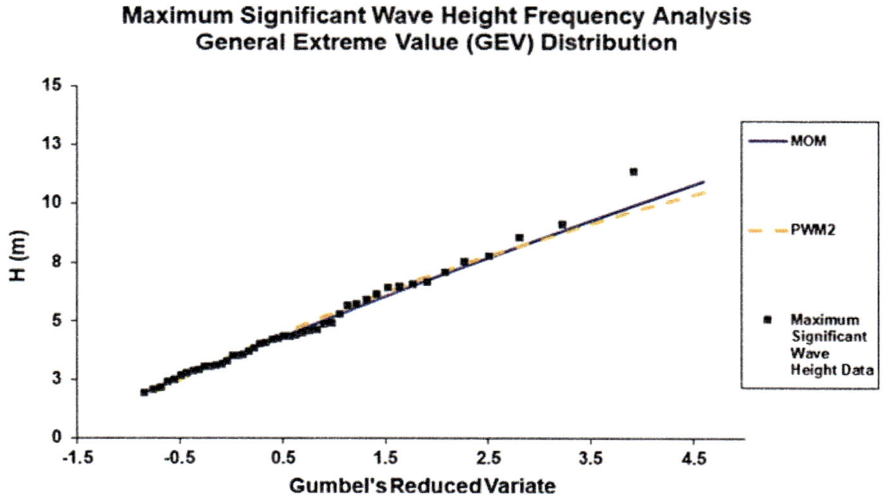

Fig. 10.29 Empirical and theoretical GEV distributions for maximum significant wave height data

10.9.4.1 MOM Method

By using an Excel® spreadsheet, the results contained in Fig. 10.32, were obtained for the MOM method applied to the GEV distribution for annual maximum wind speed data.

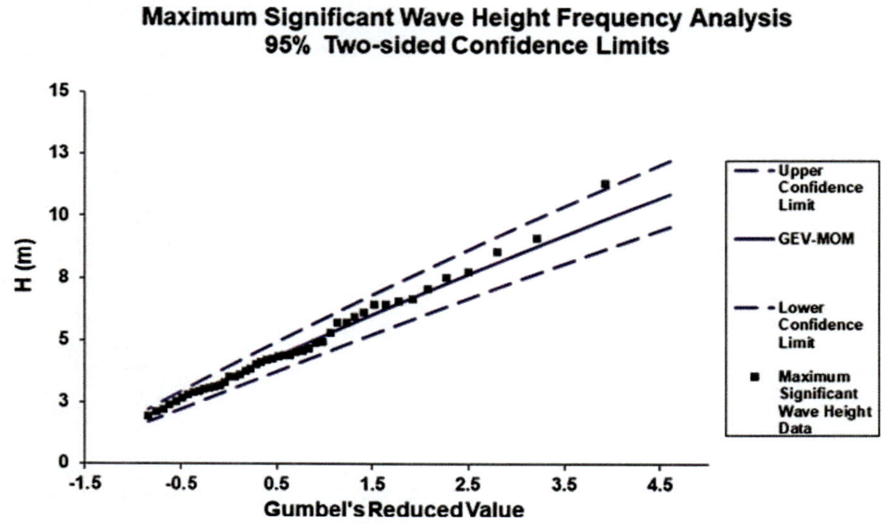

Fig. 10.30 Empirical and MOM GEV theoretical frequency curves and confidence limits for maximum significant wave height data

	A	B	C
1	ANNUAL MAXIMUM WIND SPEED FREQUENCY ANALYSIS		
2			
3	*Descriptive Statistics*		
4	Mean	54.7964	
5	Sample Standard Error	2.9320	
6	Median	47.9965	
7	Mode	#N/D	
8	Standard Deviation	20.7327	
9	Sample Variance	429.8439	
10	Kurtosis	12.5699	
11	Skewness	3.0813	
12	Range	121.5116	
13	Minimum	36.4278	
14	Maximum	157.9394	
15	Sum	2739.8217	
16	Count	50.0000	
17	Largest(1)	157.9394	
18	Smallest(1)	36.4278	
19	Mean Confidence Level(95.0%)	5.8922	

Fig. 10.31 Descriptive statistics for annual maximum wind speed

10.9.4.2 ML Method

By using an Excel® spreadsheet, the results contained in Figs. 10.33 and 10.34, were obtained for the ML method applied to the GEV distribution for annual maximum wind speed data. But no convergence was attained when applying this method to this set of data.

	A	B	C	D	E	F	G	H	I
1	ANNUAL MAXIMUM WIND SPEED FREQUENCY ANALYSIS								
2	General Extreme Value (GEV) Distribution								
3	Method: MOM								
4									
5	Estimation of Parameters and Goodness of Fit Tests								
6	$1+\beta$ =	0.8165	$1+2\beta$ =	0.6331	$1+3\beta$ =	0.4496	$\psi(1+\beta)$ =	-0.9276	
7	$\Gamma(1+\beta)$ =	1.1461	$\Gamma(1+2\beta)$ =	1.4179	$\Gamma(1+3\beta)$ =	1.9698	$\psi(1+2\beta)$ =	-1.4255	
8	$\psi(1+3\beta)$ =	-2.2357	μ_z =	-1.1461	σ_z =	0.3229	λ_1 =	164.9758	
9	λ_2 =	447.8101	A =	-18.7914	B =	64.2047	Coefficient of Skewness =		3.0813
10	Location Parameter =		45.4133	m/s	Scale Parameter =		11.7788	m/s	
11	Shape Parameter =		-0.1835	SEF =	7.17	MARD =	3.44		
12									
13	Design Values and Confidence Limits								
14–17	Return Period Tr (years)	F(x)	y	Frequency Factor K_T	$\partial K_T/\partial\gamma$	Squared Root Stand. Error S_T (m/s)	Two-sided 95% Lower Limit (m/s)	Design Values V_T (m/s)	Two-sided 95% Upper Limit (m/s)
18	2	0.5	0.6931	-0.2372	0.0046	1.5181	47	50	53
19	5	0.8	0.2231	0.5284	-0.0070	5.1671	56	66	76
20	10	0.9	0.1054	1.1302	-0.0172	7.3282	64	78	93
21	20	0.95	0.0513	1.7909	-0.0293	9.4562	73	92	110
22	50	0.98	0.0202	2.7863	-0.0490	12.3241	88	113	137
23	100	0.99	0.0101	3.6522	-0.0674	14.5443	102	131	159
24	Two-sided Confidence Limits:		z_α =	1.96					

Fig. 10.32 MOM estimation of the parameters, standard errors, quantiles, and confidence limits for annual maximum wind speed

	A	B	C	D	E	F	G
1	ANNUAL MAXIMUM WIND SPEED FREQUENCY ANALYSIS						
2	General Extreme Value (GEV) Distribution						
3	Method: ML						
4							
5	Estimation of Parameters and Goodness of Fit Tests						
6	Initial Moment Estimators:			$(1-\beta)-1$ =	-0.0005	$(1-2\beta)-1$ =	-0.0010
7	Location Parameter =	48.6631	m/s	$1-\beta$ =	0.9995	$1-2\beta$ =	0.9990
8	Scale Parameter =	14.0507	m/s	$\Gamma(1-\beta)$ =	1.0003	$\Gamma(1-2\beta)$ =	1.0006
9	Shape Parameter =	0.0005		$\psi(1-\beta)$ =	-0.5780		
10	Iterative Procedure:			p =	0.9996	q =	-1999.1555
11	i-th iteration				(i+1)-th iteration		
12	Location Parameter =	48.6631		0.9996	-0.4217	-0.3570	
13	Scale Parameter =	14.0507		-0.4217	1.8223	5.0692	
14	Shape Parameter =	0.0005		-0.3570	5.0692	-14822.8463	
15	P =	7.5353			Determinant =	-24389.1526	
16	Q =	-3.5280E+00		a =	0.6075	f =	0.0001
17	R =	1.8562E+01		b =	1.1086	g =	0.0002
18	ABS($\partial LL/\partial x_0$) =	2.5109E-01		c =	-0.0001	h =	0.2564
19	ABS($\partial LL/\partial\alpha$) =	5.7040E+02		$\delta(x_0)$ =	-3.0228E+02		
20	ABS($\partial LL/\partial\beta$) =	1.5992E+07		$\delta(\alpha)$ =	-4.6206E+02		
21	Final MV Estimators:			$\delta(\beta)$ =	-2.1588E+01		
22	Location Parameter =	48.6631	m/s	Location Parameter=		-253.6184	
23	Scale Parameter =	14.0507	m/s	Scale Parameter =		-448.0095	
24	Shape Parameter =	0.0005		ShapeParameter =		-21.5877	
25	Log-Likelihood Function=	-199.6209					
26	SEF =	9.38	MARD =	7.12	AIC =	405	

Fig. 10.33 ML estimation of the parameters for the annual maximum wind speed

10.9.4.3 PWM1 Method

By using an Excel® spreadsheet, the results contained in Fig. 10.35, were obtained by the PWM1 method applied to the GEV distribution for annual maximum wind speed data.

	A	B	C	D	E	F	G	H
28		Variance-Covariance Matrix						
29	$var(x_0)$ =	4.3772		$cov(x_0,\alpha)$ =	1.0122			
30	$var(\alpha)$ =	2.3987		$cov(x_0,\beta)$ =	0.0000			
31	$var(\beta)$ =	0.0000		$cov(\alpha,\beta)$ =	0.0001			
32								
33			Design Values and Confidence Limits					
34–37	Return Period Tr (years)	F(x)	$\partial x_T/\partial\alpha$	$\partial x_T/\partial\beta$	Squared Root Stand. Error S_T (m/s)	Two-sided 95% Lower Limit (m/s)	Design Values V_T (m/s)	Two-sided 95% Upper Limit (m/s)
38	2	0.5	0.3665	-20596.2214	#¡NUM!	#¡NUM!	54	#¡NUM!
39	5	0.8	1.4994	-84253.4256	#¡NUM!	#¡NUM!	70	#¡NUM!
40	10	0.9	2.2491	-126370.2658	#¡NUM!	#¡NUM!	80	#¡NUM!
41	20	0.95	2.9680	-166747.4780	#¡NUM!	#¡NUM!	90	#¡NUM!
42	50	0.98	3.8981	-218979.2706	#¡NUM!	#¡NUM!	103	#¡NUM!
43	100	0.99	4.5949	-258095.7244	#¡NUM!	#¡NUM!	113	#¡NUM!
44	Two-sided Confidence Limits:	z_α =	1.96					

Fig. 10.34 ML estimation of standard errors, quantiles and confidence limits for annual maximum wind speed

	A	B	C	D	E	F	G	H	I	J	K	L
1	ANNUAL MAXIMUM WIND SPEED FREQUENCY ANALYSIS											
2	General Extreme Value (GVE) Distribution											
3	Method: PWM1											
4												
5	Probability Weighted Moments											
6	b_0 =	54.7964	m/s	b_1 =	23.5702	m/s	b_2 =	15.2788	m/s	c_H =	0.2235	
7												
8	Estimation of Parameters and Goodness of Fit Tests											
9	$1+\beta$ =	2.9043	$\Gamma(1+\beta)$ =	9.7311								
10	Location Parameter =		45.4232	m/s	Scale Parameter =		-2.0444	m/s	Shape Parameter =		1.9043	
11	SEF =	13.00	MARD =	36.21			Shape parameter is outside of the range $-0.5 < \beta < 0.5$					
12												
13	Variance-Covariance Matrix											
14	w_{11} =	63.1515	w_{12} =	83.1115	w_{13} =	153.2092	w_{22} =	120.9026	w_{23} =	2.09E+02	w_{33} =	3.79E+02
15	$var(x_0)$ =	5.2787	$cov(x_0,\alpha)$ =	6.9471	$cov(x_0,\beta)$	-6.2643	$var(\alpha)$ =	10.1059	$cov(\alpha,\beta)$ =	4.1708	$var(\beta)$ =	7.5797
16												

	A	B	C	D	E	F	G	H
17			Design Values and Confidence Limits					
18–21	Return Period Tr (years)	F(x)	$\partial x_T/\partial\alpha$	$\partial x_T/\partial\beta$	Squared Root Stand. Error S_T (m/s)	Two-sided 95% Lower Limit (m/s)	Design Values V_T (m/s)	Two-sided 95% Upper Limit (m/s)
22	2	0.5	0.2638	0.0874	0.4173	44	45	46
23	5	0.8	0.4949	0.4388	0.4796	43	44	45
24	10	0.9	0.5179	0.5227	0.4865	43	44	45
25	20	0.95	0.5233	0.5506	0.4882	43	44	45
26	50	0.98	0.5248	0.5609	0.4887	43	44	45
27	100	0.99	0.5250	0.5629	0.4887	43	44	45
28	Two-sided Confidence Limits:	z_α =	1.96					

Fig. 10.35 PWM1 estimation of the parameters, standard errors, quantiles, and confidence limits for annual maximum wind speed

This set of data can not comply with the restrictions set for this method on the shape parameter of the GEV distribution.

10.9.4.4 PWM2 Method

By using an Excel® spreadsheet, the results contained in Fig. 10.36, were obtained by the PWM2 method applied to the GEV distribution annual maximum wind speed data.

In Fig. 10.37 a comparison is made between the histogram and GEV theoretical density for annual maximum wind speed data.

	A	B	C	D	E	F	G	H	I	J	K	L
1	ANNUAL MAXIMUM WIND SPEED FREQUENCY ANALYSIS											
2	General Extreme Value (GVE) Distribution											
3	Method: PWM2											
4												
5	Probability Weighted Moments											
6	$M_{(0)}$ =	54.7964	m/s	$M_{(1)}$ =	22.5305	m/s	$M_{(2)}$ =	14.2391	m/s	C_{RV} =	-1.2408	
7												
8	Estimation of Parameters and Goodness of Fit Tests											
9	Test β =	0.0005	1+β =	1.0005	Γ(1+β) =	0.9998	F(β) =	6.0467E-05				
10	Location Parameter =		48.6631	m	Scale Parameter =		14.0507	m	Shape Parameter =		0.0005	
11	SEF =	9.38	MARD =	7.12								
12												
13	Variance-Covariance Matrix											
14	w_{11} =	1.2686	w_{12} =	0.3698	w_{13} =	0.2993	w_{22} =	0.7387	w_{23} =	0.2249	w_{33} =	0.5632
15	var(x_0) =	5.0089	cov(x_0,α) =	1.4601	cov(x_0,β)	0.0841	var(α) =	2.9166	cov(α,β) =	0.0045	var(β) =	0.0113
16												
17	Design Values and Confidence Limits											
18					Squared Root	Two-sided	Design	Two-sided				
19					Stand. Error	95%	Values	95%				
20	Return Period				S_T	Lower Limit	V_T	Upper Limit				
21	Tr (years)	F(x)	$\partial x_T/\partial\alpha$	$\partial x_T/\partial\beta$	(m/s)	(m/s)	(m/s)	(m/s)				
22	2	0.5	0.3665	-0.9436	0.3555	53	54	55				
23	5	0.8	1.4994	-15.7979	0.5655	69	70	71				
24	10	0.9	2.2491	-35.5508	0.8274	79	80	82				
25	20	0.95	2.9680	-61.9168	1.1942	88	90	93				
26	50	0.98	3.8981	-106.8229	1.8407	100	103	107				
27	100	0.99	4.5949	-148.4384	2.4501	108	113	118				
28	Two-sided Confidence Limits:		z_α =	1.96								

Fig. 10.36 PWM2 estimation of the parameters, standard errors, quantiles, and confidence limits for annual maximum wind speed

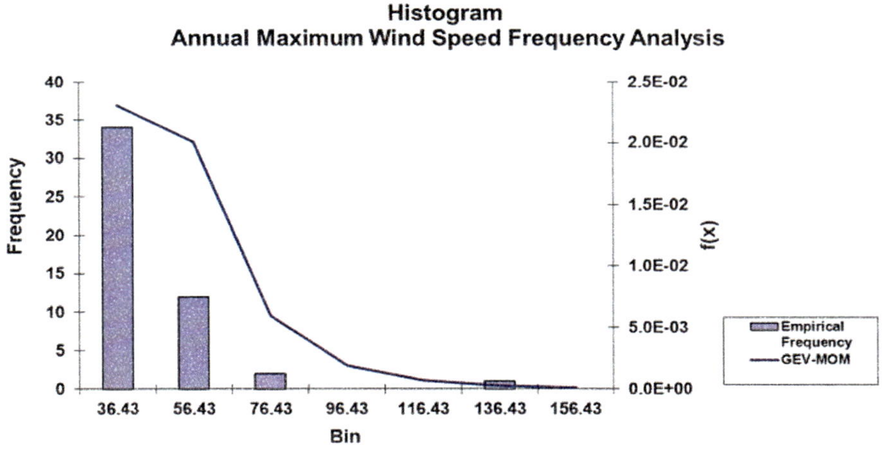

Fig. 10.37 Histogram and theoretical GEV density for annual maximum wind speed

A graphical comparison between the empirical and MOM and PWM2 theoretical frequency curves from the results provided by the three methods shown before is shown in Fig. 10.38.

A graphical depiction of the empirical and theoretical frequency curves and the confidence limits of the best fit provided, in this case that of MOM, is shown in Fig. 10.39.

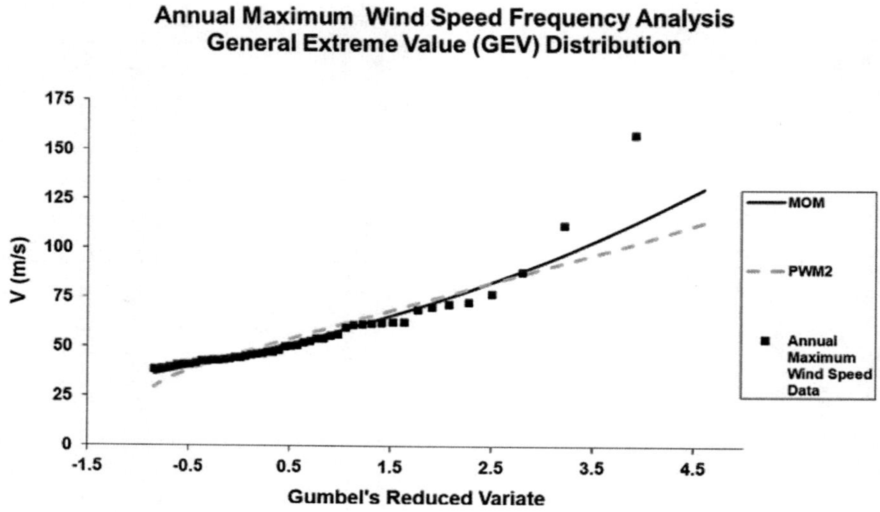

Fig. 10.38 Empirical and theoretical GEV distributions for annual maximum wind speed

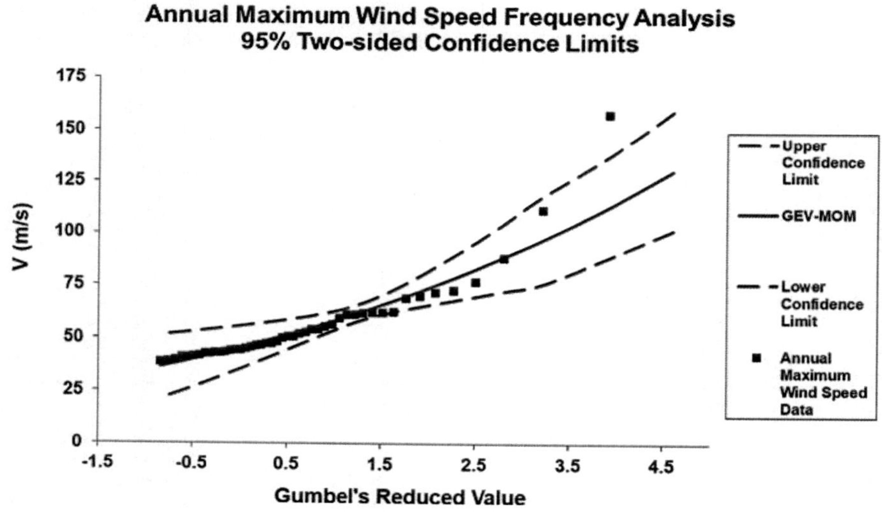

Fig. 10.39 Empirical and PWM2 GEV theoretical frequency curves and confidence limits for annual maximum wind speed

11 Log-Normal Distribution with Three Parameters for the Minima

If you want to fly, you have to give up the things that weigh you down. T. Morrison

11.1 Introduction

If $y = \text{Ln}(x - x_0)$ is normally distributed, then the random variable x has a Log-Normal with 3 parameters distribution.

Chow (1954, 1959 and 1964), Sangal and Singh (1970), Singh and Sinclair (1972) and Yevjevich (1972) have demonstrated the applicability of Log-Normal distributions with 2 and 3 parameters to solve hydrologic problems. Matalas (1963) analyzed the suitability of using several probability distributions functions for modeling low flows. He found that the Pearson type III and the extreme value type III for the minima, when using the method of moments, were the best models among those tested in such study, namely the three-parameters Log-Normal, Pearson types III and V, and the extreme value type III for the minima. Caruso (2000) analyzed 21 rivers in the Otago Region, South Island, New Zealand by performing low flow frequency analysis using Three-parameters Log-Normal, extreme value type I, extreme value type III, and general extreme value distributions, and he reported that log-Normal distribution gave overestimated values while extreme value I gave underestimated values when compared with the results produced by the extreme value type III distribution. In some cases, the general extreme value distribution was the best option. Kroll and Vogel (2002) analyzed the low flows of 1505 gauged river sites and they suggested to use either the Three-parameters Log-Normal or the Pearson type III distributions for describing low flow statistics the United States of America. Durrans and Tomic (2007) tested five methods

J. A. Raynal Villaseñor, *Frequency Analyses of Natural Extreme Events*,
Earth and Environmental Sciences Library,
https://doi.org/10.1007/978-3-030-86390-6_11

of estimation of parameters for the 3-parameters Log-Normal distribution: distributional truncation, MLE treatment of censored data, partial probability weighted moments, LL-moments, and expected moments. They found almost no difference in the results among such methods for estimation of parameters.

The LN3 distribution for the minima (LN3M), has the same mathematical form that its maxima counterpart, being the only difference the use of the exceedance probability as its probability distribution function in the former and the non-exceedance probability as its probability distribution function in the latter.

11.2 Chapter Objectives

After reading this chapter, you will know how to:

(a) Recognize the distribution and density functions of the LN3M distribution
(b) Estimate the parameters of the LN3M distribution
(c) Compute the quantiles and confidence limits of the LN3M distribution
(d) Make a graphic display of your data and the LN3M distribution
(e) Develop an application of all the above using Excel® spreadsheets.

11.3 Probability Distribution and Density Functions

The probability distribution function of the Three-parameters Log-Normal (LN3M) distribution is:

$$F(x) = \int_{-\infty}^{x} \frac{1}{(x - x_0)\sigma_y\sqrt{2\pi}} \exp\left[-\frac{\left(Ln(x - x_0) - \mu_y\right)^2}{2\sigma_y^2}\right] dx \tag{11.1}$$

where x_0, σ_y and μ_y are the location, scale, and shape parameters. The domain of variable x in this distribution is $(x - x_0) < x < \infty$. The sample values must comply with the following restriction: $(x - x_0) > 0$, the involvement of natural logarithms in all the formulation of such probability distribution puts such a condition in its use.

The probability density function for the LN3M distribution for the minima is:

$$f(x) = \frac{1}{(x - x_0)\sigma_y\sqrt{2\pi}} \exp\left[-\frac{\left(Ln(x - x_0) - \mu_y\right)^2}{2\sigma_y^2}\right] \tag{11.2}$$

11.4 Estimation of the Parameters

11.4.1 MOM Method

The MOM estimators for the parameters of the LN3M distribution for the minima are:

$$\hat{\mu}_y = \frac{1}{N}\sum_{i=1}^{N} Ln(x_i - \hat{x}_0) \tag{11.3}$$

$$\hat{\sigma}_y = \left[\frac{\sum_{i=1}^{N}\left(Ln(x_i - x_0) - \hat{\mu}_y\right)^2}{N}\right]^{1/2} \tag{11.4}$$

$$\hat{x}_0 = \mu_x\left(1 - \frac{CV_x}{CV_z}\right) \tag{11.5}$$

where

$$CV_x = \frac{\sigma_x}{\mu_x} \tag{11.6}$$

$$CV_z = \frac{\left(1 - w^{2/3}\right)}{w^{1/3}} \tag{11.7}$$

and:

$$w = \frac{1}{2}\left[-\hat{\gamma}_x + \left(\hat{\gamma}_x^2 + 4\right)^{1/2}\right] \tag{11.8}$$

where $\hat{\gamma}_x$ is the skewness coefficient of the x's, as it was defined in Chap. 2.

11.4.1.1 Example of Application of Estimation of the Parameters of the LN3M Distribution for the Minima Using the MOM Method

Find the MOM estimators for the parameters of the LN3M distribution for the minima for the one-day low flow sample data of gauging station Villalba, Mexico (1939–1991), contained in Appendix A.

The following statistics can be obtained:

$$\hat{\mu} = \frac{1}{N}\sum_{i=1}^{N} x_i = 0.3306\,\mathrm{m}^3/\mathrm{s}$$

$$\hat{\sigma} = \left[\frac{1}{N}\sum_{i=1}^{N}(x_i - \hat{\mu})^2\right]^{1/2} = 0.1465\,\mathrm{m}^3/\mathrm{s}$$

$$\hat{\gamma} = \hat{\gamma}_x = \frac{N}{(N-1)(N-2)\hat{\sigma}^3} \sum_{i=1}^{N} (x_i - \hat{\mu})^3 = 0.5853$$

$$CV = CV_x = \frac{\hat{\sigma}}{\hat{\mu}} = \frac{0.1465}{0.3306} = 0.4431$$

then:

$$w = \frac{1}{2}\left[-\hat{\gamma}_x + \left(\hat{\gamma}_x^2 + 4\right)^{1/2}\right] = \frac{1}{2}\left[-0.5853 + \left((0.5853)^2 + 4\right)^{\frac{1}{2}}\right] = 0.7493$$

$$CV_z = \frac{\left(1 - w^{2/3}\right)}{w^{1/3}} = \frac{\left(1 - (0.7493)^{0.6666}\right)}{(0.7493)^{0.3333}} = 0.1927$$

so:

$$\hat{x}_0 = \mu_x\left(1 - \frac{CV_x}{CV_z}\right) = 0.3306\left(1 - \frac{0.4431}{0.1927}\right) = -0.4294\,\mathrm{m}^3/\mathrm{s}$$

and:

$$\hat{\mu}_y = \frac{1}{N}\sum_{i=1}^{N} Ln\left(x_i - \hat{x}_0\right) = \frac{1}{51}\sum_{i=1}^{51} Ln(x_i - (-0.4294) = -0 - 2870$$

$$\hat{\sigma}_y = \left[\frac{\sum_{i=1}^{N}\left(Ln(x_i - x_0) - \hat{\mu}_y\right)^2}{N}\right]^{1/2}$$

$$= \left[\frac{\sum_{i=1}^{51}[Ln(x_i - (-(-0.4294) - (-0.2870]^2}{51}\right]^{1/2} = 0.1945$$

Finally, the MOM estimators of the parameters of the LN3M distribution for the minima for the one-day low flow sample data of gauging station Villalba, Mexico, are:

location parameter:

$$\hat{x}_0 = -0.4294\,\mathrm{m}^3/\mathrm{s}$$

scale parameter:

$$\hat{\sigma}_y = 0.1945$$

and:

shape parameter:

$$\hat{\mu}_y = -0.2870$$

11.4.2 ML Method

The likelihood function of the LN3M distribution for the minima is:

$$L\left(x,\mu_y,\sigma_y,x_0\right)=\prod_{i=1}^{N}\left\{\frac{1}{(x-x_0)\sigma_y\sqrt{2\pi}}\exp\left[-\frac{\left(Ln(x-x_0)-\mu_y\right)^2}{2\sigma_y^2}\right]\right\} \quad (11.9)$$

which simplifies to:

$$L\left(x,\mu_y,\sigma_y,x_0\right)=\left(\frac{1}{\sigma_y\sqrt{2\pi}}\right)^N\prod_{i=1}^{N}\left(\frac{1}{(x_i-x_0)}\right)\exp\left[-\sum_{i=1}^{N}\frac{\left(Ln(x_i-x_0)-\mu_y\right)^2}{2\sigma_y^2}\right] \quad (11.10)$$

The Log-Likelihood function of the LN3M distribution for the minima is, Kite (1988):

$$LL\left(x,\mu_y,\sigma_y,x_0\right)=-N\left(Ln\left(\sigma_y\right)+\frac{1}{2}Ln(2\pi)\right)$$
$$-\sum_{i=1}^{N}Ln(x_i-x_0)-\frac{1}{\left(2\sigma_y^2\right)}\sum_{i=1}^{N}\left(Ln(x_i-x_0)-\mu_y\right)^2 \quad (11.11)$$

by taking the partial derivatives of Eq. (11.11) with respect to the parameters μ_y,σ_y, and x_0 and then solving the system of equations:

$$\frac{\partial LL}{\partial\mu_y}=\frac{1}{\sigma_y^2}\sum_{i=1}^{N}\left[Ln(x_i-x_0)-\mu_y\right]=0 \quad (11.12)$$

$$\frac{\partial LL}{\partial\sigma_y}=-\frac{N}{\sigma_y}+\frac{1}{\sigma_y^3}\sum_{i=1}^{N}\left[Ln(x_i-x_0)-\mu_y\right]^2=0 \quad (11.13)$$

$$\frac{\partial LL}{\partial x_0}=\left(\mu_y-\sigma_y^2\right)\sum_{i=1}^{N}\frac{1}{(x_i-x_0)}-\sum_{i=1}^{N}\frac{Ln(x_i-x_0)}{(x_i-x_0)}=0 \quad (11.14)$$

Finally, the ML estimators of the parameters of the LN3M distribution for the minima can be estimated as:

$$\hat{\mu}_y=\frac{1}{N}\sum_{i=1}^{N}Ln\left(x_i-\hat{x}_0\right) \quad (11.15)$$

$$\hat{\sigma}_y=\left[\frac{\sum_{i=1}^{N}\left(Ln\left(x_i-\hat{x}_0\right)-\hat{\mu}_y\right)^2}{N}\right]^{1/2} \quad (11.16)$$

for the case of the location parameter x_0, the following equation must be solved by a root searching procedure, the method of bisection is recommended:

$$f(\hat{x}_0) = \left[\sum_{i=1}^{N} \frac{1}{(x_i - \hat{x}_0)}\right] \left\{ \frac{1}{N} \sum_{i=1}^{N} Ln^2(x_i - \hat{x}_0) - \left[\frac{1}{N} \sum_{i=1}^{N} Ln(x_i - \hat{x}_0)\right]^2 - \frac{1}{N} \sum_{i=1}^{N} Ln(x_i - \hat{x}_0) \right\} + \sum_{i=1}^{N} \frac{Ln(x_i - \hat{x}_0)}{(x_i - \hat{x}_0)} = 0 \quad (11.17)$$

11.4.2.1 Example of Application of Estimation of the Parameters of the LN3M Distribution for the Minima Using the ML Method

Find the ML estimators for the parameters of the LN3M distribution for the minima for the one-day low flow sample data of gauging station Villalba, Mexico (1939–1991), contained in Appendix A.

Assuming an initial value, which is less than the least value in the one-day low flow sample data, the location parameter will be:

$$\hat{x}_0 = 400\,\mathrm{m}^3/\mathrm{s}$$

then:

$$F(\hat{x}_0) = -1.1939 \times 10^{-1}$$

By using an adequate method of search of roots for Eq. (11.17), as the method of bisection, the following solution is found:

$$\hat{x}_0 = 243.4908\,\mathrm{m}^3/\mathrm{s}$$

and:

$$F(\hat{x}_0) = 4.7018 \times 10^{-7}$$

The two other parameters are found as:

$$\hat{\mu}_y = \frac{1}{N} \sum_{i=1}^{N} Ln(x_i - \hat{x}_0) = \frac{1}{51} \sum_{i=1}^{51} Ln(x_i - (-0.4294) = -0.2870$$

$$\hat{\sigma}_y = \left[\frac{\sum_{i=1}^{N}\left(Ln(x_i - x_0) - \hat{\mu}_y\right)^2}{N}\right]^{1/2}$$

$$= \left[\frac{\sum_{i=1}^{51}[Ln(x_i - (-0.4294) - (-0.2870)]^2}{51}\right]^{1/2} = 0.1945$$

Finally, the MOM estimators of the parameters of the LN3M distribution for the minima for the one-day low flow sample data of gauging station Villalba, Mexico, are:

location parameter:

$$\hat{x}_0 = -26.0294\,\mathrm{m}^3/\mathrm{s}$$

scale parameter:

$$\hat{\sigma}_y = 0.0057$$

and:

shape parameter:

$$\hat{\mu}_y = 3.2520$$

11.5 Estimation of Quantiles for the LN3M Distribution for the Minima

The quantiles for the LN3M distribution for the minima can be obtained by inverting the Normal distribution to obtain the quantiles given by the following expression, Abramowitz and Stegun (1965):

$$z_T = w - \left(\frac{c_0 + c_1 w + c_2 w^2}{1 + d_1 w + d_2 w^2 + d_3 w^3}\right) \tag{11.18}$$

where z_T is the standard normal deviate that corresponds to a certain return period, T_r, which is linked to a certain value of the standard normal distribution function:

$$F(x) = 1 - \frac{1}{T_r} \tag{11.19}$$

The coefficients c and d are as follows, Abramowitz and Stegun (1965):

$$c_0 = 2.515517;\ c_1 = 0.802853;\ c_2 = 0.010328$$

Table 11.1 Estimation of MOM quantiles for the LN3M distribution for gauging station Villalba, Mexico

Tr (years)	MOM
	Q_T (m^3/s)
2	0.32
5	0.21
10	0.16
20	0.12
50	0.07
100	0.05

$$d_1 = 1.432788;\ d_2 = 0.189269;\ d_3 = 0.001308$$

and, Abramowitz and Stegun (1965):

$$w = \sqrt{Ln\left(\frac{1}{1 - F(x)}\right)^2} \tag{11.20}$$

where F(x) is the normal standard distribution function. If F(x) < 0.5, then put 1 – F(x) instead F(x) in Eq. (11.17) and change the sign in the resulting values of z_T.

Now, the quantiles for the LN3M distribution for the minima are obtained as:

$$Q_T = \hat{x}_0 + \exp\left(\hat{\mu}_y + z_T \hat{\sigma}_y\right) \tag{11.21}$$

11.5.1 Examples of Estimation of MOM Quantiles for the LN3M Distribution

Find the MOM estimators of the quantiles 2, 5, 10, 20, 50, and 100 years of return period, for the LN3M distribution for the minima for the one-day low flow sample data of gauging station Villalba, Mexico (1939–1991), contained in Appendix A.

The estimation of MOM quantiles for the LN3M distribution for the minima is made by using the parameters estimated in the preceding section and then inserted the parameters into Eq. (11.21). Table 11.1 summarizes these results.

11.6 Goodness of Fit Test

The standard error of fit (SEF) for the LN3M distribution has the following form:

$$SEF = \left[\frac{\sum_{i=1}^{N} (xi - yi)^2}{(N - 3)}\right]^{1/2} \tag{11.22}$$

while the mean absolute relative deviation (MARD) remains the same as it was defined in Chap. 2:

$$MARD = \frac{100}{N}\sum_{i=1}^{N}\left|\frac{(x_i - y_i)}{x_i}\right| \tag{11.23}$$

where x_i are the sample historical values, y_i are the distribution function values, corresponding to the same return periods of the historical values, N is the sample size.

11.6.1 Examples of Application of the SEF and MARD to the MOM and ML Estimators of the Parameters of the LN3M Distribution

Find the values of the SEF and MARD of the MOM and ML estimators of the parameters of the LN3M distribution for the minima for the one-day low flow sample data of gauging station Villalba, Mexico (1939–1991), contained in Appendix A.

The values of SEF and MARD for the MOM and ML estimators of the parameters of the LN3M distribution for the minima for the one-day low flow sample data of gauging station Villalba, Mexico, have been obtained through the application of Eqs. (11.22) and (11.23) using the parameters obtained in previous example. The results are as follows:

For the MOM method:

$$SEF = \left[\frac{\sum_{i=1}^{N}(xi - yi)^2}{(N-3)}\right]^{1/2} = \left[\frac{754754.9619}{(51-3)}\right]^{\frac{1}{2}} = 869$$

$$MARD = \frac{100}{N}\sum_{i=1}^{N}\left|\frac{(x_i - y_i)}{x_i}\right| = \frac{(100)(9.8856)}{51} = 19$$

For the ML method:

$$SEF = \left[\frac{\sum_{i=1}^{N}(xi - yi)^2}{(N-3)}\right]^{1/2} = \left[\frac{305250.4119}{(51-3)}\right]^{\frac{1}{2}} = 552$$

$$MARD = \frac{100}{N}\sum_{i=1}^{N}\left|\frac{(x_i - y_i)}{x_i}\right| = \frac{(100)(4.0522)}{51} = 8$$

As a conclusion, it can be said the ML method showed to be the best method, when the LN3M distribution was applied to this set of one-day low flow sample data, given that the values of the SEF and MARD measures resulted to be the least in both cases when compared with the values obtained by the MOM method.

11.7 Estimation of the Confidence Limits for the LN3M Distribution

By assuming that the quantiles are normally distributed, the following form can be used to set the confidence limits on such quantiles:

$$x_l = x_T \pm z_\alpha S_T \tag{11.24}$$

where x_l is the upper or lower confidence limit, depending on the sign in the preceding formula (+) for the upper confidence limit and (−) for the lower confidence limit, z_α is the standard normal variate corresponding to a confidence level α, and S_T is the squared root of the standard error of the estimate. The evaluation procedures of the standard errors of the estimates will be described in the following subsections.

11.8 Estimation of the Standard Errors for the LN3M Distribution

11.8.1 MOM Method

The general form of the MOM estimator of the standard error of the estimate of a three-parameter distribution is, Kite (1988):

$$\begin{aligned} S_T^2 = & \left(\frac{\partial x_T}{\partial m_1'}\right)^2 \mathrm{var}(m_1') + \left(\frac{\partial x_T}{\partial m_2}\right)^2 \mathrm{var}(m_2) \\ & + \left(\frac{\partial x_T}{\partial m_3}\right)^2 \mathrm{var}(m_3) + 2\left(\frac{\partial x_T}{\partial m_1'^{s}}\right)\left(\frac{\partial x_T}{\partial m_2}\right)\mathrm{cov}(m_1', m_2) \\ & + 2\left(\frac{\partial x_T}{\partial m_1'}\right)\left(\frac{\partial x_T}{\partial m_3}\right)\mathrm{cov}(m_1', m_3) \\ & + 2\left(\frac{\partial x_T}{\partial m_2}\right)\left(\frac{\partial x_T}{\partial m_3}\right)\mathrm{cov}(m_2, m_3) \end{aligned} \tag{11.25}$$

and Eq. (11.25) can be simplified in terms of the frequency factor, K_T, as, Kite (1988):

$$S_T^2 = \frac{\mu_2}{N}\left[1 + K_T\hat{\gamma} + \frac{K_T^2}{4}(\hat{\kappa} - 1)\right] \tag{11.26}$$

and given that computing the skewness and kurtosis coefficients for the logarithms will produce values of zero and 3, respectively, also the frequency factor will be equal to z_T in the LN3M distribution for the minima, and the MOM estimator of the standard error of the estimate for the LN3M distribution for the minima is, Kite (1988):

$$S_T^2 = \frac{\sigma_y^2}{N}\left(1 + \frac{z_T^2}{2}\right) \tag{11.26}$$

where z_T is the standard normal variate corresponding to the return period T. To introduce the squared root standard error of the estimate into Eq. (11.25), the following expressions must be used for the positive value associated with the upper confidence limit and the negative value associated with the lower confidence limit, Kite (1988):

$$S_T(+) = x_T(\exp(S_T) - 1) \tag{11.27}$$

$$S_T(-) = -(x_T - x_0)(\exp(-S_T) - 1) \tag{11.28}$$

11.8.1.1 Example of Estimation of MOM Standard Errors and the Two-Sided 95% Confidence Limits for the LN3M Distribution

Find the MOM estimators of the standard errors and the two-sided 95% confidence limits for the LN3M distribution for the minima for the one-day low flow sample data of gauging station Villalba, Mexico (1939–1991), in Appendix A.

The estimation of MOM estimators of the standard errors and the two-sided 95% confidence limits for the LN3M distribution for the minima is made by using the MOM estimators of the parameters and then inserted in the Eqs. (11.26), (4.27) and (11.24), using selected values of the return intervals. Table 11.2 summarizes these results.

Table 11.2 Estimation of MOM standard errors and the two-sided 95% confidence limits for the LN3M distribution for Gauging Station Villalba, Mexico

Tr (years)	Logarithmic S_T (m^3/s)	95% Lower limit (m^3/s)	Q_T (m^3/s)	95% Upper limit (m^3/s)
2	0.0606	0.23	0.32	0.41
5	0.0705	0.12	0.21	0.30
10	0.0818	0.07	0.16	0.25
20	0.0929	0.02	0.12	0.22
50	0.1068	−0.03	0.07	0.19
100	0.1166	−0.06	0.05	0.16

Two-sided Limits: $z_T = 1.96$

11.8.2 ML Method

The general form of the ML estimator of the standard error of the estimate of a three-parameter distribution is, Kite (1988):

$$
\begin{aligned}
S_T^2 = {} & \left(\frac{\partial x_T}{\partial \alpha}\right)^2 \operatorname{var}(\alpha) + \left(\frac{\partial x_T}{\partial \beta}\right)^2 \operatorname{var}(\beta) + \left(\frac{\partial x_T}{\partial \gamma}\right)^2 \operatorname{var}(\gamma) \\
& + 2\left(\frac{\partial x_T}{\partial \alpha}\right)\left(\frac{\partial x_T}{\partial \beta}\right)\operatorname{cov}(\alpha, \beta) + 2\left(\frac{\partial x_T}{\partial \alpha}\right)\left(\frac{\partial x_T}{\partial \gamma}\right)\operatorname{cov}(\alpha, \gamma) \\
& + 2\left(\frac{\partial x_T}{\partial \beta}\right)\left(\frac{\partial x_T}{\partial \gamma}\right)\operatorname{cov}(\beta, \gamma)
\end{aligned}
\tag{11.29}
$$

and for the LN3M distribution for the minima the following relationships hold, Kite (1988):

$$\frac{\partial x_T}{\partial x_0} = 1 \tag{11.30}$$

$$\frac{\partial x_T}{\partial \sigma_y^2} = \frac{z_T \exp(\mu_y + z_T \sigma_y)}{2\sigma_y} \tag{11.31}$$

$$\frac{\partial x_T}{\partial \mu_y} = \exp(\mu_y + z_T \sigma_y) \tag{11.32}$$

So, the ML estimator of the standard error of the estimate for the LN3M distribution parameters can be expressed as, Kite (1988):

$$
\begin{aligned}
S_T^2 = {} & \operatorname{var}(x_0) + \frac{z_T^2 \exp\left[2\left(\mu_y + z_T \sigma_y\right)\right]}{4\sigma_y^2} \operatorname{var}(\sigma_y^2) \\
& + \exp\left[2\left(\mu_y + z_T \sigma_y\right)\right] \operatorname{var}(\mu_y) \\
& + \frac{z_T \exp\left[\left(\mu_y + z_T \sigma_y\right)\right]}{\sigma_y} \operatorname{cov}(x_0, \sigma_y^2) \\
& + 2\exp\left[\left(\mu_y + z_T \sigma_y\right)\right] \operatorname{cov}(x_0, \mu_y) \\
& + \frac{z_T \exp\left[2\left(\mu_y + z_T \sigma_y\right)\right]}{\sigma_y} \operatorname{cov}(\sigma_y^2, \mu_y)
\end{aligned}
\tag{11.33}
$$

where

$$\operatorname{var}(x_0) = \frac{1}{2ND} \tag{11.34}$$

$$\operatorname{var}(\mu_y) = \frac{\sigma_y^2}{ND}\left[\frac{\left(\sigma_y^2 + 1\right)}{2\sigma_y^2} \exp\left[2\left(\sigma_y^2 - \mu_y\right)\right] - \exp\left[\left(\sigma_y^2 - 2\mu_y\right)\right]\right] \tag{11.35}$$

$$\operatorname{var}(\sigma_y^2) = \frac{\sigma_y^2}{ND}\left[\left(\sigma_y^2 + 1\right)\exp\left[2\left(\sigma_y^2 - \mu_y\right)\right] - \exp\left[\left(\sigma_y^2 - 2\mu_y\right)\right]\right] \tag{11.36}$$

$$\operatorname{cov}(x_0, \mu_y) = -\frac{\exp\left(\frac{\sigma_y^2}{2} - \mu_y\right)}{2ND} \tag{11.37}$$

$$\operatorname{cov}(x_0, \sigma_y^2) = \frac{\sigma_y^2 \exp\left(\frac{\sigma_y^2}{2} - \mu_y\right)}{ND} \tag{11.38}$$

$$\operatorname{cov}(\mu_y, \sigma_y^2) = -\frac{\sigma_y^2 \exp\left(\sigma_y^2 - 2\mu_y\right)}{ND} \tag{11.39}$$

and:

$$D = \frac{\left(\sigma_y^2 + 1\right)}{2\sigma_y^2}\exp\left[2\left(\sigma_y^2 - \mu_y\right)\right] - \frac{\left(2\sigma_y^2 + 1\right)}{2\sigma_y^2}\exp\left(\sigma_y^2 - 2\mu_y\right) \tag{11.40}$$

11.8.2.1 Example of Estimation of ML Standard Errors and the Two-Sided 95% Confidence Limits for the LN3M Distribution

Find the ML estimators of the standard errors and the two-sided 95% confidence limits for the LN3M distribution for the minima for the one-day low flow sample data of gauging station Villalba, Mexico (1939–1991), contained in Appendix A. No feasible solution was found for the standard errors of the LN3M distribution.

11.9 Examples of Application for the LN3M Distribution Using Excel® Spreadsheets

11.9.1 One-Day Low Flow Frequency Analysis

By using the one-day low flow data from gauging station Villalba, Mexico (1939–1991) the following descriptive statistics were obtained and are displayed in Fig. 11.1.

11.9.1.1 MOM Method

By using an Excel® spreadsheet, the results contained in Fig. 11.2, were obtained for the MOM method applied to the LN3M distribution for the minima for gauging station Villalba, Mexico.

ONE-DAY LOW FLOW FREQUENCY ANALYSIS

Gauging Station: Villalba, Mexico (1939-1991)

Descriptive Statistics		*Descriptive Statistics of the Logarithms*	
Mean	0.3306	Mean	-1.2129
Standard Error	0.0201	Standard Error	0.0668
Median	0.3280	Median	-1.1147
Mode	0.4850	Mode	-0.7236
Standard Deviation	0.1465	Standard Deviation	0.4860
Sample Variance	0.0214	Sample Variance	0.2362
Kurtosis	0.1906	Kurtosis	-0.0416
Skewness	0.5853	Skewness	-0.5433
Range	0.6250	Range	2.1873
Minimum	0.0790	Minimum	-2.5383
Maximum	0.7040	Maximum	-0.3510
Sum	17.5214	Sum	-64.2851
Count	53.0000	Count	53.0000
Largest(1)	0.7040	Largest(1)	-0.3510
Smallest(1)	0.0790	Smallest(1)	-2.5383
Confidence Level(95.0%)	0.0404	Confidence Level(95.0%)	0.1340

Fig. 11.1 Descriptive statistics for the one-day low flow sample of Villalba, Mexico

ONE-DAY LOW FLOW FREQUENCY ANALYSIS

Gauging Station: Villalba, Mexico (1939-1991)

Three-parameters Log-Normal (LN3M) Distribution for the Minima

Method: MOM

Estimation of Parameters				
w =	0.7493	Skew(Ln(x-x0)) =		0.1209
η_z =	0.1927		CV_x =	0.4430
Location Parameter =		-0.4294	m^3/s	
Scale Parameter =		0.1945		
Shape Parameter =		-0.2870		
SEF =	0.02		MARD =	4.81

Design Values and Confidence Limits								
Tr (years)	F(x)	Standard Normal Variate z_T	Logarithmic Squared Root Stand.Error S_T	Positive Squared Root Stand.Error S_T (m^3/s)	Negative Squared Root Stand.Error S_T (m^3/s)	Two-sided 95% Lower Limit (m^3/s)	Design Values Q_T (m^3/s)	Two-sided 95% Upper Limit (m^3/s)
2	0.5	0.0000	0.0606	0.0469	0.0441	0.23	0.32	0.41
5	0.2	-0.8416	0.0705	0.0465	0.0434	0.12	0.21	0.30
10	0.1	-1.2816	0.0818	0.0498	0.0459	0.07	0.16	0.25
20	0.05	-1.6449	0.0929	0.0531	0.0484	0.02	0.12	0.22
50	0.02	-2.0537	0.1068	0.0567	0.0510	-0.03	0.07	0.19
100	0.01	-2.3263	0.1166	0.0590	0.0525	-0.06	0.05	0.16
Two-sided Confidence Limits:		z_α =	1.96					

Fig. 11.2 MOM estimators for the parameters, standard errors, quantiles, and confidence limits of the LN3M distribution for the one-day low flow sample of Villalba, Mexico

11.9.1.2 ML Method

By using an Excel® spreadsheet, the results contained in Fig. 11.3, were obtained for the ML method applied to the LN3M distribution for the one-day low flow for Villalba, Mexico.

In Fig. 11.4 a comparison is made between the histogram and LN3M theoretical density for the one-day low flow sample of Villalba, Mexico.

ONE-DAY LOW FLOW FREQUENCY ANALYSIS
Gauging Station: Villalba, Mexico (1939-1991)
Three-Parameters Log-Normal (LN3M) Distribution for the Minima
Method: ML

Estimation of Parameters						
Test Location Parameter =	-26.0294			Skew(Ln(x-x0)) =		0.5706
$F(x_0)$ =	9.2448E-08	Log-likelihood function =		27.1660	AIC =	-48.3320
Location Parameter =	-26.0294	m^3/s				
Scale Parameter =	0.0055					
Shape Parameter =	3.2720		SEF =	0.03	MARD =	8
Variance-Covariance Matrix						
$var(x_0)$ =	1.6496E-01	$cov(x_0,\mu_y)$ =	-6.2570E-03	$cov(x_0,\sigma_y^2)$ =	3.7785E-07	
$var(\mu_y)$ =	2.3733E-04	$cov(\mu_y,\sigma_y^2)$ =	1.4332E-08	$var(\sigma_y^2)$ =	8.6553E-13	
D =	5.7189E-02					
Design Values and Confidence Limits						
Return Period		Standard Normal Variate	Squared Root Stand.Error S_T	Two-sided 95% Lower Limit	Design Values Q_T	Two-sided 95% Upper Limit
Tr (years)	F(x)	z_T	(m^3/s)	(m^3/s)	(m^3/s)	(m^3/s)
2	0.5	0.0000	1.6224E-05	0.34	0.34	0.34
5	0.2	-0.8416	#¡NUM!	#¡NUM!	0.21	#¡NUM!
10	0.1	-1.2816	#¡NUM!	#¡NUM!	0.15	#¡NUM!
20	0.05	-1.6449	#¡NUM!	#¡NUM!	0.10	#¡NUM!
50	0.02	-2.0537	#¡NUM!	#¡NUM!	0.04	#¡NUM!
100	0.01	-2.3263	#¡NUM!	#¡NUM!	0.00	#¡NUM!
Two-sided Confidence Limits:		z_α =	1.96			

Fig. 11.3 ML estimators for the parameters, standard errors, quantiles, and confidence limits of the LN3M distribution for the one-day low flow sample of Villalba, Mexico

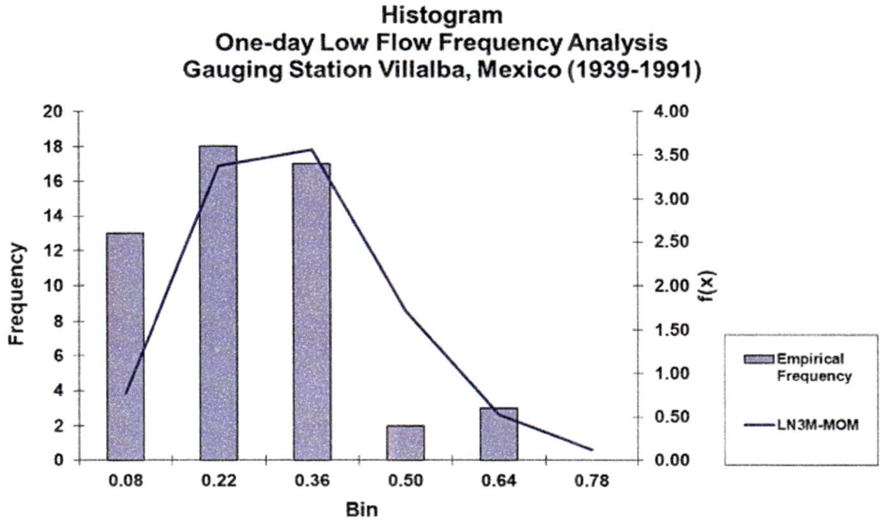

Fig. 11.4 Histogram and LN3M theoretical density of one-day low flow sample of Villalba, Mexico

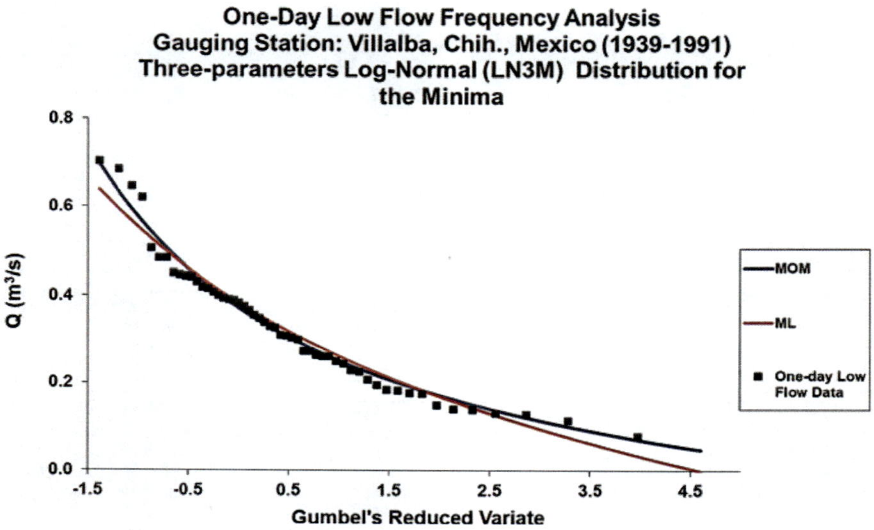

Fig. 11.5 Empirical and MOM and ML theoretical curves of LN3M distribution of one-day low flow sample of Villalba, Mexico

A graphical comparison between the empirical and theoretical frequency curves from the results provided by the three methods shown before is depicted in Fig. 11.5.

A graphical description of the empirical and theoretical frequency curves and the confidence limits of the best fit provided, in this case that is MOM, is shown in Fig. 11.6.

11.9.2 7-day Low Flow Frequency Analysis

By using the 7-day low flow data from gauging station Villalba, Mexico (1939–1986) the following descriptive statistics were obtained and are displayed in Fig. 11.7.

11.9.2.1 MOM Method

By using an Excel® spreadsheet, the results contained in Fig. 11.8, were obtained for the MOM method applied to the LN3M distribution for the minima for gauging station Villalba, Mexico.

11.9.2.2 ML Method

By using an Excel® spreadsheet, the results contained in Fig. 11.9, were obtained for the ML method applied to the LN3M distribution for the 7-day low flow for Villalba, Mexico.

In Fig. 11.10 a comparison is made between the histogram and LN3M theoretical density for the 7-day low flow sample of Villalba, Mexico.

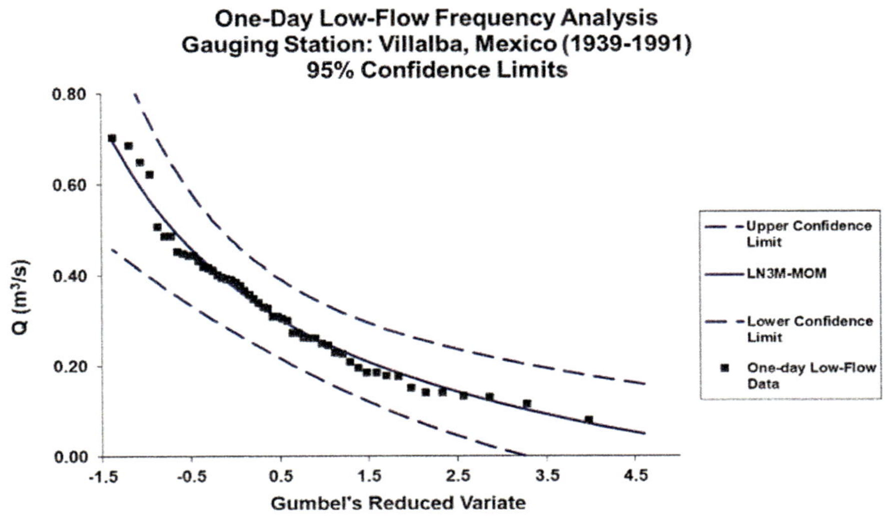

Fig. 11.6 Empirical and MOM LN3M theoretical frequency curves and confidence limits of one-day low flow sample of Villalba, Mexico

	A	B	C	D
1	7-DAY LOW FLOW FREQUENCY ANALYSIS			
2	Gauging Station: Villalba, Mexico (1939-1991)			
3				
4	*Descriptive Statistics*		*Descriptive Statistics of the Logarithms*	
5	Mean	2.8483	Mean	0.9451
6	Standard Error	0.1813	Standard Error	0.0679
7	Median	2.7125	Median	0.9979
8	Mode	#N/D	Mode	#N/D
9	Standard Deviation	1.2561	Standard Deviation	0.4705
10	Sample Variance	1.5779	Sample Variance	0.2214
11	Kurtosis	-0.2429	Kurtosis	-0.2698
12	Skewness	0.5782	Skewness	-0.3803
13	Range	5.2120	Range	2.0789
14	Minimum	0.7450	Minimum	-0.2944
15	Maximum	5.9570	Maximum	1.7846
16	Sum	136.7200	Sum	45.3646
17	Count	48.0000	Count	48.0000
18	Largest(1)	5.9570	Largest(1)	1.7846
19	Smallest(1)	0.7450	Smallest(1)	-0.2944
20	Confidence Level(95.0%)	0.3647	Confidence Level(95.0%)	0.1366

Fig. 11.7 Descriptive statistics for the 7-day low flow sample of Villalba, Mexico

7-DAY LOW FLOW FREQUENCY ANALYSIS
Gauging Station: Villalba, Mexico (1939-1986)
Three-parameters Log-Normal (LN3M) Distribution for the Minima
Method: MOM

Estimation of Parameters

	A	B	C	D	E
7	w =	0.7518	Skew(Ln(x-x0)) =		0.2075
8	n_z =	0.1904		CV_x =	0.4410
9	Location Parameter =		-3.7477	m³/s	
10	Scale Parameter =		0.1859		
11	Shape Parameter =		1.8659		
12	SEF =	0.17		MARD =	4.61

Design Values and Confidence Limits

Tr (years)	F(x)	Standard Normal Variate z_T	Logarithmic Squared Root Stand.Error S_T	Positive Squared Root Stand.Error S_T (m³/s)	Negative Squared Root Stand.Error S_T (m³/s)	Two-sided 95% Lower Limit (m³/s)	Design Values Q_T (m³/s)	Two-sided 95% Upper Limit (m³/s)
2	0.5	0.0000	0.0622	0.4149	0.3899	1.95	2.71	3.53
5	0.2	-0.8416	0.0724	0.4150	0.3860	1.02	1.78	2.59
10	0.1	-1.2816	0.0840	0.4461	0.4102	0.54	1.34	2.22
20	0.05	-1.6449	0.0955	0.4767	0.4333	0.16	1.01	1.95
50	0.02	-2.0537	0.1097	0.5116	0.4584	-0.24	0.66	1.67
100	0.01	-2.3263	0.1198	0.5337	0.4734	-0.48	0.45	1.49
Two-sided Confidence Limits:		z_α =	1.96					

Fig. 11.8 MOM estimators for the parameters, standard errors, quantiles, and confidence limits of the LN3M distribution for the 7-day low flow sample of Villalba, Mexico

7-DAY LOW FLOW FREQUENCY ANALYSIS
Gauging Station: Villalba, Mexico (1939-1986)
Three-Parameters Log-Normal (LN3M) Distribution for the Minima
Method: ML

Estimation of Parameters

	A	B	C	D	E	F	G
7	Test Location Parameter =	-25.0000			Skew(Ln(x-x0)) =		0.4872
8	$F(x_0)$ =	3.5585E-05	Log-likelihood function =		-1616.0112	AIC =	3238.0224
9	Location Parameter =	-25.0000	m³/s				
10	Scale Parameter =	0.1847		No convergence was attained			
11	Shape Parameter =	1.8692		SEF =	21.97	MARD =	925

Variance-Covariance Matrix

	A	B	C	D	E	F
14	$var(x_0)$ =	1.1577E-02	$cov(x_0,\mu_y)$ =	-1.8164E-03	$cov(x_0,\sigma_y^2)$ =	1.2398E-04
15	$var(\mu_y)$ =	2.8549E-04	$cov(\mu_y,\sigma_y^2)$ =	1.9452E-05	$var(\sigma_y^2)$ =	1.3622E-06
16	D =	8.9980E-01				

Design Values and Confidence Limits

Return Period Tr (years)	F(x)	Standard Normal Variate z_T	Squared Root Stand.Error S_T (m³/s)	Two-sided 95% Lower Limit (m³/s)	Design Values Q_T (m³/s)	Two-sided 95% Upper Limit (m³/s)
2	0.5	0.0000	4.9677E-03	-18.53	-18.52	-18.51
5	0.2	-0.8416	#¡NUM!	#¡NUM!	-19.45	#¡NUM!
10	0.1	-1.2816	#¡NUM!	#¡NUM!	-19.88	#¡NUM!
20	0.05	-1.6449	#¡NUM!	#¡NUM!	-20.22	#¡NUM!
50	0.02	-2.0537	#¡NUM!	#¡NUM!	-20.56	#¡NUM!
100	0.01	-2.3263	#¡NUM!	#¡NUM!	-20.78	#¡NUM!
Two-sided Confidence Limits:		z_α =	1.96			

Fig. 11.9 ML estimators for the parameters, standard errors, quantiles, and confidence limits of the LN3M distribution for the 7-day low flow sample of Villalba, Mexico

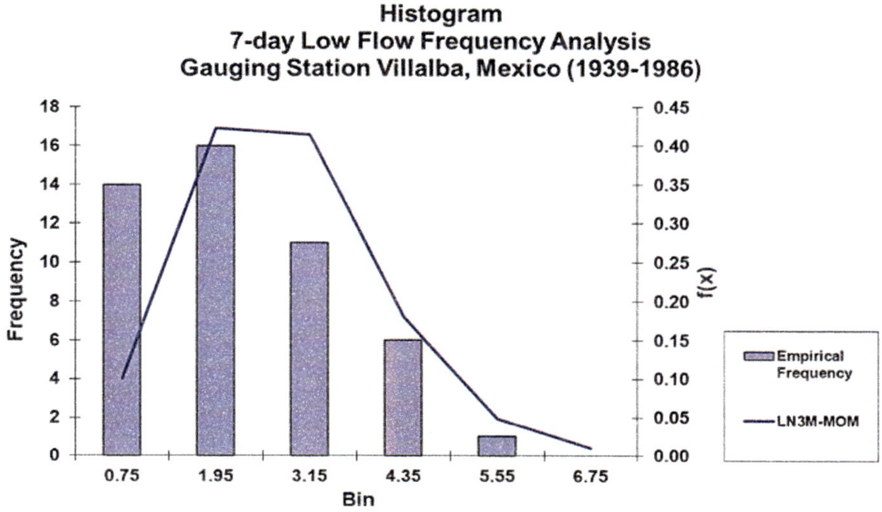

Fig. 11.10 Histogram and LN3M theoretical density of 7-day low flow sample of Villalba, Mexico

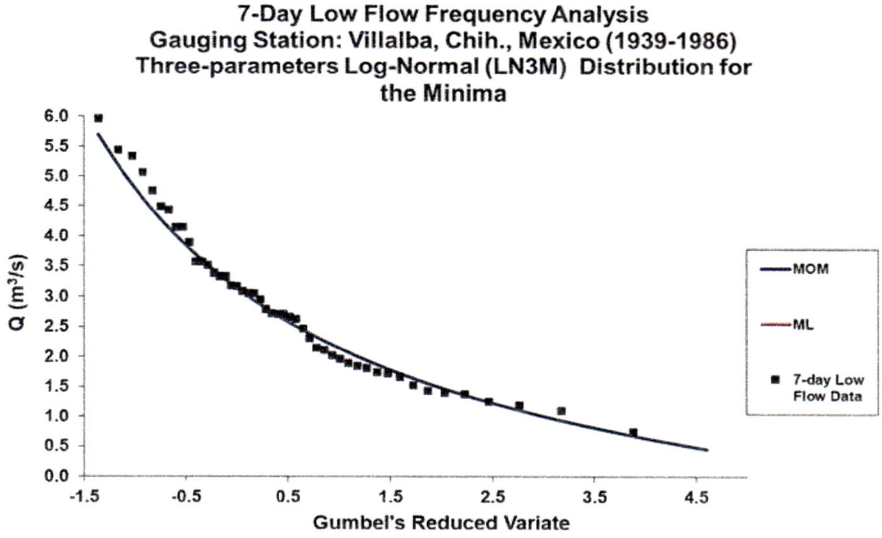

Fig. 11.11 Empirical and MOM and theoretical curves of LN3MOM distribution of 7-day low flow sample of Villalba, Mexico

A graphical comparison between the empirical and theoretical frequency curves from the results provided by the three methods shown before is depicted in Fig. 11.11.

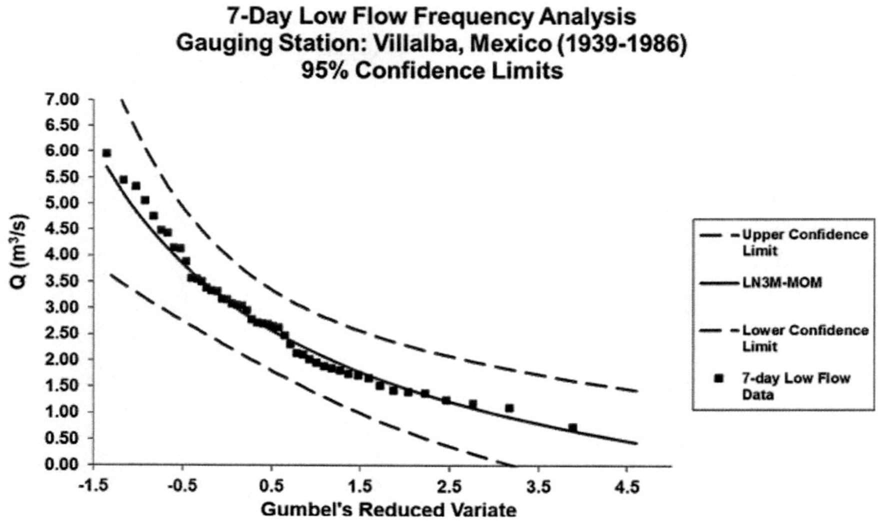

Fig. 11.12 Empirical and MOM LN3M theoretical frequency curves and confidence limits of 7-day low flow sample of Villalba, Mexico

A graphical description of the empirical and theoretical frequency curves and the confidence limits of the best fit provided, in this case that is MOM, is shown in Fig. 11.12.

11.9.3 Earthquake Epicenter Distance Frequency Analysis

By using the earthquake epicenter distance data, Castillo (1988), the following descriptive statistics were obtained and are shown in Fig. 11.13.

11.9.3.1 MOM Method

By using an Excel© spreadsheet, the results contained in Fig. 11.14, were obtained by the MOM method applied to the LN3M distribution for the earthquake epicenter distance data.

11.9.3.2 ML Method

By using an Excel© spreadsheet, the results contained in Fig. 11.15, were obtained by the ML method applied to the LN3M distribution for the earthquake epicenter distance data.

In Fig. 11.16 a comparison is made between the histogram and LN3M- MOM theoretical density for the earthquake epicenter distance sample data.

A graphical comparison between the empirical and theoretical frequency curves from the results provided by the two methods shown before is shown in Fig. 11.17.

	A	B	C	D
1	EARTHQUAKE EPICENTER DISTANCE FREQUENCY ANALYSIS			
2				
3	*Descriptive Statistics*		*Descriptive Statistics of the Logarithms*	
4	Mean	227.2069	Mean	5.3282
5	Standard Error	11.6856	Standard Error	0.0619
6	Median	248.1883	Median	5.5142
7	Mode	93.6438	Mode	4.5395
8	Standard Deviation	90.5161	Standard Deviation	0.4798
9	Sample Variance	8193.1719	Sample Variance	0.2302
10	Kurtosis	-1.1039	Kurtosis	1.1900
11	Skewness	-0.0732	Skewness	-0.9978
12	Range	347.5440	Range	2.3449
13	Minimum	36.8461	Minimum	3.6067
14	Maximum	384.3901	Maximum	5.9517
15	Sum	13632.4134	Sum	319.6947
16	Count	60.0000	Count	60.0000
17	Largest(1)	384.3901	Largest(1)	5.9517
18	Smallest(1)	36.8461	Smallest(1)	3.6067
19	Confidence Level(95.0%)	23.3828	Confidence Level(95.0%)	0.1240

Fig. 11.13 Descriptive statistics for the earthquake epicenter distance sample data

	A	B	C	D	E	F	G	H	I
1	EARTHQUAKE EPICENTER DISTANCE FREQUENCY ANALYSIS								
2	Three-parameters Log-Normal (LN3M) Distribution for the Minima								
3	Method: MOM								
4									
5	Estimation of Parameters								
6	w =	1.0373		Skew(Ln(x-x0)) =		#¡NUM!			
7	n_z =	-0.0244			CV_x =	0.3984			
8	Location Parameter =		3936.7550	km					
9	Scale Parameter =		#¡NUM!						
10	Shape Parameter =		#¡NUM!						
11	SEF =	#¡NUM!		MARD =	#¡NUM!				
12									
13	Design Values and Confidence Limits								
14–18	Tr (years)	F(x)	Standard Normal Variate z_T	Logarithmic Squared Root Stand.Error S_T	Positive Squared Root Stand.Error S_T (km)	Negative Squared Root Stand.Error S_T (km)	Two-sided 95% Lower Limit (km)	Design Values D_T (km)	Two-sided 95% Upper Limit (km)
19	2	0.5	0.0000	#¡NUM!	#¡NUM!	#¡NUM!	#¡NUM!	#¡NUM!	#¡NUM!
20	5	0.2	-0.8416	#¡NUM!	#¡NUM!	#¡NUM!	#¡NUM!	#¡NUM!	#¡NUM!
21	10	0.1	-1.2816	#¡NUM!	#¡NUM!	#¡NUM!	#¡NUM!	#¡NUM!	#¡NUM!
22	20	0.05	-1.6449	#¡NUM!	#¡NUM!	#¡NUM!	#¡NUM!	#¡NUM!	#¡NUM!
23	50	0.02	-2.0537	#¡NUM!	#¡NUM!	#¡NUM!	#¡NUM!	#¡NUM!	#¡NUM!
24	100	0.01	-2.3263	#¡NUM!	#¡NUM!	#¡NUM!	#¡NUM!	#¡NUM!	#¡NUM!
25	Two-sided Confidence Limits:			z_α =	1.96				

Fig. 11.14 MOM estimators for the parameters, standard errors, quantiles, and confidence limits of the LN3M distribution for the earthquake epicenter distance sample data

A graphical depiction of the empirical and theoretical frequency curves and the confidence limits of the best fit provided, in this case that of MOM, is shown in Fig. 11.18.

	A	B	C	D	E	F	G
1	EARTHQUAKE EPICENTER DISTANCE FREQUENCY ANALYSIS						
2	Three-parameters Log-Normal (LN3M) Distribution for the Minima						
3	Method: ML						
4							
5	Estimation of Parameters						
6	Test Location Parameter =	-150000.0000			Skew(Ln(x-x0)) =		0.1276
7	$F(x_0)$ =	5.6567E-03	Log-likelihood function =		-286.6305	AIC =	579.2611
8	Location Parameter =	-150000.0000	km	SEF =	56.45	MARD =	33
9	Scale Parameter =	0.0004					
10	Shape Parameter =	11.9202					
11							
12	Variance-Covariance Matrix						
13	$var(x_0)$ =	5.1515E+06	$cov(x_0,\mu_y)$ =	-3.4282E+01	$cov(x_0,\sigma_y^2)$ =	1.2209E-05	
14	$var(\mu_y)$ =	2.2813E-04	$cov(\mu_y,\sigma_y^2)$ =	8.1247E-11	$var(\sigma_y^2)$ =	2.8936E-17	
15	D =	1.8313E-09					
16							
17	Design Values and Confidence Limits						
18			Standard	Squared Root	Two-sided	Design	Two-sided
19	Return Period		Normal	Stand.Error	95%	Values	95%
20			Variate	S_T	Lower Limit	D_T	Upper Limit
21	Tr (years)	F(x)	z_T	(km)	(km)	(km)	(km)
22	2	0.5	0.0000	5.4421E-04	271.44	271.44	271.44
23	5	0.2	-0.8416	#¡NUM!	#¡NUM!	218.08	#¡NUM!
24	10	0.1	-1.2816	#¡NUM!	#¡NUM!	190.20	#¡NUM!
25	20	0.05	-1.6449	#¡NUM!	#¡NUM!	167.18	#¡NUM!
26	50	0.02	-2.0537	#¡NUM!	#¡NUM!	141.27	#¡NUM!
27	100	0.01	-2.3263	#¡NUM!	#¡NUM!	124.00	#¡NUM!
28	Two-sided Confidence Limits:		z_α =	1.96			

Fig. 11.15 Maximum likelihood estimators for the parameters, standard errors, quantiles, and confidence limits of the LN3M distribution for the earthquake epicenter distance sample data

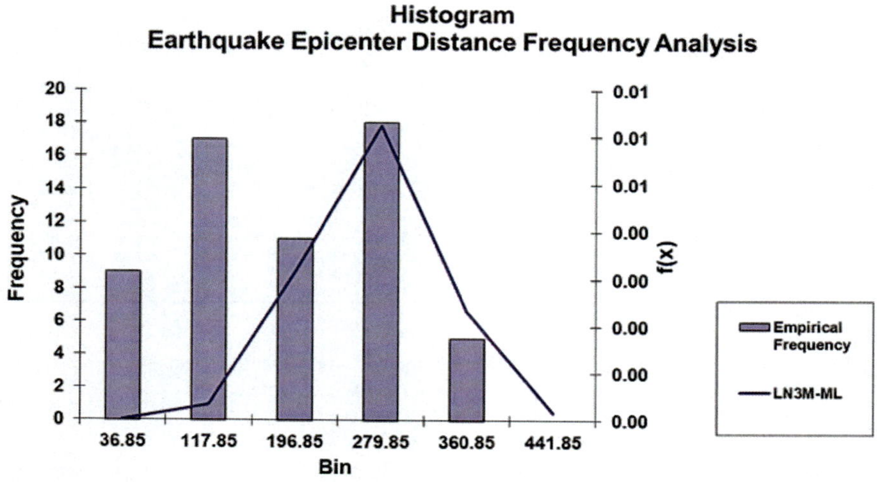

Fig. 11.16 Histogram and MOM-LN3M theoretical density for the earthquake epicenter distance sample data

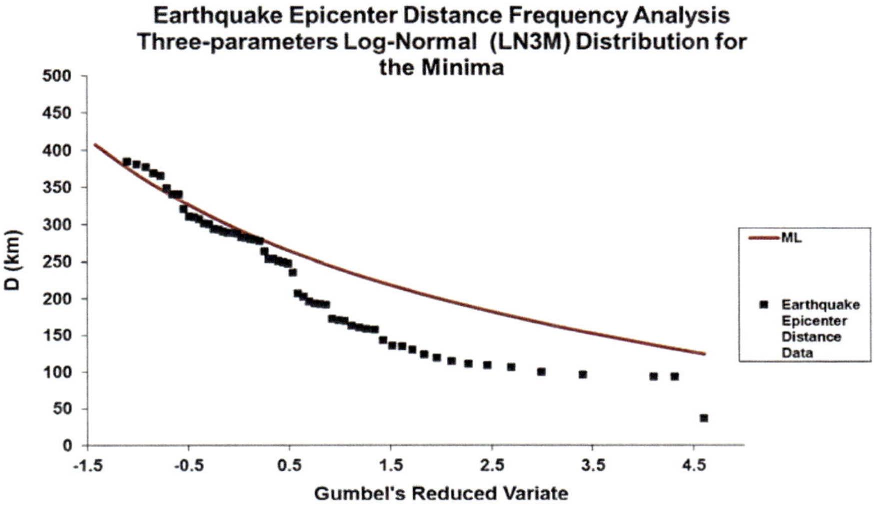

Fig. 11.17 Empirical and ML theoretical curves of LN3M distribution of earthquake epicenter distance sample data

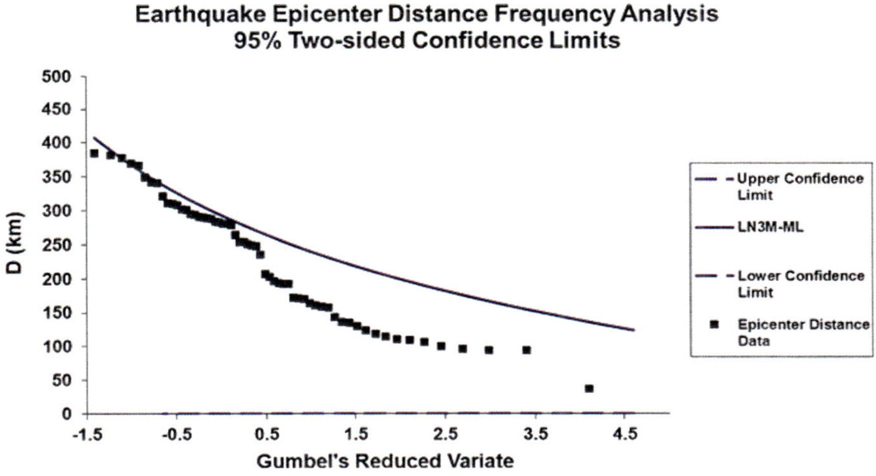

Fig. 11.18 Empirical and ML-LN3M theoretical frequency curves and confidence limits of earthquake epicenter distance sample data

Pearson Type III Distribution for the Minima

12

Start with what is right rather than what is acceptable. F. Kafka.

12.1 Introduction

The probability distribution function of the Pearson type III (PIIIM) distribution it is a gamma distribution with three parameters.

Matalas (1963) analyzed the suitability of using several probability distributions functions for modeling low flows. He found that the Pearson type III and the extreme value type III for the minima, when using the method of moments, were the best models among those tested in such study, namely the 3 parameters Log-Normal, Pearson types III and V, and the extreme value type III for the minima. Kroll and Vogel (2002) analyzed the low flows of 1505 gauged river sites and they suggested to use either the three-parameters Log-Normal or the Pearson type III distributions for describing low flow statistics in the United States of America.

The PIIIM distribution for the minima has the same mathematical form that its maxima counterpart, being the only difference the use of the exceedance probability as its probability distribution function in the former and the non-exceedance probability as its probability distribution function in the latter.

12.2 Chapter Objectives

After reading this chapter, you will know how to:

(1) Recognize the distribution and density functions of the PIIIM distribution

J. A. Raynal Villaseñor, *Frequency Analyses of Natural Extreme Events*,
Earth and Environmental Sciences Library,
https://doi.org/10.1007/978-3-030-86390-6_12

(2) Estimate the parameters of the PIIIM distribution
(3) Compute the quantiles and confidence limits of the PIIIM distribution
(4) Make a graphic display of your data and the PIIIM distribution
(5) Develop an application of all the above using Excel® spreadsheets

12.3 Probability Distribution and Density Functions

The probability distribution function of the Pearson (PIIIM) distribution for the minima is:

$$F(x) = \frac{1}{\alpha\Gamma(\beta)} \int_{x_0}^{x} \left(\frac{x - x_0}{\alpha}\right)^{\beta-1} \exp\left(-\frac{(x - x_0)}{\alpha}\right) dx \quad (12.1)$$

where x_0,α and β are the location, scale, and shape parameters, respectively, and $\Gamma(.)$ is the complete gamma function, as it was defined in Chap. 6. Such parameters are restricted to be greater than zero, that is $\alpha > 0$ and $\beta > 0$. The domain of variable x in this distribution is $x_0 \leq x < \infty$.

The probability density function for the PIIIM distribution is:

$$f(x) = \frac{1}{\alpha\Gamma(\beta)} \left(\frac{x - x_0}{\alpha}\right)^{\beta-1} \exp\left(-\frac{(x - x_0)}{\alpha}\right) \quad (12.2)$$

The PIII distribution for the minima (PIIIM), has the same mathematical form that its maxima counterpart, being the only difference the use of the exceedance probability as its probability distribution function in the former and the non-exceedance probability as its probability distribution function in the latter.

12.4 Estimation of the Parameters

12.4.1 MOM Method

The mean, variance and coefficients of skewness and kurtosis of the PIIIM distribution are, Kite(1988):

$$\mu = \alpha\beta + x_0 \quad (12.3)$$

$$\sigma^2 = \alpha^2\beta \quad (12.4)$$

$$\gamma = \frac{2\alpha}{|\alpha|\sqrt{\beta}} \quad (12.5)$$

$$\kappa = 3\left(1 + \frac{\gamma^2}{2}\right) \quad (12.6)$$

Other useful relationships are, Kite (1988):

$$\lambda_1 = \frac{\mu_5}{\mu_2^{2.5}} = \gamma\left(10 + 3\gamma^2\right) \tag{12.7}$$

$$\lambda_2 = \frac{\mu_6}{\mu_2^3} = 5\left(3 + \frac{13\gamma^2}{2} + \frac{3\gamma^4}{2}\right) \tag{12.8}$$

So, these previous equations would serve to estimate the parameters of the PIIIM distribution by the MOM method as:

$$\hat{x}_0 = \hat{\mu} - \hat{\sigma}\sqrt{\hat{\beta}} \tag{12.9}$$

$$\hat{\alpha} = \frac{\hat{\sigma}}{\sqrt{\hat{\beta}}} \tag{12.10}$$

$$\hat{\beta} = \left(\frac{2}{\gamma}\right)^2 \tag{12.11}$$

In the above equations the mean, the variance and/or the standard deviation and the coefficient of skewness can be estimated by the procedures outlined in Chap. 2. In the case of the skewness coefficient, the following formula allows to correct the bias in such coefficient:

$$\hat{\gamma}_{PIII} = \frac{\frac{1}{N}\sum_{i=1}^{N}\left(x_i - \hat{\mu}\right)^3}{\hat{\sigma}^3}\left[\frac{\left(N(N-1)\right)^{\frac{1}{2}}}{(N-2)}\right]\left[1 + \frac{8.5}{N}\right] \tag{12.12}$$

12.4.1.1 Example of Application of Estimation of the Parameters of the PIIIM Distribution Using the MOM Method

Find the MOM estimators for the parameters of the PIIIM distribution for the one-day low flow sample data of gauging station Villalba, Mexico (1939–1991), contained Appendix A.

The following statistics can be obtained:

$$\hat{\mu} = \frac{1}{N}\sum_{i=1}^{N} x_i = 0.3306\ \mathrm{m}^3/s$$

$$\hat{\sigma} = \left[\frac{1}{N}\sum_{i=1}^{N}\left(x_i - \hat{\mu}\right)^2\right]^{1/2} = 0.1465\ \mathrm{m}^3/s$$

$$\hat{\gamma} = \frac{N}{(N-1)(N-2)\hat{\sigma}^3}\sum_{i=1}^{N}\left(x_i - \hat{\mu}\right)^3 = 0.6201$$

then:

$$\hat{\beta} = \left(\frac{2}{\gamma^2}\right)^2 = \left(\frac{2}{0.6201}\right)^2 = 8.1843$$

$$\hat{x}_0 = \hat{\mu} - \hat{\sigma}\sqrt{\hat{\beta}} = 0.3306 - (0.1464)(8.1843)^{0.5} = -0.0884$$

$$\hat{\alpha} = \frac{\hat{\sigma}}{\sqrt{\hat{\beta}}} = \frac{0.1465}{(8.1843)^{0.5}} = 0.0512$$

Finally, the MOM estimators of the parameters of the PIIIM distribution for the one-day low flow sample data of gauging station Villalba, Mexico are:
location parameter:

$$\hat{x}_0 = -0.0884 m^3/s$$

scale parameter:

$$\hat{\alpha} = 0.0512 m^3/s$$

and:
shape parameter:

$$\hat{\beta} = 8.1843$$

12.4.2 ML Method

The likelihood function of the PIIIM distribution is:

$$L(\alpha, \beta, x_0) = \prod_{i=1}^{N}\left\{\frac{1}{\alpha\Gamma(\beta)}\left(\frac{x - x_0}{\alpha}\right)^{\beta-1} \exp\left(-\frac{(x - x_0}{\alpha}\right)\right\} \tag{12.13}$$

So, the Log-likelihood function of the PIIIM distribution is:

$$LL(\alpha, \beta, x_0) = -NLn(\Gamma(\beta)) - \frac{1}{\alpha}\sum_{i=1}^{N}(x_i - x_0) + (\beta - 1)\sum_{i=1}^{N} Ln(x_i - x_0) - N\beta Ln(\alpha) \tag{12.14}$$

Now, taking the first-order partial derivatives of this equation with respect to parameters α, β and x_0 and set them equal to zero:

$$\frac{\partial LL}{\partial \alpha} = -\frac{N\beta}{\alpha} + \frac{1}{\alpha^2}\sum_{i=1}^{N}(x_i - x_0) = 0 \tag{12.15}$$

$$\frac{\partial LL}{\partial \beta} = -NLn(\alpha) - \frac{N\Gamma'(\beta)}{\Gamma(\beta)} + \sum_{i=1}^{N} Ln(x_i - x_0) = 0 \tag{12.16}$$

$$\frac{\partial LL}{\partial x_0} = \frac{N}{\alpha} - (\beta - 1)\sum_{i=1}^{N}\left[\frac{1}{(x_i - x_0)}\right] = 0 \tag{12.17}$$

The solution of the system of equations conformed by Eqs. (12.15) to (12.17) provides the ML estimators for the parameters of the PIIIM distribution:

$$\hat{\alpha} = \frac{1}{N}\sum_{i=1}^{N}(x_i - \hat{x}_0) - N\left[\sum_{i=1}^{N}\frac{1}{(x_i - \hat{x}_0)}\right]^{-1} \tag{12.18}$$

$$\hat{\beta} = \left\{1 - N^2\left[\sum_{i=1}^{N}(x_i - \hat{x}_0)\sum_{i=1}^{N}\frac{1}{(x_i - \hat{x}_0)}\right]^{-1}\right\}^{-1} \tag{12.19}$$

$$F(\hat{x}_0) = -N\psi(\hat{\beta}) + \sum_{i=1}^{N} Ln(x_i - \hat{x}_0) - NLn(\hat{\alpha}) = 0 \tag{12.20}$$

Equation (12.20) must be solved by a procedure of finding the roots of such an equation, again the bisection method will give good results in solving this equation.

Kite (1988) proposed the following procedure to solve the system of equation formed by Eqs. (12.18) to (12.20), the results are:

$$A = \frac{1}{N}\sum_{i=1}^{N}(x_i - \hat{x}_0) \tag{12.21}$$

$$B = \left[\prod_{i=1}^{N}(x_i - \hat{x}_0)\right]^{\frac{1}{N}} \tag{12.22}$$

and if:

$$C = Ln(A) - Ln(B) \tag{12.23}$$

and then:

$$C = Ln\left(\hat{\beta}\right) - \frac{\Gamma'\left(\hat{\beta}\right)}{\Gamma(\beta)} \tag{12.24}$$

An approximation for the solution for $\hat{\beta}$ is given by, Greenwood and Durand (1960):

$$\hat{\beta} = \frac{1}{C}\left(0.500876 + 0.1648852C - 0.054427C^2\right) \tag{12.25}$$

for $0 \leq C \leq 0.5772$, and:

$$\hat{\beta} = \frac{8.898919 + 9.05995C + 0.9775373C^2}{C\left(17.7928 + 11.966847C + C^2\right)} \tag{12.26}$$

for $0.5772 \leq C \leq 17.0$.

This approach will not be used in this procedure, instead the solution of Eqs. (12.18) to (12.20) will be preferred.

As a word of warning, it must be said that the ML method may not always be applicable for the PIIIM distribution, this was specifically observed for small values of the sample skewness coefficient, Matalas and Wallis (1973). Furthermore, if $\beta < 1$ the ML solution is not possible, Kite (1988). This situation is coupled with the fact that the sample skewness coefficient must be less than two ($\hat{\gamma} < 2$), see Eq. (12.1). If the sample skewness coefficient is negative, then the PIIIM distribution becomes upper bounded.

12.4.2.1 Example of Application of Estimation of the Parameters of the PIIIM Distribution Using the ML Method

Find the ML estimators for the parameters of the PIIIM distribution for the one-day low flow sample data of gauging station Villalba, Mexico (1939–1991), Appendix A.

The following statistics can be obtained:

$$\hat{\mu} = \frac{1}{N}\sum_{i=1}^{N} x_i = 0.3306 \text{ m}^3/s$$

$$\hat{\sigma} = \left[\frac{1}{N}\sum_{i=1}^{N}(x_i - \hat{\mu})^2\right]^{1/2} = 0.1465 \text{ m}^3/s$$

$$\hat{\gamma} = \frac{N}{(N-1)(N-2)\hat{\sigma}^3}\sum_{i=1}^{N}(x_i - \hat{\mu})^3 = 0.6201$$

then by finding the roots of Eq. (12.20):

$$\hat{x}_0 = -0.0139 m^3/s$$

$$\hat{\alpha} = \frac{1}{N}\sum_{i=1}^{N}(x_i - \hat{x}_0) - N\left[\sum_{i=1}^{N}\frac{1}{(x_i - \hat{x}_0)}\right]^{-1} = 0.0642 m^3/s$$

$$\hat{\beta} = \left\{1 - N^2\left[\sum_{i=1}^{N}(x_i - \hat{x}_0)\sum_{i=1}^{N}\frac{1}{(x_i - \hat{x}_0)}\right]^{-1}\right\}^{-1} = 5.3705$$

Finally, the ML estimators of the parameters of the PIIIM distribution for the one-day low flow sample data of gauging station Villalba, Mexico are:

location parameter:

$$\hat{x}_0 = -0.0139 m^3/s$$

scale parameter:

$$\hat{\alpha} = 0.0642 m^3/s$$

and:

shape parameter:

$$\hat{\beta} = 5.3705$$

12.5 Estimation of Quantiles for the PIIIM Distribution

By using the following reduced value, Kite (1988):

$$y = \frac{x - x_0}{\alpha} \tag{12.27}$$

into Eq. (12.) produces the following expression, Kite (1988):

$$F(y) = \frac{1}{\Gamma(\beta)} \int_0^{y_0} y^{\beta-1} \exp(-y) dy \tag{12.28}$$

and Abramowitz and Stegun (1965) have shown that:

$$F(y) = F(\chi^2 | v) \tag{12.29}$$

where $F(\chi^2/v)$ is the chi-square distribution with 2β degrees of freedom and $\chi^2 = 2\,y$.

So, the reduced event magnitude, y_0, may be obtained as, Kite (1988):

$$y_0 = \frac{\chi^2}{2} \tag{12.30}$$

and the expected event of magnitude associated with a given return period T_r is, Kite (1988):

$$x_T = \frac{\alpha \chi^2}{2} + x_0 \tag{12.31}$$

but using the Wilson-Hilferty approximation, Kendall and Stuart (1963):

$$\left[\left(\frac{\chi^2}{v}\right)^{\frac{1}{3}} + \frac{2}{9v} - 1\right]\left(\frac{9v}{2}\right)^{\frac{1}{2}} \sim N(0, 1) \tag{12.32}$$

and this expression is valid for $v > 30$. So, the values of χ^2 can be approximated as:

$$\chi^2 \approx v\left[1 - \frac{2}{9v} + z_T\sqrt{\frac{2}{9v}}\right]^3 \tag{12.33}$$

where z_T is the standard normal (N(0,1)) variate corresponding a certain return period, T_r, and from Eqs. (12.31) and (12.33) one may obtain:

$$x_T = \hat{\alpha}\hat{\beta}\left[1 - \frac{1}{9\hat{\beta}} + z_T\sqrt{\frac{1}{9\hat{\beta}}}\right]^3 + \hat{x}_0 \tag{12.34}$$

Table 12.1 Estimation of MOM and ML Quantiles for the PIIIM distribution for Gauging station Villalba, Mexico

Tr (Years)	MOM	ML
	Q_T (m^3/s)	Q_T (m^3/s)
2	0.3138	0.3096
5	0.2054	0.2039
10	0.1574	0.1591
20	0.1219	0.1272
25	0.1123	0.0962
50	0.0863	0.0783
100	0.0650	0.0483

12.5.1 Examples of Estimation of MOM and ML Quantiles for the PIIIM Distribution

Find the MOM and ML estimators of the quantiles 2, 5, 10, 20, 50, and 100 years of return period, for the PIIIM distribution for the one-day low flow sample data of gauging station Villalba, Mexico (1942–1992), contained Appendix A.

The estimation of MOM and ML quantiles for the PIIIM distribution is made by using the parameters estimated in the preceding sections and then inserting the parameters into Eq. (12.34). Table 12.1 summarizes these results.

12.6 Goodness of Fit Test

The standard error of fit (SEF) for the PIIIM has the following form:

$$SEF = \left[\frac{\sum_{i=1}^{N} (xi - yi)^2}{(N-3)} \right]^{1/2} \tag{12.35}$$

while the mean absolute relative deviation (MARD) remains the same as it was defined in Chap. 2:

$$MARD = \frac{100}{N} \sum_{i=1}^{N} \left| \frac{(x_i - y_i)}{x_i} \right| \tag{12.36}$$

where x_i are the sample historical values, y_i are the distribution function values, corresponding to the same return periods of the historical values, N is the sample size.

12.6.1 Examples of Application of the SEF and MARD to the MOM and ML Estimators of the Parameters of the PIIIM Distribution

Find the values of the SEF and MARD of the MOM and ML estimators of the parameters of the PIIIM distribution for the one-day low flow sample data of gauging station Villalba, Mexico (1939–1991), Appendix A.

The values of SEF and MARD for the MOM and ML estimators of the parameters of the PIIIM distribution for the one-day low flow sample data of gauging station Villalba, Mexico have been obtained through the application of Eqs. (12.35) and (12.36) using the parameters obtained in previous examples. The results are as follows:

(a) MOM Method

$$SEF = \left[\frac{\sum_{i=1}^{N}(xi - yi)^2}{(N-3)}\right]^{1/2} = \left[\frac{0.0319}{(53-3)}\right]^{\frac{1}{2}} = 0.03$$

$$MARD = \frac{100}{N}\sum_{i=1}^{N}\left|\frac{(x_i - y_i)}{x_i}\right| = \frac{(100)(2.9354)}{53} = 5.54$$

(b) ML Method

$$SEF = \left[\frac{\sum_{i=1}^{N}(xi - yi)^2}{(N-3)}\right]^{1/2} = \left[\frac{0.0301}{(53-3)}\right]^{\frac{1}{2}} = 0.02$$

$$MARD = \frac{100}{N}\sum_{i=1}^{N}\left|\frac{(x_i - y_i)}{x_i}\right| = \frac{(100)(3.2699)}{53} = 6.17$$

12.7 Estimation of Confidence Limits for the PIIIM Distribution

By assuming that the quantiles are normally distributed, the following form can be used to set the confidence limits on such quantiles:

$$x_l = x_T \pm z_\alpha S_T \tag{12.37}$$

where x_l is the upper or lower confidence limit, depending on the sign in the preceding formula (+) for the upper confidence limit and (-) for the lower confidence limit, z_α is the standard normal variate corresponding to a confidence level α, and S_T is squared root of the standard error of the estimate. The evaluation procedures of the standard errors of the estimates will be described in the following subsections.

12.8 Estimation of Standard Errors for the PIIIM Distribution

12.8.1 MOM Method

The general form of the MOM estimator of the standard error of the estimate of a three-parameter distribution is, Kite (1988):

$$
\begin{aligned}
S_T^2 = & \left(\frac{\partial x_T}{\partial m_1'}\right)^2 \mathrm{var}(m_1') + \left(\frac{\partial x_T}{\partial m_2}\right)^2 \mathrm{var}(m_2) + \left(\frac{\partial x_T}{\partial m_3}\right)^2 \mathrm{var}(m_3) \\
& + 2\left(\frac{\partial x_T}{\partial m_1'}\right)\left(\frac{\partial x_T}{\partial m_2}\right)\mathrm{cov}(m_1', m_2) + 2\left(\frac{\partial x_T}{\partial m_1'}\right)\left(\frac{\partial x_T}{\partial m_3}\right)\mathrm{cov}(m_1', m_3) \\
& + 2\left(\frac{\partial x_T}{\partial m_2}\right)\left(\frac{\partial x_T}{\partial m_3}\right)\mathrm{cov}(m_2, m_3)
\end{aligned} \tag{12.38}
$$

and Eq. (12.38) can be simplified in terms of the frequency factor, K_T, as, Kite (1988):

$$
\begin{aligned}
S_T^2 = \frac{\mu_2}{N}\Bigg\{ & 1 + K_T\gamma + \frac{K_T^2}{4}(\kappa - 1) + \frac{\partial K_T}{\partial \gamma}\left[2\kappa - 3\gamma^2 - 6 + K_T\left(\lambda_1 - \frac{6\gamma\kappa}{4} - \frac{10\gamma}{4}\right)\right] \\
& + \left(\frac{\partial K_T}{\partial \gamma}\right)^2\left[\lambda_2 - 3\gamma\lambda_1 - 6\gamma + \frac{9\gamma^2\kappa}{4} + \frac{35\gamma^2}{4} + 9\right]\Bigg\}
\end{aligned} \tag{12.39}
$$

$$
\hat{\lambda}_1 = \frac{\frac{\left[\sum_{i=1}^{N}(x_i - \hat{\mu})^5\right]}{N}}{\left(\sigma^2\right)^{2.5}} \tag{12.40}
$$

$$
\hat{\lambda}_2 = \frac{\frac{\left[\sum_{i=1}^{N}(x_i - \hat{\mu})^6\right]}{N}}{\left(\sigma^2\right)^{3}} \tag{12.41}
$$

The previous equation may be further simplified by using the relationships given in Eqs. (12.6) to (12.8), Kite (1988):

$$
S_T^2 = \frac{\mu_2}{N}
$$
$$
\left\{1 + K_T\gamma + \frac{K_T^2}{2}\left(\frac{3\gamma^2}{4} + 1\right) + 3K_T\frac{\partial K_T}{\partial \gamma}\left(\gamma + \frac{\gamma^3}{4}\right) + 3\left(\frac{\partial K_T}{\partial \gamma}\right)^2\left(2 + 3\gamma^2 + \frac{5\gamma^4}{8}\right)\right\} \tag{12.42}
$$

The partial derivative of the frequency factor with respect to skewness coefficient may be evaluated as, Kite (1988):

$$
\frac{\partial K_T}{\partial \gamma} \approx \frac{z_T^2 - 1}{6} + \frac{4\left(z_T^3 - 6z_T\right)\gamma}{6^3} - \frac{3\left(z_T^2 - 1\right)\gamma^2}{6^3} + \frac{4z_T\gamma^3}{6^4} - \frac{10\gamma^4}{6^6} \tag{12.43}
$$

Table 12.2 Estimation of MOM Standard Errors and the Two-sided 95% confidence limits for the PIIIM distribution for Gauging station Villalba, Mexico

Return period Tr (years)	Squared root stand.error S_T	Two-sided 95% lower limit (m^3/s)	Design values Q_T (m^3/s)	Two-sided 95% upper limit (m^3/s)
2	0.0225	0.2697	0.3138	0.3578
5	0.0189	0.1683	0.2054	0.2424
10	0.0198	0.1185	0.1574	0.1962
20	0.0247	0.0735	0.1219	0.1704
25	0.0269	0.0596	0.1123	0.1649
50	0.0345	0.0187	0.0863	0.1539
100	0.0429	-0.0190	0.0650	0.1490

Two-sided Limits: $z_\alpha = 1.96$.

and the frequency factor for the PIIIM distribution is, Kite (1988):

$$K_T = z_T + \left(z_T^2 - 1\right)\left(\frac{\hat{\gamma}}{6}\right) + \frac{\left(z_T^3 - 6z_T\right)}{3}\left(\frac{\hat{\gamma}}{6}\right)^2 - \left(z_T^2 - 1\right)\left(\frac{\hat{\gamma}}{6}\right)^3 + z_T\left(\frac{\hat{\gamma}}{6}\right)^4 - \frac{1}{3}\left(\frac{\hat{\gamma}}{6}\right)^5 \quad (12.44)$$

By inserting these results into Eq. (12.42), the moment estimator of the standard error of the estimate for the PIIIM distribution may be obtained.

12.8.1.1 Example of Estimation of MOM Standard Errors and the Two-Sided 95% Confidence Limits for the PIIIM Distribution

Find the MOM estimators of the standard errors and the two-sided 95% confidence limits for the PIIIM distribution for the one-day low flow sample data of gauging station Villalba, Mexico (1939–1991), Appendix A.

The estimation of MOM estimators of the standard errors and the two-sided 95% confidence limits for the PIIIM distribution is made by using the moment estimators of the parameters and then inserted in the Eqs. (12.42) to (12.44), using selected values of the return intervals. Table 12.2 summarizes these results.

12.8.2 ML Method

The general equation for the ML estimator of the standard error of the estimate of a three-parameter distribution is, Kite (1988):

$$S_T^2 = \left(\frac{\partial x_T}{\partial \alpha}\right)^2 \mathrm{var}(\alpha) + \left(\frac{\partial x_T}{\partial \beta}\right)^2 \mathrm{var}(\beta) + \left(\frac{\partial x_T}{\partial \gamma}\right)^2 \mathrm{var}(\gamma)$$

$$+2\left(\frac{\partial x_T}{\partial \alpha}\right)\left(\frac{\partial x_T}{\partial \beta}\right)\text{cov}(\alpha, \beta) + 2\left(\frac{\partial x_T}{\partial \alpha}\right)\left(\frac{\partial x_T}{\partial \gamma}\right)\text{cov}(\alpha, \gamma)$$
$$+2\left(\frac{\partial x_T}{\partial \beta}\right)\left(\frac{\partial x_T}{\partial \gamma}\right)\text{cov}(\beta, \gamma) \quad (12.45)$$

and the elements of the Fisher's information matrix, are:

$$-\frac{\partial^2 LL}{\partial \alpha^2} = \frac{2\sum_{i=1}^{N}(x_i - x_0)}{\alpha^3} - \frac{N\beta}{\alpha^2} \quad (12.46)$$

$$-\frac{\partial^2 LL}{\partial \beta^2} = N\psi'(\beta) \quad (12.47)$$

$$-\frac{\partial^2 LL}{\partial x_0^2} = (\beta - 1)\sum_{i=1}^{N}\frac{1}{(x_i - x_0)^2} \quad (12.48)$$

$$-\frac{\partial^2 LL}{\partial \alpha \partial \beta} = \frac{N}{\alpha} \quad (12.49)$$

$$-\frac{\partial^2 LL}{\partial \alpha \partial x_0} = \frac{N}{\alpha^2} \quad (12.50)$$

$$-\frac{\partial^2 LL}{\partial \beta \partial x_0} = \frac{N}{\alpha(\beta - 1)} \quad (12.51)$$

but the Eqs. (12.45) and (12.46) can be simplified by using the following result, Kite (1988):

$$\sum_{i=1}^{N}(x_i - x_0)^r = \frac{N\alpha^r\Gamma(\beta + r)}{\Gamma(\beta)} \quad (12.52)$$

so:

$$-\frac{\partial^2 LL}{\partial \alpha^2} = \frac{N\beta}{\alpha^2} \quad (12.53)$$

$$-\frac{\partial^2 LL}{\partial x_0^2} = \frac{N}{\alpha^2(\beta - 2)} \quad (12.54)$$

and the Fisher's information matrix is constructed as:

$$[I] = N\begin{bmatrix} \frac{\beta}{\alpha^2} & \frac{1}{\alpha} & \frac{1}{\alpha^2} \\ \frac{1}{\alpha} & \psi'(\beta) & \frac{1}{\alpha(\beta-1)} \\ \frac{1}{\alpha^2} & \frac{1}{\alpha(\beta-1)} & \frac{1}{\alpha^2(\beta-2)} \end{bmatrix} \quad (12.55)$$

Finally, the variances and covariances of the parameters of the PIIIM distribution are found as:

$$\mathrm{var}(\alpha) = \frac{1}{ND\alpha^2}\left[\frac{\psi'(\beta)}{(\beta-2)} - \frac{1}{(\beta-1)^2}\right] \tag{12.56}$$

$$\mathrm{var}(\beta) = \frac{2}{ND\alpha^4(\beta-2)} \tag{12.57}$$

$$\mathrm{var}(x_0) = \frac{1}{ND\alpha^2}[\beta\psi'(\beta) - 1] \tag{12.58}$$

$$\mathrm{cov}(\alpha, \beta) = -\frac{1}{ND\alpha^3}\left[\frac{1}{(\beta-2)} - \frac{1}{(\beta-1)}\right] \tag{12.59}$$

$$\mathrm{cov}(\alpha, x_0) = \frac{1}{ND\alpha^2}\left[\frac{1}{(\beta-1)} - \psi'(\beta)\right] \tag{12.60}$$

$$\mathrm{cov}(\beta, x_0) = -\frac{1}{ND\alpha^3}\left[\frac{\beta}{(\beta-1)} - 1\right] \tag{12.61}$$

and D is the determinant of the Fisher's information matrix with the following form:

$$D = \frac{1}{(\beta-2)\alpha^4}\left[2\psi'(\beta) - \frac{(2\beta-3)}{(\beta-1)^2}\right] \tag{12.62}$$

The first order partial derivatives of x_T with respect to the parameters are evaluated as, Kite (1988):

$$\frac{\partial x_T}{\partial \alpha} = \left[\beta^{\frac{1}{3}} - \frac{1}{9\beta^{\frac{2}{3}}} + \frac{z_T}{3\beta^{\frac{1}{6}}}\right]^3 \tag{12.63}$$

$$\frac{\partial x_T}{\partial \beta} = 3\alpha\left[\beta^{\frac{1}{3}} - \frac{1}{9\beta^{\frac{2}{3}}} + \frac{z_T}{3\beta^{\frac{1}{6}}}\right]^2\left[\frac{1}{3\beta^{\frac{2}{3}}} + \frac{2}{27\beta^{\frac{5}{3}}} - \frac{z_T}{18\beta^{\frac{7}{6}}}\right] \tag{12.64}$$

$$\frac{\partial x_T}{\partial x_0} = 1 \tag{12.65}$$

The substitution of the results of Eqs. (12.46) to (12.65) into Eq. (12.45) will provide the value of the ML estimator of the standard error of the estimate of a PIIIM distribution.

Table 12.3 Estimation of ML Standard Errors and the two-sided 95% confidence limits for the PIIIM distribution for Gauging station Villalba, Mexico

Return Period Tr (years)	Squared root Stand. Error S_T	Two-sided 95% Lower Limit (m^3/s)	Design Values Q_T (m^3/s)	Two-sided 95% Upper Limit (m^3/s)
2	0.0205	0.2758	0.3096	0.3435
5	0.0178	0.1745	0.2039	0.2333
10	0.0173	0.1306	0.1591	0.1876
20	0.0178	0.0979	0.1272	0.1565
25	0.0199	0.0635	0.0962	0.1290
50	0.0223	0.0416	0.0783	0.1151
100	0.0293	−0.0001	0.0483	0.0967

Two-sided Limits: $z_\alpha = 1.96$.

12.8.2.1 Example of Estimation of ML Standard Errors and the Two-Sided 95% Confidence Limits for the PIIIM Distribution

Find the ML estimators of the standard errors and the two-sided 95% confidence limits for the PIIIM distribution for the one-day low flow sample data of gauging station Villalba, Mexico (1939–1991), contained in Appendix A.

The estimation of ML estimators of the standard errors and the two-sided 95% confidence limits for the PIIIM distribution is made by using the ML estimators of the parameters and then inserted in the Eqs. (12.54) to (12.63) and then into Eq. (12.43), using selected values of the return intervals. Table 12.3 summarizes these results.

12.9 Examples of Application for the PIIIM Distribution Using Excel© Spreadsheets

12.9.1 One-Day Low Flow Frequency Analysis

By using the one-day low flow data from gauging station Villalba, Mexico (1939–1991), the following descriptive statistics were obtained and are shown in Fig. 12.1.

12.9.1.1 MOM Method

By using an Excel© spreadsheet, the results contained in Fig. 12.2, were obtained for the MOM method applied to the PIIIM distribution to one-day low flow from gauging station Villalba, Mexico (1939–1991).

ONE-DAY LOW FLOW FREQUENCY ANALYSIS
Gauging Station: Villalba, Mexico (1939-1991)

Descriptive Statistics		*Descriptive Statistics of the Logarithms*	
Mean	0.3306	Mean	-1.2129
Standard Error	0.0201	Standard Error	0.0668
Median	0.3280	Median	-1.1147
Mode	0.4850	Mode	-0.7236
Standard Deviation	0.1465	Standard Deviation	0.4860
Sample Variance	0.0214	Sample Variance	0.2362
Kurtosis	0.1906	Kurtosis	-0.0416
Skewness	0.5853	Skewness	-0.5433
Range	0.6250	Range	2.1873
Minimum	0.0790	Minimum	-2.5383
Maximum	0.7040	Maximum	-0.3510
Sum	17.5214	Sum	-64.2851
Count	53.0000	Count	53.0000
Largest(1)	0.7040	Largest(1)	-0.3510
Smallest(1)	0.0790	Smallest(1)	-2.5383
Confidence Level(95.0%)	0.0404	Confidence Level(95.0%)	0.1340

Fig. 12.1 Descriptive statistics for the one-day low flow sample data at Villalba, Mexico

LOW FLOW FREQUENCY ANALYSIS
Gauging Station: Villalba, Mexico (1939-1991)
Pearson Type III (PIIIM) Distribution for the Minima
Method: MOM

Estimation of Parameters and Goodness of Fit Tests				
Skewness Coef. (Gamma Correction) =			0.6991	
Location Parameter =		-0.0884	m^3/s	
Scale Parameter =		0.0512	m^3/s	
Shape Parameter =		8.1843		
$\Gamma(\beta)$ =	7323.9768			
SEF =	0.02		MARD =	5

Design Values and Confidence Limits

Return Period Tr (years)	$\Pi(x)$	Standard Normal Variate z_T	$\partial K_T/\partial\gamma$	Frequency Factor K_T	Squared root Stand. Error S_T (m^3/s)	Two-sided 95% Lower Limit (m^3/s)	Design Values Q_T (m^3/s)	Two-sided 95% Upper Limit (m^3/s)
2	0.5	0.0000	-0.1599	-0.1149	0.0225	0.2697	0.3138	0.3578
5	0.2	-0.8416	0.0101	-0.8552	0.0189	0.1683	0.2053	0.2424
10	0.1	-1.2816	0.1736	-1.1827	0.0198	0.1185	0.1574	0.1963
20	0.05	-1.6449	0.3410	-1.4246	0.0247	0.0735	0.1219	0.1704
25	0.04	-1.7507	0.3948	-1.4904	0.0269	0.0597	0.1123	0.1649
50	0.02	-2.0537	0.5596	-1.6677	0.0345	0.0188	0.0863	0.1539
100	0.01	-2.3263	0.7206	-1.8135	0.0428	-0.0190	0.0650	0.1490
Two-sided Confidence Limits:		z_α =	1.96					

Fig. 12.2 MOM estimators for the parameters, standard errors, Quantiles, and confidence limits of the PIIIM distribution for the one-day low flow sample data at Villalba, Mexico

	A	B	C	D	E	F	G	H	I
2	Gauging Station: Villalba, Mexico (1939-1991)								
3	Pearson Type III (PIIIM) Distribution for the Minima								
4	Method: ML								
5									
6	Estimation of Parameters and Goodness of Fit Tests								
7	Test Location Parameter =		-0.3887		$F(x_o)$ =	-2.0274E-02			
8	Skewness Coef. (Gamma Correction) =			0.6991	<	2			
9	Location Parameter =		-0.3887	m^3/s	$\psi(\beta)$ =	3.2316			
10	Scale Parameter =		0.0279	m^3/s	$\Gamma(\beta)$ =	1.5102E+24			
11	Shape Parameter =		25.8196	Log-Likelihood Function =		-8.0043E+25			
12	SEF =	0.02	MARD =	5.00	AIC =	1.6009E+26			
13									
14	Variance-Covariance Matrix								
15	$var(x_0)$ =	3.1389E-01	$cov(\alpha,x_0)$ =	1.2815E-02	$cov(\beta,x_0)$ =	-2.3145E+01	Determinant =	1.5193E+00	
16	$var(\alpha)$ =	5.5263E-04	$cov(\alpha,\beta)$ =	-9.7167E-01	$var(\beta)$ =	1.7314E+03	$\psi'(\beta)$ =	0.0395	
17									
18	Design Values and Confidence Limits								
19–22	Return Period Tr (years)	$\Pi(x)$	Standard Normal Variate z_T	$\partial x_T/\partial\alpha$	$\partial x_T/\partial\beta$	Squared root Stand. Error S_T (m^3/s)	Two-sided 95% Lower Limit (m^3/s)	Design Values Q_T (m^3/s)	Two-sided 95% Upper Limit (m^3/s)
23	2	0.5	0.0000	25.4793	0.0279	0.0208	0.2807	0.3213	0.3620
24	5	0.2	-0.8416	21.4703	0.0255	0.0203	0.1699	0.2097	0.2494
25	10	0.1	-1.2816	19.5532	0.0243	0.0220	0.1132	0.1563	0.1993
26	20	0.05	-1.6449	18.0589	0.0233	0.0254	0.0649	0.1146	0.1644
27	50	0.02	-2.0537	16.4704	0.0222	0.0318	0.0080	0.0704	0.1328
28	100	0.01	-2.3263	15.4649	0.0215	0.0376	-0.0314	0.0424	0.1161
29	Two-sided Confidence Limits:			z_α =	1.96				

Fig. 12.3 ML estimators of the parameters, standard errors, Quantiles and confidence limits of the PIIIM distribution for the one-day low flow sample data at Villalba, Mexico

12.9.1.2 ML Method

By using an Excel© spreadsheet, the results contained in Fig. 12.3, were obtained for the method of ML applied to the PIIIM distribution to one-day low flow from gauging station Villalba, Mexico (1939–1991).

In Fig. 12.4 a comparison is made between the histogram and PIIIM theoretical density for the one-day low flow sample of gauging station Villalba, Mexico.

A graphical comparison between the empirical and theoretical frequency curves from the results provided by the two methods mentioned before is shown in Fig. 12.5.

A graphical description of the empirical and theoretical frequency curves and the confidence limits of the best fit provided, in this case that of MOM, is shown in Fig. 12.6.

12.9.2 7-day Low Flow Frequency Analysis

By using the 7-day low flow data from gauging station Villalba, Mexico (1939–1986), the following descriptive statistics were obtained and are shown in Fig. 12.7.

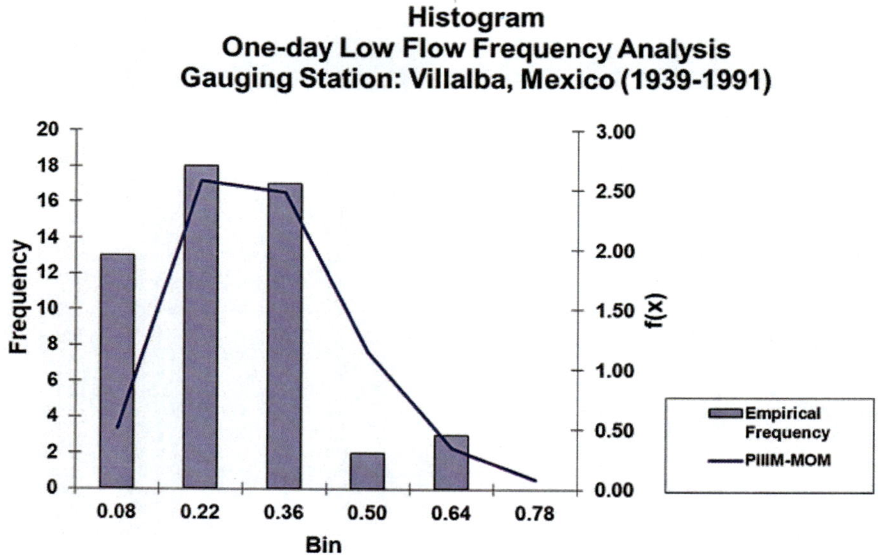

Fig. 12.4 Histogram and MOM-PIIIM theoretical density for the one-day low flow sample of Villalba, Mexico

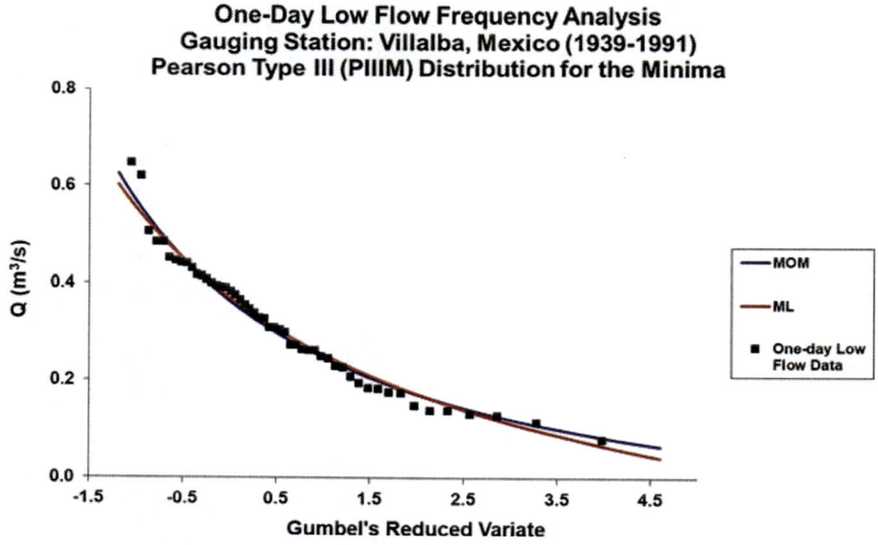

Fig. 12.5 Empirical and MOM-ML theoretical curves of PIIIM distribution of one-day low flow sample of Villalba, Mexico

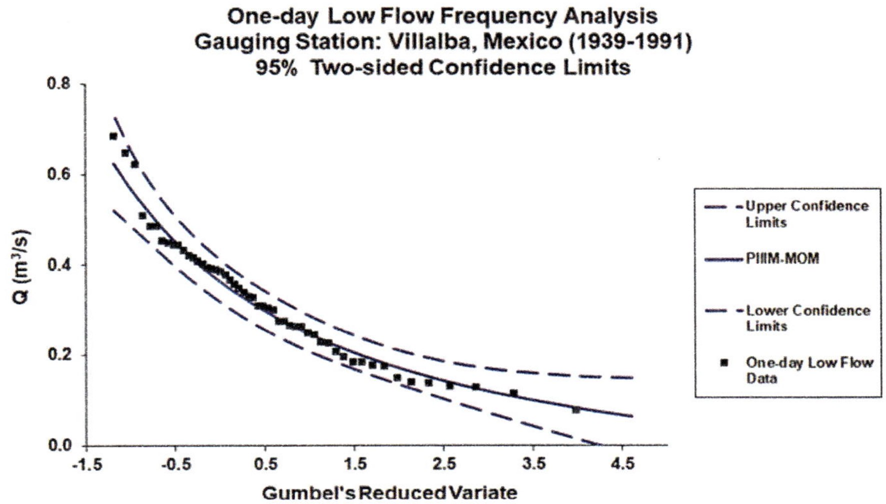

Fig. 12.6 Empirical and MOM-PIIIM theoretical frequency curves and confidence limits of one-day low flow sample of Villalba, Mexico

	A	B	C	D
1	7-DAY LOW FLOW FREQUENCY ANALYSIS			
2	Gauging Station: Villalba, Mexico (1939-1986)			
3				
4	*Descriptive Statistics*		*Descriptive Statistics of the Logarithms*	
5	Mean	2.8483	Mean	0.9451
6	Standard Error	0.1813	Standard Error	0.0679
7	Median	2.7125	Median	0.9979
8	Mode	#N/D	Mode	#N/D
9	Standard Deviation	1.2561	Standard Deviation	0.4705
10	Sample Variance	1.5779	Sample Variance	0.2214
11	Kurtosis	-0.2429	Kurtosis	-0.2698
12	Skewness	0.5782	Skewness	-0.3803
13	Range	5.2120	Range	2.0789
14	Minimum	0.7450	Minimum	-0.2944
15	Maximum	5.9570	Maximum	1.7846
16	Sum	136.7200	Sum	45.3646
17	Count	48.0000	Count	48.0000
18	Largest(1)	5.9570	Largest(1)	1.7846
19	Smallest(1)	0.7450	Smallest(1)	-0.2944
20	Confidence Level(95.0%)	0.3647	Confidence Level(95.0%)	0.1366

Fig. 12.7 Descriptive statistics for the 7-Day low flow sample data at Villalba, Mexico

7-DAY LOW FLOW FREQUENCY ANALYSIS
Gauging Station: Villalba, Mexico (1939-1986)
Pearson Type III (PIIIM) Distribution for the Minima
Method: MOM

Estimation of Parameters and Goodness of Fit Tests			
Skewness Coef. (Gamma Correction) =		0.7028	
Location Parameter =	-0.7265	m^3/s	
Scale Parameter =	0.4414	m^3/s	
Shape Parameter =	8.0989		
$\Gamma(\beta)$ =	6155.4596		
SEF =	0.14	MARD =	3.97

Return Period Tr (years)	$\Pi(x)$	Standard Normal Variate z_T	$\partial K_T/\partial\gamma$	Frequency Factor K_T	Squared root Stand. Error S_T (m^3/s)	Two-sided 95% Lower Limit (m^3/s)	Design Values Q_T (m^3/s)	Two-sided 95% Upper Limit (m^3/s)
2	0.5	0.0000	-0.1599	-0.1155	0.2027	2.3058	2.7032	3.1006
5	0.2	-0.8416	0.0104	-0.8551	0.1701	1.4408	1.7742	2.1076
10	0.1	-1.2816	0.1739	-1.1821	0.1785	1.0136	1.3635	1.7134
20	0.05	-1.6449	0.3413	-1.4234	0.2226	0.6241	1.0604	1.4966
25	0.04	-1.7507	0.3949	-1.4890	0.2419	0.5038	0.9780	1.4521
50	0.02	-2.0537	0.5596	-1.6657	0.3107	0.1470	0.7560	1.3650
100	0.01	-2.3263	0.7203	-1.8109	0.3864	-0.1837	0.5736	1.3309
Two-sided Confidence Limits:		z_α =	1.96					

Fig. 12.8 MOM estimators for the parameters, standard errors, Quantiles, and confidence limits of the PIIIM distribution for the 7-Day low flow sample data at Villalba, Mexico

12.9.2.1 MOM Method

By using an Excel© spreadsheet, the results contained in Fig. 12.8, were obtained for the MOM method applied to the PIIIM distribution to 7-day low flow from gauging station Villalba, Mexico (1939–1986).

12.9.2.2 ML Method

By using an Excel© spreadsheet, the results contained in Fig. 12.9, were obtained for the method of ML applied to the PIIIM distribution to 7-day low flow from gauging station Villalba, Mexico (1939–1986).

In Fig. 12.10 a comparison is made between the histogram and PIIIM theoretical density for the 7-day low flow sample of Villalba, Mexico.

A graphical comparison between the empirical and theoretical frequency curves from the results provided by the two methods mentioned before is shown in Fig. 12.11.

A graphical description of the empirical and theoretical frequency curves and the confidence limits of the best fit provided, in this case that of MOM, is shown in Fig. 12.12.

12.9.3 Earthquake Epicenter Distance Frequency Analysis

By using the earthquake epicenter distance data, Castillo (1988), the following descriptive statistics were obtained and are shown in Fig. 12.13.

7-DAY LOW FLOW FREQUENCY ANALYSIS
Gauging Station: Villalba, Mexico (1939-1986)
Pearson Type III (PIIIM) Distribution for the Minima
Method: ML

Estimation of Parameters and Goodness of Fit Tests				
Test Location Parameter =	-0.3888		$F(x_0)$ =	-1.3956E-08
Skewness Coef. (Gamma Correction) =	0.7028		<	2
Location Parameter =	-0.3888	m^3/s	$\psi(\beta)$ =	1.8607
Scale Parameter =	0.4668	m^3/s	$\Gamma(\beta)$ =	-1.4406E+20
Shape Parameter =	6.9217		Log-Likelihood Function =	6.9151E+21
SEF = 0.16	MARD =	4.55	AIC =	-1.3830E+22

Variance-Covariance Matrix							
$var(x_0)$ =	1.0596E+00	$cov(\alpha, x_0)$ =	1.7064E-01	$cov(\beta, x_0)$ =	-5.0637E+00	Determinant =	6.8300E-03
$var(\alpha)$ =	4.2822E-02	$cov(\alpha, \beta)$ =	-9.3179E-01	$var(\beta)$ =	2.6103E+01	$\psi'(\beta)$ =	0.1554

Return Period Tr (years)	$\Pi(x)$	Standard Normal Variate z_T	$\partial x_T/\partial \alpha$	$\partial x_T/\partial \beta$	Squared root Stand. Error S_T (m^3/s)	Two-sided 95% Lower Limit (m^3/s)	Design Values Q_T (m^3/s)	Two-sided 95% Upper Limit (m^3/s)
2	0.5	0.0000	6.5924	0.4664	0.6288	1.4567	2.6892	3.9217
5	0.2	-0.8416	4.6726	0.3900	0.4623	0.8869	1.7930	2.6991
10	0.1	-1.2816	3.8372	0.3508	0.3882	0.6422	1.4030	2.1638
20	0.05	-1.6449	3.2276	0.3190	0.3378	0.4563	1.1184	1.7805
50	0.02	-2.0537	2.6229	0.2841	0.3006	0.2470	0.8361	1.4252
100	0.01	-2.3263	2.2649	0.2615	0.2919	0.0969	0.6690	1.2411

Two-sided Confidence Limits: z_α = 1.96

Fig. 12.9 ML estimators of the parameters, standard errors, Quantiles and confidence limits of the PIIIM distribution for the 7-day low flow sample data at Villalba, Mexico

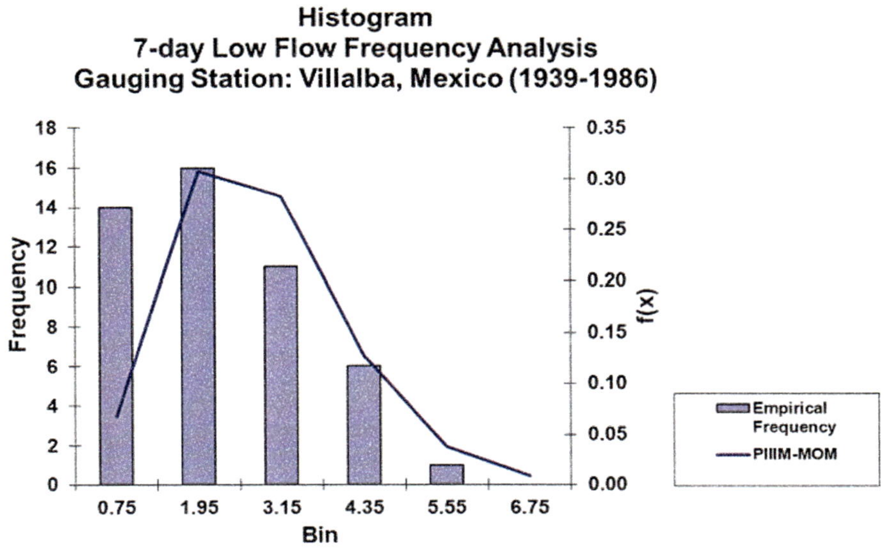

Fig. 12.10 Histogram and MOM-PIIIM theoretical density for the 7-Day low flow sample of Villalba, Mexico

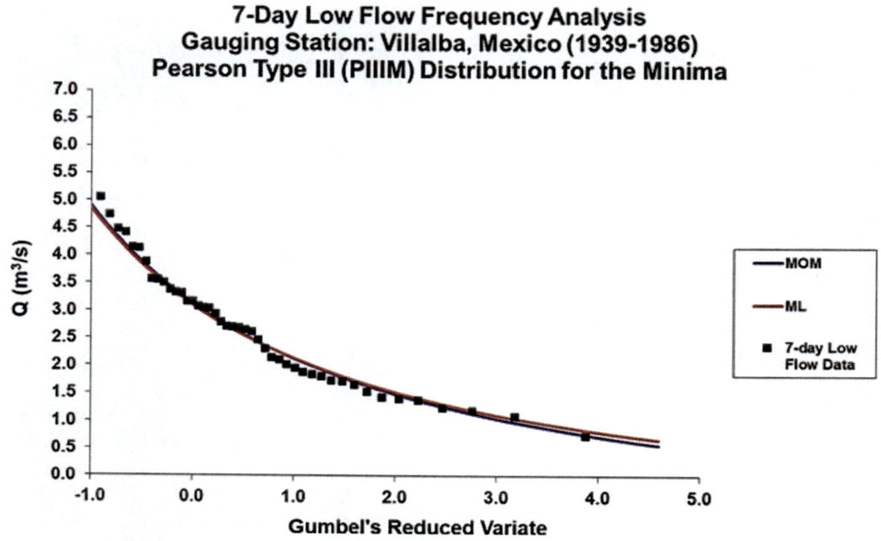

Fig. 12.11 Empirical and MOM-ML theoretical curves of PIIIM distribution of 7-day low flow sample of Villalba, Mexico

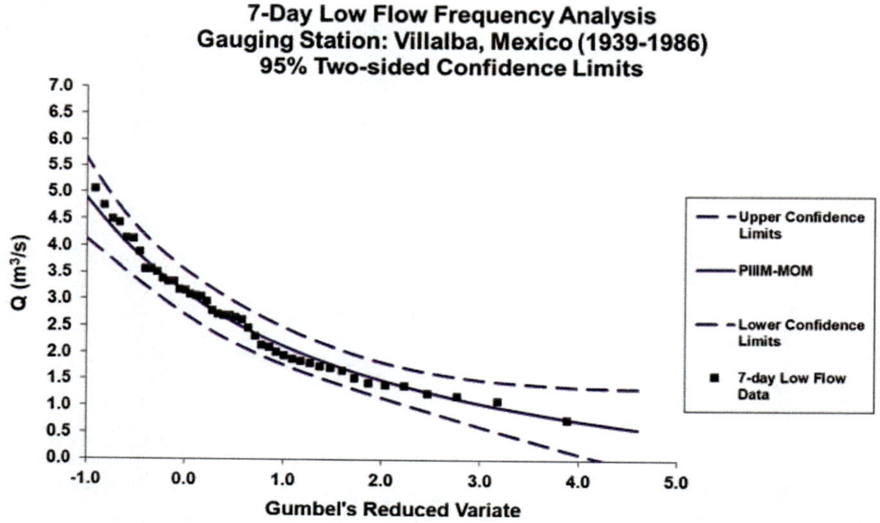

Fig. 12.12 Empirical and MOM-PIIIM theoretical frequency curves and confidence limits of 7-day low flow sample of Villalba, Mexico

	A	B	C	D
1	EARTHQUAKE EPICENTER DISTANCE FREQUENCY ANALYSIS			
2				
3	*Descriptive Statistics*		*Descriptive Statistics of the Logarithms*	
4	Mean	227.2069	Mean	5.3282
5	Standard Error	11.6856	Standard Error	0.0619
6	Median	248.1883	Median	5.5142
7	Mode	93.6438	Mode	4.5395
8	Standard Deviation	90.5161	Standard Deviation	0.4798
9	Sample Variance	8193.1719	Sample Variance	0.2302
10	Kurtosis	-1.1039	Kurtosis	1.1900
11	Skewness	-0.0732	Skewness	-0.9978
12	Range	347.5440	Range	2.3449
13	Minimum	36.8461	Minimum	3.6067
14	Maximum	384.3901	Maximum	5.9517
15	Sum	13632.4134	Sum	319.6947
16	Count	60.0000	Count	60.0000
17	Largest(1)	384.3901	Largest(1)	5.9517
18	Smallest(1)	36.8461	Smallest(1)	3.6067
19	Confidence Level(95.0%)	23.3828	Confidence Level(95.0%)	0.1240

Fig. 12.13 Descriptive statistics for the earthquake epicenter distance sample data

	A	B	C	D	E	F	G	H	I
1	EARTHQUAKE EPICENTER DISTANCE FREQUENCY ANALYSIS								
2	Pearson Type III (PIIIM) Distribution for the Minima								
3	Method: MOM								
4									
5	Estimation of Parameters and Goodness of Fit Tests								
6	Skewness Coef. (Gamma Correction) =			-0.0857					
7	Location Parameter =		-1883.9991	km					
8	Scale Parameter =		3.8808	km					
9	Shape Parameter =		544.0128						
10	Γ(β) =	1.3294E+19							
11	SEF =	17.17		MARD =	7.69				
12									
13	Design Values and Confidence Limits								
14			Standard			Squared root	Two-sided	Design	Two-sided
15			Normal		Frequency	Stand. Error	95%	Values	95%
16	Return Period		Variate		Factor	S_T	Lower Limit	D_T	Upper Limit
17	Tr (years)	Π(x)	z_T	$\partial K_T/\partial \gamma$	K_T	(km)	(km)	(km)	(km)
18	2	0.5	0.0000	-0.1666	0.0143	12.6280	201.1628	225.9136	250.6644
19	5	0.2	-0.8416	-0.0557	-0.8372	13.9784	123.2792	150.6768	178.0744
20	10	0.1	-1.2816	0.0981	-1.2903	16.6854	79.3676	112.0710	144.7744
21	20	0.05	-1.6449	0.2755	-1.6689	20.6480	40.0902	80.5603	121.0305
22	25	0.04	-1.7507	0.3358	-1.7798	22.1202	28.0881	71.4438	114.7995
23	50	0.02	-2.0537	0.5302	-2.0995	27.1249	-7.6710	45.4938	98.6587
24	100	0.01	-2.3263	0.7327	-2.3893	32.6001	-41.5468	22.3493	86.2455
25	Two-sided Confidence Limits:		z_α =	1.96					

Fig. 12.14 MOM estimators for the parameters and standard errors, quantiles, and confidence limits of the PIIIM distribution for the earthquake epicenter distance sample data

12.9.3.1 MOM Method

By using an Excel© spreadsheet, the results contained in Fig. 12.14, were obtained by the MOM method applied to the PIIIM distribution for the earthquake epicenter distance data.

	A	B	C	D	E	F	G	H	I
1	EARTHQUAKE EPICENTER DISTANCE FREQUENCY ANALYSIS								
2	Pearson Type III (PIIIM) Distribution for the Minima								
3	Method: ML								
4									
5	Estimation of Parameters and Goodness of Fit Tests								
6	Test Location Parameter =		-1883.9991		$F(x_o)$ =	6.6000E-05			
7	Skewness Coef. (Gamma Correction) =			-0.0857	<	2			
8	Location Parameter =		-1883.9991	km	$\psi(\beta)$ =	6.3102			
9	Scale Parameter =		3.8339	km	$\Gamma(\beta)$ =	4.5358E+73			
10	Shape Parameter =		550.6696	Log-Likelihood Function =		-2.7215E+75			
11	SEF =	17.27	MARD =	7.83	AIC =	5.4429E+75			
12									
13	Variance-Covariance Matrix								
14	$var(x_0)$ =	6.0843E+07	$cov(\alpha,x_0)$ =	1.1076E+05	$cov(\beta,x_0)$ =	-3.1778E+07	Determinant =	1.6932E-14	
15	$var(\alpha)$ =	2.0211E+02	$cov(\alpha,\beta)$ =	-5.7918E+04	$var(\beta)$ =	1.6608E+07	$\psi'(\beta)$ =	0.0018	
16									
17	Design Values and Confidence Limits								
18–21	Return Period Tr (years)	$\Pi(x)$	Standard Normal Variate z_T	$\partial x_T/\partial\alpha$	$\partial x_T/\partial\beta$	Squared root Stand. Error S_T (km)	Two-sided 95% Lower Limit (m^3/s)	Design Values Q_T (m^3/s)	Two-sided 95% Upper Limit (m^3/s)
22	2	0.5	0.0000	549.9889	3.8339	15.1013	196.3307	225.9292	255.5277
23	5	0.2	-0.8416	530.4824	3.7650	14.6702	122.3890	151.1427	179.8964
24	10	0.1	-1.2816	520.4721	3.7291	15.3555	82.6668	112.7636	142.8605
25	20	0.05	-1.6449	512.3010	3.6994	17.4641	47.2061	81.4357	115.6653
26	50	0.02	-2.0537	503.2073	3.6659	22.0332	3.3850	46.5701	89.7553
27	100	0.01	-2.3263	497.2049	3.6437	26.3629	-28.1144	23.5568	75.2281
28	Two-sided Confidence Limits:		z_α =	1.96					

Fig. 12.15 ML estimators for the parameters, estimators of the standard errors, quantiles, and confidence limits of the PIIIM distribution for the earthquake epicenter distance sample data

12.9.3.2 ML Method

By using an Excel© spreadsheet, the results contained in Fig. 12.15, were obtained by the ML method applied to the PIIIM distribution for the earthquake epicenter distance data.

In Fig. 12.16 a comparison is made between the histogram and MOM-PIIIM theoretical density for the earthquake epicenter distance sample data.

A graphical comparison between the empirical and theoretical frequency curves from the results provided by the two methods shown before is shown in Fig. 12.17.

A graphical depiction of the empirical and theoretical frequency curves and the confidence limits of the best fit provided, in this case that of MOM, is shown in Fig. 12.18.

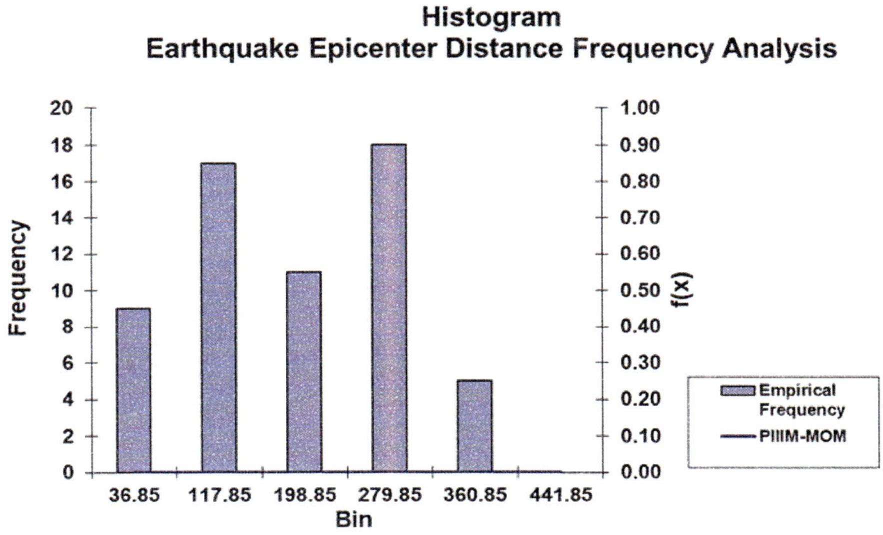

Fig. 12.16 Histogram and MOM-PIIIM theoretical density for the earthquake epicenter distance sample data

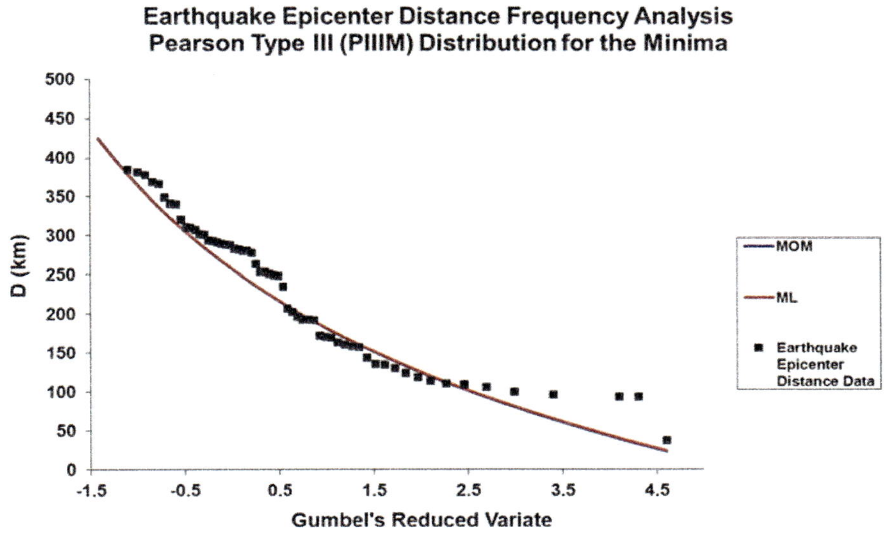

Fig. 12.17 Empirical and MOM-ML theoretical curves of PIIIM distribution of earthquake epicenter distance sample data

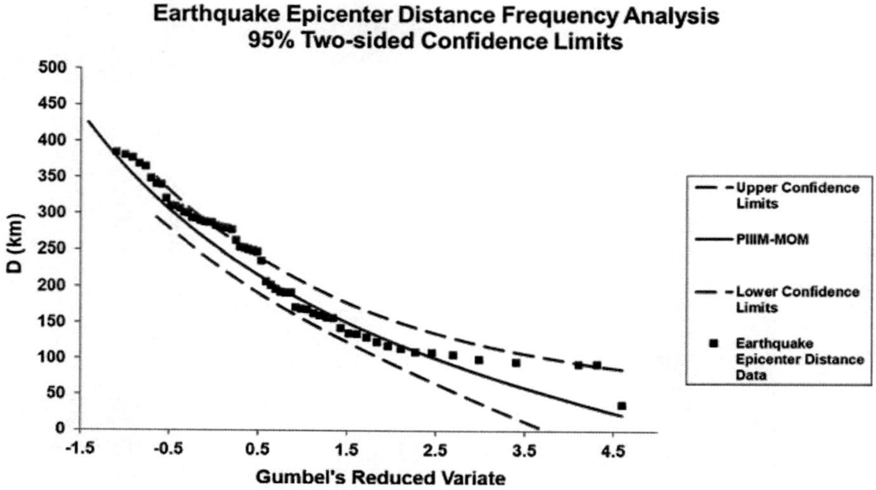

Fig. 12.18 Empirical and MOM-PIIIM theoretical frequency curves and confidence limits of earthquake epicenter distance sample data

Extreme Value Type III Distribution for the Minima

13

Don't follow the path. Go where there is no path and begin the trail. R. Bridges

13.1 Introduction

The extreme value type III (EVIII) distribution is one of the three particular solutions, independently found by Fisher-Tippett (1928) and Fréchet (1927), to the Stability Postulate that all the extremes must comply with. The EVIII distribution, also known as Weibull's distribution, has been studied extensively by Gumbel (1958), he provided the means to use this distribution in statistics and in engineering practice and since then the EVIII distribution is a widely used member of the family of extreme value distributions in engineering practice. Caruso (2000) analyzed 21 rivers in the Otago Region, South Island, New Zealand by performing low flow frequency analysis using Three-parameters Log-Normal, extreme value type I, extreme value type III, and general extreme value distributions, and he reported that log-Normal distribution gave overestimated values, while extreme value type I gave underestimated values when compared with the results produced by the extreme value type III distribution. In some cases, the general extreme value distribution was the best option.

J. A. Raynal Villaseñor, *Frequency Analyses of Natural Extreme Events*, Earth and Environmental Sciences Library,
https://doi.org/10.1007/978-3-030-86390-6_13

13.2 Chapter Objectives

After reading this chapter, you will know how to:

(a) Recognize the distribution and density functions of the EVIIIM distribution
(b) Estimate the parameters of the EVIIIM distribution
(c) Compute the quantiles and confidence limits of the EVIIIM distribution
(d) Make a graphic display of your data and the EVIIIM distribution
(e) Develop an application of all the above using Excel® spreadsheets.

13.3 Probability Distribution and Density Functions

The probability distribution function of the extreme value type III (EVIIIM) distribution for the minima is, Gumbel (1958):

$$\Pi(x) = \exp\left[-\left(\frac{x-\varepsilon}{v-\varepsilon}\right)^{k}\right] \tag{13.1}$$

where ε, k and v, are the location, scale, and shape parameters, respectively. $\Pi(x)$ is the probability distribution function of the random variable x and for the case of the minima frequency analysis, and is equal to the exceedance probability, $Pr(X > x)$. There are additional conditions: $k > 0$, $(v - \varepsilon) > 0$, $P(\varepsilon) = 1$ and $P(v) = 1/\mathrm{e}$. Usually, for the case of drought frequency analysis, ε is also called the smallest possible drought, v is called the characteristic drought and k is a parameter greater than zero. The domain of variable x in EVIIIM distribution is $\varepsilon \leq x < \infty$.

The probability density function for the EVIIIM distribution is, Gumbel (1958):

$$\pi(x) = \frac{k}{(v-\varepsilon)}\left(\frac{x-\varepsilon}{v-\varepsilon}\right)^{k-1} \exp\left[-\left(\frac{x-\varepsilon}{v-\varepsilon}\right)^{k}\right] \tag{13.2}$$

where $\pi(x)$ is the probability density function of random variable x. A graphical representation for the EVIIIM densities, for different values of κ, is depicted in Fig. 13.1

13.4 Estimation of the Parameters

13.4.1 MOM Method

The MOM method is based on the procedure devised to obtain the moments of inertia in statics. Fisher-Tippet (1928) adapted the MOM method to be used in statistics by considering the probability density function as the body to which the moments of inertia must be computed.

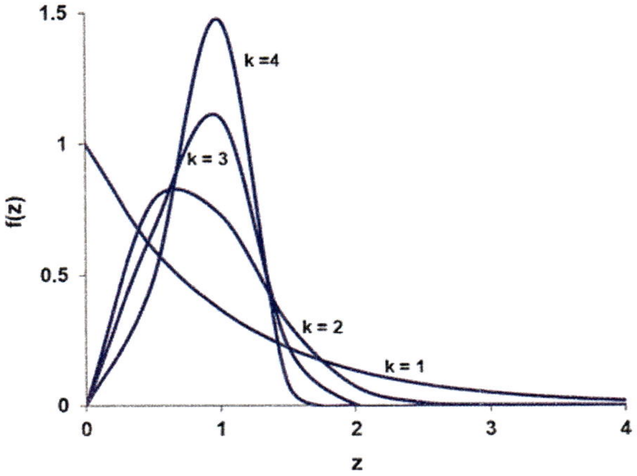

Fig. 13.1 EVIIIM densities for different values of scale parameter (κ)

The population mean and variance of the EVIIIM distribution are as follows, Gumbel (1958):

$$\mu = \varepsilon + (v - \varepsilon)\Gamma\left(1 + \frac{1}{k}\right) \tag{13.3}$$

$$\sigma^2 = (v - \varepsilon)^2\left[\Gamma\left(1 + \frac{2}{k}\right) - \Gamma^2\left(1 + \frac{1}{k}\right)\right] \tag{13.4}$$

and the third central moments are, Gumbel (1958):

$$\mu_3 = (v - \varepsilon)^3\left[\Gamma\left(1 + \frac{3}{k}\right) - 3\Gamma\left(1 + \frac{2}{k}\right)\Gamma\left(1 + \frac{1}{k}\right) + 2\Gamma^3\left(1 + \frac{1}{k}\right)\right] \tag{13.5}$$

so, the coefficient of skewness is, Gumbel (1958):

$$\gamma = \frac{\left[\Gamma(1 + \frac{3}{k}) - 3\Gamma(1 + \frac{2}{k})\Gamma(1 + \frac{1}{k}) + 2\Gamma^3(1 + \frac{1}{k})\right]}{\left[\Gamma(1 + \frac{2}{k}) - \Gamma^2(1 + \frac{1}{k})\right]^{\frac{3}{2}}} \tag{13.6}$$

and the fourth, fifth and sixth central moments are, Kite (1988):

$$\begin{aligned}\mu_4 = (v - \varepsilon)^4\Bigg[&\Gamma\left(1 + \frac{4}{k}\right) - 4\Gamma\left(1 + \frac{3}{k}\right)\Gamma\left(1 + \frac{1}{k}\right)\\ &+6\Gamma\left(1 + \frac{2}{k}\right)\Gamma^2\left(1 + \frac{1}{k}\right) - 3\Gamma^4\left(1 + \frac{1}{k}\right)\Bigg]\end{aligned} \tag{13.7}$$

$$\mu_5 = (v - \varepsilon)^5\left[\Gamma\left(1 + \frac{5}{k}\right) - 5\Gamma\left(1 + \frac{4}{k}\right)\Gamma\left(1 + \frac{1}{k}\right)\right.$$

$$+10\Gamma\left(1+\frac{3}{k}\right)\Gamma^2\left(1+\frac{1}{k}\right)-10\Gamma\left(1+\frac{2}{k}\right)\Gamma^3\left(1+\frac{1}{k}\right)$$
$$+4\Gamma^5\left(1+\frac{1}{k}\right)\bigg] \tag{13.8}$$

$$\mu_6=(v-\varepsilon)^6\left[\Gamma\left(1+\frac{6}{k}\right)-6\Gamma\left(1+\frac{5}{k}\right)\Gamma\left(1+\frac{1}{k}\right)+15\Gamma\left(1+\frac{4}{k}\right)\Gamma^2\left(1+\frac{1}{k}\right)\right.$$
$$\left.-20\Gamma\left(1+\frac{3}{k}\right)\Gamma^3\left(1+\frac{1}{k}\right)+15\Gamma\left(1+\frac{2}{k}\right)\Gamma^4\left(1+\frac{1}{k}\right)-5\Gamma^6\left(1+\frac{1}{k}\right)\right] \tag{13.9}$$

For the EVIIIM distribution, using the MOM method, the estimators of the parameters are obtained by first equating the population moments with the sample moments, and then simultaneously solving the resulting system of equations:

$$\mu=\overline{x} \tag{13.10}$$

$$\sigma^2=\hat{\sigma}^2=s^2 \tag{13.11}$$

$$\gamma=\hat{\gamma}=g \tag{13.12}$$

then using Eqs. (13.3), (13.4) and (13.6) in the left-hand side of Eqs. (13.10)–(13.12), respectively, and the expressions for the sample mean, sample variance, and sample skewness given in Chap. 2, in the right-hand side of Eqs. (13.10)–(13.12), respectively, we obtain the following set of expressions:

$$\varepsilon+(v-\varepsilon)\Gamma\left(1+\frac{1}{k}\right)=\frac{1}{N}\sum_{i=1}^{N}x_i \tag{13.13}$$

$$(v-\varepsilon)^2\left[\Gamma\left(1+\frac{2}{k}\right)-\Gamma^2\left(1+\frac{1}{k}\right)\right]=\frac{1}{N}\sum_{i=1}^{N}(x-\overline{x})^2 \tag{13.14}$$

$$\left[\Gamma\left(1+\frac{3}{k}\right)-3\Gamma\left(1+\frac{2}{k}\right)\Gamma\left(1+\frac{1}{k}\right)+2\Gamma^3\left(1+\frac{1}{k}\right)\right]B_k^3$$
$$=\frac{N}{(N-1)(N-2)\hat{\sigma}^3}\sum_{i=1}^{N}(x_i-\hat{\mu})^3 \tag{13.15}$$

The solution to the system of equations formed by Eqs. (13.13)–(13.15), provide the estimators of the MOM method for the EVIIIM distribution, Kite (1988):

$$\hat{\varepsilon}=\hat{v}-B_k\hat{\sigma} \tag{13.16}$$

$$\hat{\upsilon} = \overline{\mu} + A_k \hat{\sigma} \tag{13.17}$$

$$\hat{k} = \frac{1}{a_0 + a_1\hat{\gamma} + a_2\hat{\gamma}^2 + a_3\hat{\gamma}^3 + a_4\hat{\gamma}^4 + a_5\hat{\gamma}^5 + a_6\hat{\gamma}^6} \tag{13.18}$$

where $\hat{\mu}$ or $\overline{x}$ is the sample mean, $\hat{\sigma}$ or s is the sample standard deviation, and $\hat{\gamma}$ or g is the sample skewness coefficient and:

$$\begin{aligned} a_0 &= 0.277597 a_1 = 0.323127 a_2 = 0.061656 a_3 = -0.020235 \\ a_4 &= -0.007321 a_5 = 0.005578 a_6 = -0.001094 \end{aligned} \tag{13.19}$$

and this polynomial is valid for the range $-1.04 \leq \gamma \leq 2.0$ and it has a multiple correlation coefficient of 0.999999. The other auxiliary coefficients are, Kite (1988):

$$A_k = \left[1 - \Gamma\left(1 + \frac{1}{k}\right)\right] B_k \tag{13.20}$$

$$B_k = \left[\Gamma\left(1 + \frac{2}{k}\right) - \Gamma^2\left(1 + \frac{1}{k}\right)\right]^{-\frac{1}{2}} \tag{13.21}$$

where Γ(.) is the complete Gamma function defined in Chap. 2.

13.4.1.1 Example of Application of Estimation of the Parameters of the EVIIIM Distribution Using the MOM Method

Find the MOM estimators for the parameters of the EVIIIM distribution for the one-day low flow sample data of gauging station Villalba, Mexico (1939–1991), contained in Appendix A.

The following statistics have already been obtained:

$$\hat{\mu} = \overline{x} = \frac{1}{N} \sum_{i=1}^{N} x_i = 0.3306 \, \mathrm{m^3/s}$$

$$\hat{\sigma} = s = \left[\frac{1}{N} \sum_{i=1}^{N} (x_i - \hat{\mu})^2\right]^{1/2} = 0.1465 \, \mathrm{m^3/s}$$

$$\hat{\gamma} = \frac{N}{(N-1)(N-2)\hat{\sigma}^3} \sum_{i=1}^{N} (x_i - \hat{\mu})^3 = 0.6201$$

So, the moment estimators of the parameters for the EVIIIM distribution are as follows:

$$\hat{k} = \frac{1}{a_0 + a_1\hat{\gamma} + a_2\hat{\gamma}^2 + a_3\hat{\gamma}^3 + a_4\hat{\gamma}^4 + a_5\hat{\gamma}^5 + a_6\hat{\gamma}^6} = 2.0693$$

$$B_k = \left[\Gamma\left(1 + \frac{2}{k}\right) - \Gamma^2\left(1 + \frac{1}{k}\right)\right]^{-\frac{1}{2}} = \left[0.9863 - (0.8858)^2\right]^{-\frac{1}{2}} = 2.2269$$

$$A_k = \left[1 - \Gamma\left(1 + \frac{1}{k}\right)\right] B_k = (1 - 0.8858)(2.2269) = 0.2543$$

$$\hat{\varepsilon} = \overline{\mu} + A_k \hat{\sigma} = 0.3306 + (0.2543)(0.1465) = 0.0417\,\mathrm{m}^3/\mathrm{s}$$

$$\hat{\upsilon} = \varepsilon - B_k \hat{\sigma} = 0.0417 - (2.2269)(0.1465) = 0.3678$$

Finally, the moment estimators of the parameters of the EVIIIM distribution for the one-day low flow sample data of gauging station Villalba, Mexico are:

location parameter:

$$\hat{\varepsilon} = 0.0417\,\mathrm{m}^3/\mathrm{s}$$

scale parameter:

$$\hat{k} = 2.0693$$

and: shape parameter:

$$\hat{\upsilon} = 0.3678$$

13.4.2 ML Method

The likelihood function for the EVIIIM distribution is as follows, Gumbel (1958):

$$L(x, \varepsilon, k, \upsilon) = \prod_{i=1}^{N} \left\{ \frac{k}{(\upsilon - \varepsilon)} \left[-\frac{(x - \varepsilon)}{(\upsilon - \varepsilon)} \right]^{k-1} \exp\left[-\left[\frac{(x - \varepsilon)}{(\upsilon - \varepsilon)} \right]^k \right] \right\} \tag{13.22}$$

By taking the natural logarithm of the previous equation, the log-likelihood function is obtained as, Gumbel (1958):

$$LL(x, \varepsilon, k, \upsilon) = \sum_{i=1}^{N} Ln \left\{ \frac{k}{(\upsilon - \varepsilon)} \left[\frac{(x_i - \varepsilon)}{(\upsilon - \varepsilon)} \right]^{k-1} \exp\left[-\left[\frac{(x_i - \varepsilon)}{(\upsilon - \varepsilon)} \right]^k \right] \right\} \tag{13.23}$$

and the log-likelihood function for the EVIIIM distribution is finally obtained as, Gumbel (1958):

$$LL(x, \varepsilon, k, \upsilon) = NLn(k) - NkLn(\upsilon - \varepsilon)$$

$$+ (k-1)\sum_{i=1}^{N} Ln[(x_i-\varepsilon)] - \sum_{i=1}^{N}\left[\frac{(x_i-\varepsilon)}{(v-\varepsilon)}\right]^k \tag{13.24}$$

Now, the classical approach to the ML method requires the computation of the first-order partial derivatives of the log-likelihood function with respect to each of its parameters, equating them equal to zero and then solving the resulting system of equations. So, the first-order partial derivatives are obtained as follows, Kite (1988):

$$\frac{\partial LL(x,\varepsilon,k,v)}{\partial \varepsilon} = \frac{Nk}{(v-\varepsilon)} - (k-1)\sum_{i=1}^{N}\frac{1}{(x_i-\varepsilon)}$$

$$-\frac{k}{(v-\varepsilon)^{k+1}}\sum_{i=1}^{N}(x_i-\varepsilon)^k + \frac{k}{(v-\varepsilon)^k}\sum_{i=1}^{N}(x_i-\varepsilon)^{k-1} = 0 \tag{13.25}$$

$$\frac{\partial LL(x,\varepsilon,k,v)}{\partial k} = N\left[\frac{1}{k} - Ln(v-\varepsilon)\right]$$

$$+\sum_{i=1}^{N} Ln(x_i-\varepsilon) - \sum_{i=1}^{N}\left\{\left[\frac{(x_i-\varepsilon)}{(v-\varepsilon)}\right]^k Ln\left[\frac{(x_i-\varepsilon)}{(v-\varepsilon)}\right]\right\} = 0 \tag{13.26}$$

$$\frac{\partial LL(x,\varepsilon,k,v)}{\partial v} = -\frac{Nk}{(v-\varepsilon)} + k\sum_{i=1}^{N}\frac{(x_i-\varepsilon)^k}{(v-\varepsilon)^{k+1}} = 0 \tag{13.27}$$

Further simplification of Eqs. (13.25)–(13.27) yields the following results, Kite (1998):

$$(k-1)\sum_{i=1}^{N}\frac{1}{(x_i-\varepsilon)} - \frac{Nk\sum_{i=1}^{N}(x_i-\varepsilon)^{k-1}}{\sum_{i=1}^{N}(x_i-\varepsilon)^k} = 0 \tag{13.28}$$

$$N + k\sum_{i=1}^{N} Ln(x_i-\varepsilon) - \frac{Nk\sum_{i=1}^{N}(x_i-\varepsilon)^k Ln[(x_i-\varepsilon)]}{\sum_{i=1}^{N}(x_i-\varepsilon)^k} = 0 \tag{13.29}$$

$$v = \varepsilon + \left[\frac{1}{N}\sum_{i=1}^{N}(x_i-\varepsilon)^k\right]^{\frac{1}{k}} \tag{13.30}$$

The exact solution to the system formed by Eqs. (13.28)–(13.30) is not known, so an iterative procedure is needed to evaluate the ML estimators of the parameters of the EVIIIM distribution. The iterative procedure consists in solving simultaneously Eqs. (13.28) and (13.29), and then compute Eq. (13.30) with the parameters previously obtained, Kite (1988).

13.4.2.1 Example of Application of Estimation of the Parameters of the EVIIIM Distribution Using the ML Method

Find the ML estimators for the parameters of the EVIIIM distribution for the one-day low flow sample data of gauging station Villalba, Mexico (1939–1991), contained in Appendix A.

Using as starting values of the iterative procedure the MOM estimators computed in the preceding section, the following results have been obtained and they are the ML estimators of the parameters of the EVIIIM distribution for the one-day low flow sample data of gauging station Villalba, Mexico:

Location parameter:

$$\hat{\varepsilon} = 0.0263\,\mathrm{m}^3/\mathrm{s}$$

Scale parameter:

$$\hat{k} = 2.3341$$

and:

Shape parameter:

$$\hat{\upsilon} = 0.2500$$

13.5 Estimation of Quantiles for the EVIIIM Distribution

The quantiles for the EVIIIM distribution are obtained by using the inverse form of the EVIIIM distribution function:

$$x_T = \hat{\varepsilon} + \left(\hat{\upsilon} - \hat{\varepsilon}\right)[-Ln(\Pi(x))]^{\frac{1}{k}} \tag{13.31}$$

where x_T is the quantile value for a certain value of the distribution function $\Pi(x)$.

The term Q_T is more frequently used in engineering instead of x_T and this also applies to T_r instead of $\Pi(x)$; so, the following expression is widely used when it is needed to relate a low flow event Q_T to a specific return period T_r:

$$Q_T = \hat{\varepsilon} + \left(\hat{\upsilon} - \hat{\varepsilon}\right)\left[-Ln\left(\frac{1}{T_r}\right)\right]^{\frac{1}{k}} \tag{13.32}$$

where Q_T is a design value corresponding to a specific return period T_r.

Table 13.1 Estimation of MOM and ML quantiles for the EVIIIM distribution for Gauging Station Villalba, Mexico

Tr (Years)	MOM	ML
	Q_T (m^3/s)	Q_T (m^3/s)
2	0.3149	0.2175
5	0.1997	0.1439
10	0.1516	0.1116
20	0.1193	0.0889
25	0.1112	0.0831
50	0.0912	0.0683
100	0.0770	0.0575

13.5.1 Examples of Estimation of MOM and ML Quantiles for the EVIIIM Distribution

Find the MOM and ML estimators of the quantiles 2, 5, 10, 20, 50, and 100 years of return period, for the EVIIIM distribution for the one-day low flow sample data of gauging station Villalba, Mexico (1939–1991), contained in Appendix A.

The estimation of MOM and ML estimators of the quantiles for the EVIIIM distribution is made by using the parameters estimated in the preceding sections and then inserted the three parameters at a time in Eq. (13.32). Table 13.1 summarizes these results.

13.6 Goodness of Fit Test

The standard error of fit (SEF) for the EVIIIM distribution has the following form:

$$SEF = \left[\frac{\sum_{i=1}^{N} (xi - yi)^2}{(N - 3)} \right]^{1/2} \tag{13.33}$$

while the mean absolute relative deviation (MARD) remains the same as it was defined in Chap. 2:

$$MARD = \frac{100}{N} \sum_{i=1}^{N} \left| \frac{(x_i - y_i)}{x_i} \right| \tag{13.34}$$

where x_i are the sample historical values of the sample, y_i are the distribution function values, corresponding to the same return periods of the historical values, N is the sample size.

Table 13.2 SEF and MARD measures for MOM and ML estimators of the parameters of the EVIIIM distribution for the one-day low flow sample data of gauging station Villalba, Mexico

Goodness of fit test	MOM	ML
SEF	**0.02**	0.13
MARD	**4.21**	30.42

13.6.1 Examples of Application of the SEF and MARD to the MOM and ML Estimators of the Parameters of the EVIIIM Distribution

Find the values of the SEF and MARD of the MOM and ML estimators of the parameters of the EVIIIM distribution for the one-day low flow sample data of gauging station Villalba, Mexico (1939–1991), contained in Appendix A.

The values of SEF and MARD for the MOM and ML estimators of the parameters of the EVIIIM distribution for the one-day low flow sample data of gauging station Villalba, Mexico, have been obtained through the application of Eqs. (13.33) and (13.34) using the parameters obtained in previous examples. Table 13.2 contains a summary of both measures of goodness of fit tests.

For the one-day low flow sample data at gauging station Villalba, Mexico, the best choice according to SEF and MARD measures is the MOM method.

13.7 Estimation of Confidence Limits for the EVIIIM Distribution

By assuming that the quantiles are normally distributed, the following form can be used to set the confidence limits on such quantiles:

$$x_l = x_T \pm z_\alpha S_T \tag{13.35}$$

where x_l is the upper or lower confidence limit, depending on the sign in the preceding formula (+) for the upper confidence limit and (−) for the lower confidence limit, z_α is the standard normal variate corresponding to a confidence level α, and S_T is squared root of the standard error of the estimate. The evaluation procedures of the standard errors of the estimates will be described in the following subsections.

13.8 Estimation of Standard Errors for the EVIIIM Distribution

13.8.1 MOM Method

The general form of the MOM estimator of the standard error of the estimate of a three-parameter distribution is, Kite (1988):

$$\begin{aligned} S_T^2 = &\left(\frac{\partial x_T}{\partial m_1'}\right)^2 \mathrm{var}(m_1') + \left(\frac{\partial x_T}{\partial m_2}\right)^2 \mathrm{var}(m_2) + \left(\frac{\partial x_T}{\partial m_3}\right)^2 \mathrm{var}(m_3) \\ &+ 2\left(\frac{\partial x_T}{\partial m_1'}\right)\left(\frac{\partial x_T}{\partial m_2}\right)\mathrm{cov}(m_1', m_2) + 2\left(\frac{\partial x_T}{\partial m_1'}\right)\left(\frac{\partial x_T}{\partial m_3}\right)\mathrm{cov}(m_1', m_3) \\ &+ 2\left(\frac{\partial x_T}{\partial m_2}\right)\left(\frac{\partial x_T}{\partial m_3}\right)\mathrm{cov}(m_2, m_3) \end{aligned} \tag{13.36}$$

and Eq. (13.36) can be simplified in terms of the frequency factor, K_T, as, Kite (1988):

$$\begin{aligned} S_T^2 = \frac{\mu_2}{N}\Bigg\{ &1 + K_T\hat{\gamma} + \frac{K_T^2}{4}(\hat{\kappa} - 1) \\ &+ \frac{\partial K_T}{\partial \hat{\gamma}}\left[2\hat{\kappa} - 3\hat{\gamma}^2 - 6 + K_T\left(\hat{\lambda}_1 - \frac{6\hat{\gamma}\hat{\kappa}}{4} - \frac{10\hat{\gamma}}{4}\right)\right] \\ &+ \left(\frac{\partial K_T}{\partial \gamma}\right)^2\left[\hat{\lambda}_2 - 3\hat{\gamma}\hat{\lambda}_1 - 6\hat{\gamma} + \frac{9\hat{\gamma}^2\hat{\kappa}}{4} + \frac{35\hat{\gamma}^2}{4} + 9\right]\Bigg\} \end{aligned} \tag{13.37}$$

and:

$$\hat{\lambda}_1 = \frac{\frac{\left[\sum_{i=1}^{N}(x_i - \hat{\mu})^5\right]}{N}}{\left(\sigma^2\right)^{2.5}} \tag{13.38}$$

$$\hat{\lambda}_2 = \frac{\frac{\left[\sum_{i=1}^{N}(x_i - \hat{\mu})^6\right]}{N}}{\left(\sigma^2\right)^{3}} \tag{13.39}$$

where K_T is the frequency factor given by, Kite (1988):

$$K_T = A_K + B_K\left[\left(-Ln\left(1 - \frac{1}{T_r}\right)\right)^{\frac{1}{\hat{k}}} - 1\right] \tag{13.40}$$

and A_k and B_k were given in Eqs. (13.20) and (13.21).

The first-order partial derivative of the frequency factor, K_T, with respect to the skewness coefficient, γ, can be evaluated by means of the chain rule for differentiation, Kite (1988):

$$\frac{\partial K_T}{\partial \gamma} = \left(\frac{\partial K_T}{\partial (1/k)}\right)\left(\frac{\partial (1/k)}{\partial \gamma}\right) \tag{13.41}$$

and:

$$\frac{\partial K_T}{\partial (1/k)} = \left\{\frac{\left[y^{\frac{1}{k}} Ln(y) - G_1 P_1\right] - \left[y^{\frac{1}{k}} - G_1\right]\left[G_2 - G_1^2\right]^{-1}\left[G_2 P_2 - G_1^2 P_1\right]}{\left[G_2 - G_1^2\right]^{\frac{1}{2}}}\right\} \tag{13.42}$$

$$\frac{\partial \gamma}{\partial (1/k)} = 3\{[G_2 - G_1^2][G_3 P_3 - 3G_1 G_2(P_1 + 2P_2 + 2G_1^3 P_1)] - [G_3 - 3G_1 G_2 + 2G_1^3][G_2 P_2 - 2G_1^2 P_1]\}/[G_2 - G_1^2]^{2.5} \tag{13.43}$$

where

$$y = -Ln\left(1 - \frac{1}{T_r}\right) \tag{13.44}$$

$$G_r = \Gamma\left(1 + \frac{r}{k}\right) \tag{13.45}$$

$$P_r = \psi\left(1 + \frac{r}{k}\right) \tag{13.46}$$

and $\Gamma(.)$ and $\psi(.)$ are the complete Gamma and Digamma functions, previously defined.

Then, the first-order partial derivative of the frequency factor, K_T, with respect to the skewness coefficient, γ, is, Kite (1988):

$$\frac{\partial K_T}{\partial \gamma} = \frac{\left[G_2 - G_1^2\right]^2\left\{\left[y^{\frac{1}{k}} Ln(y) - G_1 P_1\right] - \left[y^{\frac{1}{k}} - G_1\right]\left[G_2 - G_1^2\right]^{-1}\left[G_2 P_2 - G_1^2 P_1\right]\right\}}{3\left\{\left[G_2 - G_1^2\right]\left[G_3 P_3 + G_1 G_2(P_1 + 2P_2) + 2G_1^3 P_1\right] - \left[G_3 - 3G_1 G_2 + 2G_1^3\right]\left[G_2 P_2 - G_1^2 P_1\right]\right\}} \tag{13.47}$$

The substitution of Eqs. (13.38)–(13.47) into Eq. (13.37) will provide the moment estimator of the standard error of the variate for the EVIIIM distribution.

Table 13.3 Estimation of MOM standard errors and the two-sided 95% confidence limits for the EVIIIM distribution for Gauging Station **Villalba, Mexico**

Tr (Years)	S_T (m^3/s)	95% Lower limit (m^3/s)	Q_T (m^3/s)	95% Upper limit (m^3/s)
2	0.0195	0.2768	0.3149	0.3530
5	0.0233	0.1539	0.1997	0.2454
10	0.0240	0.1046	0.1516	0.1986
20	0.0238	0.0727	0.1193	0.1659
25	0.0236	0.0650	0.1112	0.1574
50	0.0226	0.0469	0.0912	0.1355
100	0.0212	0.0355	0.0770	0.1185

Two-sided Limits: $z_\alpha = 1.96$

13.8.1.1 Example of Estimation of MOM Standard Errors and the Two-Sided 95% Confidence Limits for the EVIIIM Distribution

Find the MOM estimators of the standard errors and the two-sided 95% confidence limits for the EVIIIM distribution for the one-day low flow sample data of gauging station Villalba, Mexico (1939–1991), Appendix A.

The estimation of MOM estimators of the standard errors and the two-sided 95% confidence limits for the EVIIIM distribution is made by using the MOM estimators of the parameters and then inserted in the Eqs. (13.38)–(13.45) and then insert these results into Eq. (13.37), using selected values of the return intervals. Table 13.3 summarizes these results.

13.8.2 ML Method

The general form of the ML estimator of the standard error of the estimate of a three-parameter distribution is, Kite (1988):

$$\begin{aligned} S_T^2 = {} & \left(\frac{\partial x_T}{\partial \alpha}\right)^2 \text{var}(\alpha) + \left(\frac{\partial x_T}{\partial \beta}\right)^2 \text{var}(\beta) + \left(\frac{\partial x_T}{\partial \gamma}\right)^2 \text{var}(\gamma) \\ & + 2\left(\frac{\partial x_T}{\partial \alpha}\right)\left(\frac{\partial x_T}{\partial \beta}\right)\text{cov}(\alpha, \beta) + 2\left(\frac{\partial x_T}{\partial \alpha}\right)\left(\frac{\partial x_T}{\partial \gamma}\right)\text{cov}(\alpha, \gamma) \\ & + 2\left(\frac{\partial x_T}{\partial \beta}\right)\left(\frac{\partial x_T}{\partial \gamma}\right)\text{cov}(\beta, \gamma) \end{aligned} \tag{13.48}$$

and the elements of the Fisher's information matrix, are, Kite (1988):

$$-\frac{\partial^2 LL}{\partial \varepsilon^2} = -\frac{Nk}{(v-\varepsilon)^2} + (k-1)\sum_{i=1}^{N}\frac{1}{(x_i-\varepsilon)^2} + \frac{k(k+1)}{(v-\varepsilon)^{(k+2)}}\sum_{i=1}^{N}(x_i-\varepsilon)^k$$

$$-\frac{k^2}{(v-\varepsilon)^{k+1}}\sum_{i=1}^{N}\frac{1}{(x_i-\varepsilon)^{k-1}}+\frac{k(k-1)}{(v-\varepsilon)^k}\sum_{i=1}^{N}\frac{1}{(x_i-\varepsilon)^{k-2}} \quad (13.49)$$

$$-\frac{\partial^2 LL}{\partial k^2}=\frac{N}{k^2}+\frac{1}{(v-\varepsilon)^k}\left[\sum_{i=1}^{N}(x_i-\varepsilon)^k Ln^2(x_i-\varepsilon)\right.$$
$$\left.-Ln^2(v-\varepsilon)\sum_{i=1}^{N}(x_i-\varepsilon)^k\right] \quad (13.50)$$

$$-\frac{\partial^2 LL}{\partial v^2}=-\frac{Nk}{(v-\varepsilon)^2}+\frac{k(k+1)}{(v-\varepsilon)^{k+2}}\sum_{i=1}^{N}(x_i-\varepsilon)^k \quad (13.51)$$

$$-\frac{\partial^2 LL}{\partial\varepsilon\partial k}=-\frac{N}{(v-\varepsilon)}+\sum_{i=1}^{N}\frac{1}{(x_i-\varepsilon)}$$
$$-\frac{1}{(v-\varepsilon)}\left\{\sum_{i=1}^{N}(x_i-\varepsilon)^{k-1}[1+kLn(x_i-\varepsilon)]\right.$$
$$\left.-kLn(v-\varepsilon)\sum_{i=1}^{N}(x_i-\varepsilon)^{k-1}\right\}$$
$$-\frac{1}{(v-\varepsilon)^{k+1}}[1-kLn(v-\varepsilon)]\sum_{i=1}^{N}(x_i-\varepsilon)^k$$
$$+k(v-\varepsilon)^{k-1}\sum_{i=1}^{N}(x_i-\varepsilon)^k Ln(x_i-\varepsilon) \quad (13.52)$$

$$-\frac{\partial^2 LL}{\partial\varepsilon\partial v}=\frac{Nk}{(v-\varepsilon)^2}+\frac{k^2}{(v-\varepsilon)^{k+1}}\sum_{i=1}^{N}(x_i-\varepsilon)^{k-1}-\frac{k(k+1)}{(v-\varepsilon)^{k+2}}\sum_{i=1}^{N}(x_i-\varepsilon)^k \quad (13.53)$$

$$-\frac{\partial^2 LL}{\partial k\partial v}=\frac{N}{(v-\varepsilon)}-\frac{1}{(v-\varepsilon)^{k+1}}$$
$$\left\{[Ln(v-\varepsilon)-1]\sum_{i=1}^{N}(x_i-\varepsilon)^k+\sum_{i=1}^{N}(x_i-\varepsilon)^k Ln(x_i-\varepsilon)\right\} \quad (13.54)$$

Finally, the variances and covariances of the parameters of the EVIIIM distribution are found, for k > 2 as, Cohen and Witten (1988):

$$\mathrm{var}(\alpha)=\frac{(v-\varepsilon)^2}{N}\varphi_{11} \quad (13.55)$$

$$\mathrm{var}(k)=\frac{k^2}{N}\varphi_{22} \quad (13.56)$$

$$\operatorname{var}(v) = \frac{(v - \varepsilon)^2}{N}\varphi_{33} \tag{13.57}$$

$$\operatorname{cov}(\varepsilon, k) = \frac{(v - \varepsilon)}{N}\varphi_{12} \tag{13.58}$$

$$\operatorname{cov}(\varepsilon, v) = \frac{(v - \varepsilon)^2}{N}\varphi_{13} \tag{13.59}$$

$$\operatorname{cov}(k, v) = \frac{(v - \varepsilon)}{N}\varphi_{23} \tag{13.60}$$

where

$$\varphi_{11} = \frac{1.64493}{M} \tag{13.61}$$

$$\varphi_{22} = \frac{C - \Gamma^2\left(2 - \frac{1}{k}\right)}{M} \tag{13.62}$$

$$\varphi_{33} = \frac{1.82367C - J^2}{k^2 M} \tag{13.63}$$

$$\varphi_{12} = -\left[\frac{J + 0.42278\Gamma\left(2 - \frac{1}{k}\right)}{M}\right] \tag{13.64}$$

$$\varphi_{13} = -\left[\frac{0.42278J + 1.82367\Gamma\left(2 - \frac{1}{k}\right)}{k^2 M}\right] \tag{13.65}$$

$$\varphi_{23} = \left[\frac{0.42278C + J\Gamma\left(2 - \frac{1}{k}\right)}{M}\right] \tag{13.66}$$

and:

$$M = 1.82367C - 0.84556J\Gamma\left(2 - \frac{1}{k}\right) - 0.17874C - 1.82367\Gamma^2\left(2 - \frac{1}{k}\right) - J^2 \tag{13.67}$$

$$C = \left[\Gamma\left(1 - \frac{2}{k}\right) + k\Gamma\left(2 - \frac{2}{k}\right)\right]\left(\frac{k-1}{k^2}\right) \tag{13.68}$$

$$J = \Gamma\left(1 - \frac{1}{k}\right) + \Gamma\left(2 - \frac{1}{k}\right)\left[1 + \psi\left(2 - \frac{1}{k}\right)\right] \tag{13.69}$$

and from Eq. (13.38):

$$\frac{\partial x_T}{\partial \varepsilon} = 1 - y^{\frac{1}{k}} \tag{13.70}$$

$$\frac{\partial x_T}{\partial k} = -\frac{1}{k^2}(v-\varepsilon)y^{\frac{1}{k}}Ln(y) \tag{13.71}$$

$$\frac{\partial x_T}{\partial v} = y^{\frac{1}{k}} \tag{13.72}$$

So, Eq. (13.46) becomes to:

$$S_T^2 = \left(1-y^{\frac{1}{k}}\right)^2 \text{var}(\varepsilon) + \left[-\frac{1}{k^2}(v-\varepsilon)y^{\frac{1}{k}}Ln(y)\right]^2 \text{var}(k) + y^{\frac{2}{k}}\text{var}(v)$$
$$+2\left(y^{\frac{1}{k}}-y^{\frac{2}{k}}\right)\left[-\frac{1}{k^2}(v-\varepsilon)Ln(y)\right]\text{cov}(\varepsilon,k) + 2\left(y^{\frac{1}{k}}-y^{\frac{2}{k}}\right)\text{cov}(\varepsilon,v)$$
$$+2y^{\frac{2}{k}}\left[-\frac{1}{k^2}(v-\varepsilon)Ln(y)\right]\text{cov}(k,v) \tag{13.73}$$

13.8.2.1 Example of Estimation of ML Standard Errors and the Two-Sided 95% Confidence Limits for the EVIIIM Distribution

Find the ML estimators of the standard errors and the two-sided 95% confidence limits for the EVIIIM distribution for the one-day low flow sample data of gauging station Villalba, Mexico (1939–1991), contained in Appendix A.

The estimation of ML estimators of the standard errors and the two-sided 95% confidence limits for the EVIIIM distribution is made by using the ML estimators of the parameters and then inserted in the Eqs. (13.55)–(13.72) and into Eq. (13.73), using selected values of the return intervals. Table 13.4 summarizes these results.

Table 13.4 Estimation of ML standard errors and the two-sided 95% confidence limits for the EVIIIM distribution for Gauging Station Villalba, Mexico

Tr (Years)	S_T (m^3/s)	95% Lower Limit (m^3/s)	Q_T (m^3/s)	95% Upper Limit (m^3/s)
2	0.0251	0.1683	0.2175	0.2667
5	0.0318	0.0816	0.1439	0.2062
10	0.0384	0.0364	0.1116	0.1868
20	0.0451	0.0006	0.0889	0.1773
25	0.0472	−0.0094	0.0831	0.1756
50	0.0537	−0.0368	0.0683	0.1735
100	0.0599	−0.0600	0.0575	0.1749

Two-sided Limits: $z_\alpha = 1.96$

13.9 Examples of Application for the EVIIIM Distribution Using Excel® Spreadsheets

13.9.1 One-Day Low Flow Frequency Analysis

By using the one-day low flow data from gauging station Villalba, Mexico (1939–1991) the following descriptive statistics were obtained and are shown in Fig. 13.2.

13.9.1.1 MOM Method

By using an Excel® spreadsheet, the results contained in Fig. 13.3, were obtained by the MOM method applied to the EVIIIM distribution for the one-day low flow data at gauging station Villalba, Mexico.

13.9.1.2 ML Method

By using an Excel® spreadsheet, the results contained in Figs. 13.4 and 13.5, were obtained by the ML method applied to the EVIIIM distribution for gauging station Villalba, Mexico.

In Fig. 13.6 a comparison is made between the histogram and MOM-EVIIIM theoretical density for the One-Day Low Flow Sample of Villalba, Mexico.

A graphical comparison between the empirical and theoretical frequency curves from the results provided by the two methods mentioned before is shown in Fig. 13.7.

A graphical description of the empirical and theoretical frequency curves and the confidence limits of the best fit provided, in this case that of MOM, is shown in Fig. 13.8.

Fig. 13.2 Descriptive statistics for the one-day low flow sample data at Villalba, Mexico

	A	B	C
1	ONE-DAY LOW FLOW FREQUENCY ANALYSIS		
2	Gauging Station: Villalba, Mexico (1939-1991)		
3			
4	*Descriptive Statistics*		
5	Mean	0.3306	
6	Standard Error	0.0201	
7	Median	0.3280	
8	Mode	0.4850	
9	Standard Deviation	0.1465	
10	Sample Variance	0.0214	
11	Kurtosis	0.1906	
12	Skewness	0.5853	
13	Range	0.6250	
14	Minimum	0.0790	
15	Maximum	0.7040	
16	Sum	17.5214	
17	Count	53.0000	
18	Largest(1)	0.7040	
19	Smallest(1)	0.0790	
20	Confidence Level(95.0%)	0.0404	

	A	B	C	D	E	F	G	H	I
1	ONE-DAY LOW-FLOW FREQUENCY ANALYSIS								
2	Gauging Station: Villalba, Mexico (1939-1991)								
3	Extreme Value Type III (EVIIIM) Distribution for the Minima								
4	Method: MOM								
5									
6	Estimation of Parameters and Goodness of Fit Tests								
7	1+1/k =	1.4833	1+2/k =	1.9665	1+3/k =	2.4498	λ_1 =	0.0354	
8	Γ(1+1/k) =	0.8858	Γ(1+2/k) =	0.9863	Γ(1+3/k) =	1.2840	λ2 =	0.0425	
9	ψ(1+1/k) =	0.0207	ψ(1+2/k) =	0.4010	ψ(1+3/k) =	0.6782			
10	μ_z =	0.8858	σ_z =	0.4491	A_k =	0.2543	B_k =	2.2269	
11	Location Parameter =	0.0417	m^3/s						
12	Scale Parameter =	2.0693							
13	Shape Parameter =	0.3678							
14	SEF =	0.02		MARD =	4.21				
15									
16	Design Values and Confidence Limits								
17	Return Period	Exceedance		Frequency		Squared Root	Two-sided	Design	Two-sided
18		Probability		Factor		Stand. Error	95%	Values	95%
19	Tr	Π(x)	y	K_T	$\partial K_T/\partial \gamma$	S_T	Lower Limit	Q_T	Upper Limit
20	(years)					(m^3/s)	(m^3/s)	(m^3/s)	(m^3/s)
21	2	0.5	0.6931	-0.1072	-5.2051E-05	0.0195	0.2768	0.3149	0.3530
22	5	0.2	1.6094	0.8301	-2.6688E-05	0.0233	0.1539	0.1997	0.2454
23	10	0.1	2.3026	1.3597	1.8051E-05	0.0240	0.1046	0.1516	0.1986
24	20	0.05	2.9957	1.8116	7.0091E-05	0.0238	0.0727	0.1193	0.1659
25	25	0.04	3.2189	1.9453	8.7680E-05	0.0236	0.0650	0.1112	0.1574
26	50	0.02	3.9120	2.3325	1.4382E-04	0.0226	0.0469	0.0912	0.1355
27	100	0.01	4.6052	2.6856	2.0129E-04	0.0212	0.0355	0.0770	0.1185
28	Two-sided Confidence Limits:		z_α =	1.96					

Fig. 13.3 MOM estimators for the parameters, standard errors, quantiles, and confidence limits of the EVIIIM distribution for the one-day low flow sample data at Villalba, Mexico

	A	B	C	D	E	F	G	H	I
1	ONE-DAY LOW FLOW FREQUENCY ANALYSIS								
2	Gauging Station: Villalba, Mexico (1939-1991)								
3	Extreme Value Type III (EVIIIM) Distribution for the Minima								
4	Method: ML								
5									
6	Estimation of Parameters and Goodness of Fit Tests								
7	Initial Moment Estimators:			1-1/k =	0.5716	ϕ_{11}=	1.9084		
8	Location Parameter =	0.0417	m^3/s	Γ(1−1/k) =	1.5582	ϕ_{22}=	1.5563		
9	Scale Parameter =	2.0693	m^3/s	ψ(1−1/k) =	-1.6482	ϕ_{33}=	0.7580		
10	Shape Parameter =	0.3678		1-2/k =	0.1431	ϕ_{12}=	-1.1066		
11	Iterative Procedure:			Γ(1−2/k) =	6.5341	ϕ_{13}=	-0.3979		
12	Location Parameter =	0.0263		ψ(1−2/k) =	-7.3494	ϕ_{23}=	1.6436		
13	Scale Parameter =	2.3341		2-1/k =	1.5716	2-2/k =	1.1431		
14	Shape Parameter =	0.2500		(2-1/k)-1 =	0.5716	(2-2/k)-1 =	0.1431		
15	$F_1(\varepsilon,k)$ =	-1.0146E+00		Γ(2−1/k) =	0.8906	Γ(2−2/k) =	0.9353		
16	$F_2(\varepsilon,k)$ =	-4.1522E+00		ψ(2−1/k) =	0.1014	ψ(2−1/k) =	-0.3636		
17	Final MV Estimators:			A =	1.1014	J =	0.5773		
18	Location Parameter =	0.0263		C =	2.1347	M =	0.8619		
19	Scale Parameter =	2.3341		Log-likelihood function =		82.3039			
20	Shape Parameter =	0.2500							
21	SEF =	0.13	MARD =	30.42	AIC =	-158.6077			
22									
23	Variance-Covariance Matrix								
24	var(ε) =	0.0018		cov(ε,k) =	-0.0047				
25	var(k) =	0.1600		cov(ε,v)	-0.0004				
26	var(v) =	0.0007		cov(k,v) =	0.0069				

Fig. 13.4 ML estimators of the parameters of the EVIIIM distribution for the one-day low flow sample data at Villalba, Mexico

	A	B	C	D	E	F	G	H	I
28	Design Values and Confidence Limits								
29–32	Return Period Tr (years)	Exceedance Probability $\Pi(x)$	$\partial x_T/\partial \varepsilon$	$\partial x_T/\partial k$	$\partial x_T/\partial v$	Squared Root Stand. Error S_T (m^3/s)	Two-sided 95% Lower Limit (m^3/s)	Design Values Q_T (m^3/s)	Two-sided 95% Upper Limit (m^3/s)
33	2	0.5	0.1453	0.0129	0.8547	0.0251	0.1683	0.2175	0.2667
34	5	0.2	-0.2262	-0.0240	1.2262	0.0318	0.0816	0.1439	0.2062
35	10	0.1	-0.4295	-0.0490	1.4295	0.0384	0.0364	0.1116	0.1868
36	20	0.05	-0.6001	-0.0721	1.6001	0.0451	0.0006	0.0889	0.1773
37	25	0.04	-0.6501	-0.0792	1.6501	0.0472	-0.0094	0.0831	0.1756
38	50	0.02	-0.7939	-0.1005	1.7939	0.0537	-0.0368	0.0683	0.1735
39	100	0.01	-0.9238	-0.1206	1.9238	0.0599	-0.0600	0.0575	0.1749
40	Two-sided Confidence Limits:	z_α =	1.96						

Fig. 13.5 ML standard errors, quantiles, and confidence limits or the parameters of the EVIIIM distribution for the one-day low flow sample data at Villalba, Mexico

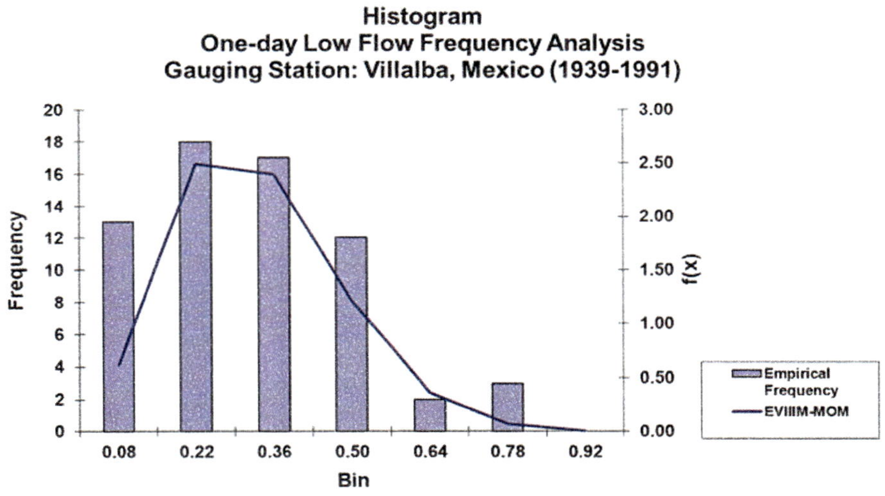

Fig. 13.6 Histogram and MOM-EVIIIM theoretical density for the one-day low flow sample of Villalba, Mexico

13.9.2 7-Day Low Flow Frequency Analysis

By using the 7-day low flow data from gauging station Villalba, Mexico (1939–1986) the following descriptive statistics were obtained and are shown in Fig. 13.9.

13.9.2.1 MOM Method

By using an Excel® spreadsheet, the results contained in Fig. 13.10, were obtained by the MOM method applied to the EVIIIM distribution for the 7-day low flow data at gauging station Villalba, Mexico.

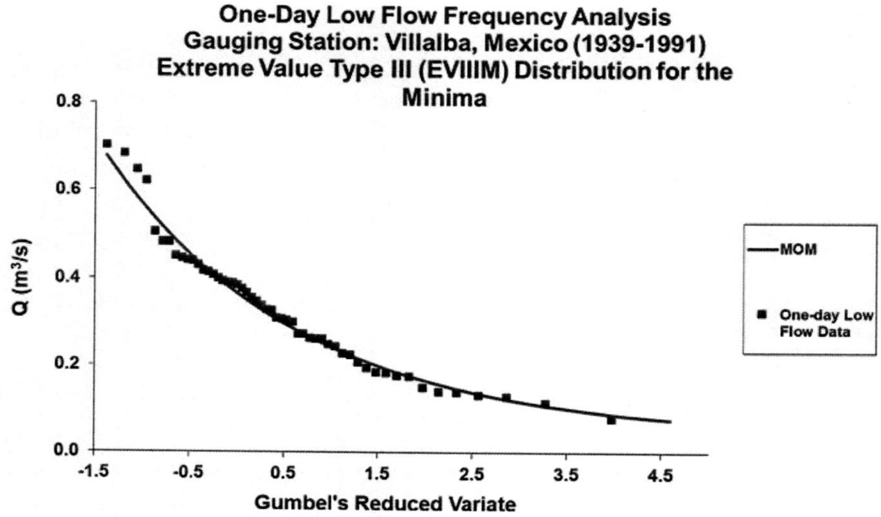

Fig. 13.7 Empirical and MOM-ML theoretical curves of EVIIIM distribution of one-day low flow sample of Villalba, Mexico

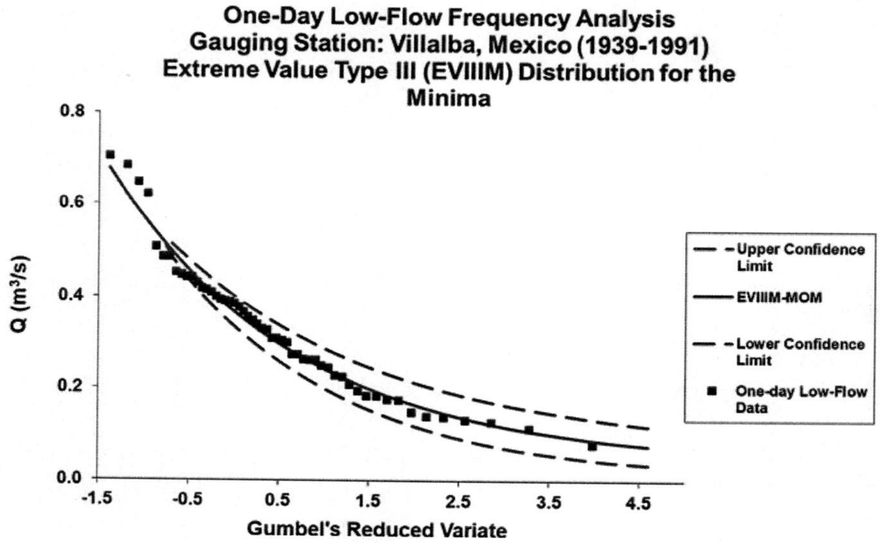

Fig. 13.8 Empirical and MOM-EVIIIM theoretical frequency curves and confidence limits of one-day low flow sample of Villalba, Mexico

	A	B	C
1	7-DAY LOW FLOW FREQUENCY ANALYSIS		
2	Gauging Station: Villalba, Mexico (1939-1986)		
3			
4	*Descriptive Statistics*		
5	Mean	2.8483	
6	Standard Error	0.1813	
7	Median	2.7125	
8	Mode	#N/D	
9	Standard Deviation	1.2561	
10	Sample Variance	1.5779	
11	Kurtosis	-0.2429	
12	Skewness	0.5782	
13	Range	5.2120	
14	Minimum	0.7450	
15	Maximum	5.9570	
16	Sum	136.7200	
17	Count	48.0000	
18	Largest(1)	5.9570	
19	Smallest(1)	0.7450	
20	Confidence Level(95.0%)	0.3647	

Fig. 13.9 Descriptive statistics for the 7-day low flow sample data at Villalba, Mexico

	A	B	C	D	E	F	G	H	I
1	7-DAY LOW-FLOW FREQUENCY ANALYSIS								
2	Gauging Station: Villalba, Mexico (1939-1986)								
3	Extreme Value Type III (EVIIIM) Distribution for the Minima								
4	Method: MOM								
5									
6	Estimation of Parameters and Goodness of Fit Tests								
7	1+1/k =	1.4806	1+2/k =	1.9613	1+3/k =	2.4419	λ_1 =	5.8804	
8	Γ(1+1/k) =	0.8858	Γ(1+2/k) =	0.9842	Γ(1+3/k) =	1.2772	λ2 =	19.4983	
9	ψ(1+1/k) =	0.0182	ψ(1+2/k) =	0.3975	ψ(1+3/k) =	0.6743			
10	μ_z =	0.8858	σ_z =	0.4468	A_k =	0.2557	B_k =	2.2379	
11	Location Parameter =	0.3583	m^3/s						
12	Scale Parameter =	2.0806							
13	Shape Parameter =	3.1695							
14	SEF =	0.13		MARD =	3.58				
15									
16	Design Values and Confidence Limits								
17	Return Period	Exceedance		Frequency		Squared Root	Two-sided	Design	Two-sided
18		Probability		Factor		Stand. Error	95%	Values	95%
19	Tr	Π(x)	y	K_T	$\partial K_T/\partial y$	S_T	Lower Limit	Q_T	Upper Limit
20	(years)					(m^3/s)	(m^3/s)	(m^3/s)	(m^3/s)
21	2	0.5	0.6931	-0.1058	-4.9689E-05	0.1754	2.3717	2.7154	3.0592
22	5	0.2	1.6094	0.8308	-2.5203E-05	0.2040	1.3255	1.7254	2.1252
23	10	0.1	2.3026	1.3592	1.7519E-05	0.1996	0.9202	1.3114	1.7026
24	20	0.05	2.9957	1.8098	6.7110E-05	0.1839	0.6722	1.0327	1.3931
25	25	0.04	3.2189	1.9430	8.3858E-05	0.1768	0.6161	0.9626	1.3090
26	50	0.02	3.9120	2.3287	1.3728E-04	0.1475	0.5001	0.7892	1.0784
27	100	0.01	4.6052	2.6803	1.9192E-04	0.1023	0.4658	0.6664	0.8670
28	Two-sided Confidence Limits:		z_α =	1.96					

Fig. 13.10 MOM estimators for the parameters, standard errors, quantiles, and confidence limits of the EVIIIM distribution for the 7-day low flow sample data at Villalba, Mexico

13.9.2.2 ML Method

By using an Excel® spreadsheet, the results contained in Figs. 13.11 and 13.12, were obtained by the ML method applied to the EVIIIM distribution for the 7-day low flow data at gauging station Villalba, Mexico.

7-DAY LOW FLOW FREQUENCY ANALYSIS
Gauging Station: Villalba, Mexico (1939-1986)
Extreme Value Type III (EVIIIM) Distribution for the Minima
Method: ML

Estimation of Parameters and Goodness of Fit Tests						
Initial Moment Estimators:			1-1/k =	0.5348	ϕ_{11}=	0.4698
Location Parameter =	0.3583	m³/s	Γ(1−1/k) =	1.6601	ϕ_{22}=	0.9057
Scale Parameter =	2.0806	m³/s	ψ(1−1/k) =	-1.8012	ϕ_{33}=	0.4150
Shape Parameter =	3.1695		1-2/k =	0.0697	ϕ_{12}=	-0.3103
Iterative Procedure:			Γ(1−2/k) =	13.8429	ϕ_{13}=	-0.1186
Location Parameter =	0.4368		ψ(1−2/k) =	-14.8234	ϕ_{23}=	0.6584
Scale Parameter =	2.1498		2-1/k =	1.5348	2-2/k =	1.0697
Shape Parameter =	3.1695		(2-1/k)-1 =	0.5348	(2-2/k)-1 =	0.0697
$F_1(\varepsilon,k)$ =	-3.8588E-07		Γ(2−1/k) =	0.8879	Γ(2−2/k) =	0.9643
$F_2(\varepsilon,k)$ =	-3.7750E+00		ψ(2−1/k) =	0.0686	ψ(2−1/k) =	-0.4681
Final MV Estimators:			A =	1.0686	J =	0.7113
Location Parameter =	0.4368		C =	3.9597	M =	3.5017
Scale Parameter =	2.1498		Log-likelihood function =		-115.8227	
Shape Parameter =	3.1695		No convergence was attained			
SEF =	0.17	MARD =	4.80	AIC =	237.6455	

Variance-Covariance Matrix			
var(ε) =	0.0731	cov(ε,k) =	-0.0177
var(k) =	0.0872	cov(ε,v)	-0.0185
var(v) =	0.0646	cov(k,v) =	0.0375

Fig. 13.11 ML estimators of the parameters of the EVIIIM distribution for the 7-day low flow sample data at Villalba, Mexico

Design Values and Confidence Limits

Return Period Tr (years)	Exceedance Probability Π(x)	$\partial x_T/\partial \varepsilon$	$\partial x_T/\partial k$	$\partial x_T/\partial v$	Squared Root Stand. Error S_T (m³/s)	Two-sided 95% Lower Limit (m³/s)	Design Values Q_T (m³/s)	Two-sided 95% Upper Limit (m³/s)
2	0.5	0.1567	0.1828	0.8433	0.2372	2.2762	2.7412	3.2061
5	0.2	-0.2478	-0.3511	1.2478	0.3021	1.2049	1.7969	2.3890
10	0.1	-0.4740	-0.7269	1.4740	0.3688	0.6732	1.3961	2.1191
20	0.05	-0.6659	-1.0808	1.6659	0.4404	0.2600	1.1232	1.9864
25	0.04	-0.7225	-1.1907	1.7225	0.4638	0.1450	1.0540	1.9630
50	0.02	-0.8861	-1.5213	1.8861	0.5365	-0.1697	0.8818	1.9332
100	0.01	-1.0348	-1.8375	2.0348	0.6084	-0.4341	0.7584	1.9509
Two-sided Confidence Limits:	z_α =	1.96						

Fig. 13.12 ML standard errors, quantiles and confidence limits or the parameters of the EVIIIM distribution for the 7-day low flow sample data at Villalba, Mexico

In Fig. 13.13 a comparison is made between the histogram and MOM-EVIIIM theoretical density for the 7-day low flow sample of Villalba, Mexico.

A graphical comparison between the empirical and theoretical frequency curves from the results provided by the two methods mentioned before is shown in Fig. 13.14.

A graphical description of the empirical and theoretical frequency curves and the confidence limits of the best fit provided, in this case that of MOM, is shown in Fig. 13.15.

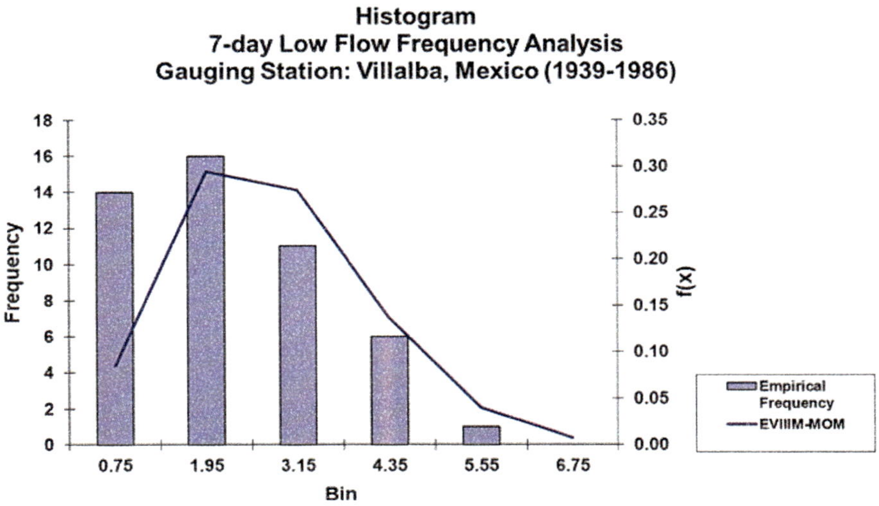

Fig. 13.13 Histogram and MOM-EVIIIM theoretical density for the 7-day low flow sample of Villalba, Mexico

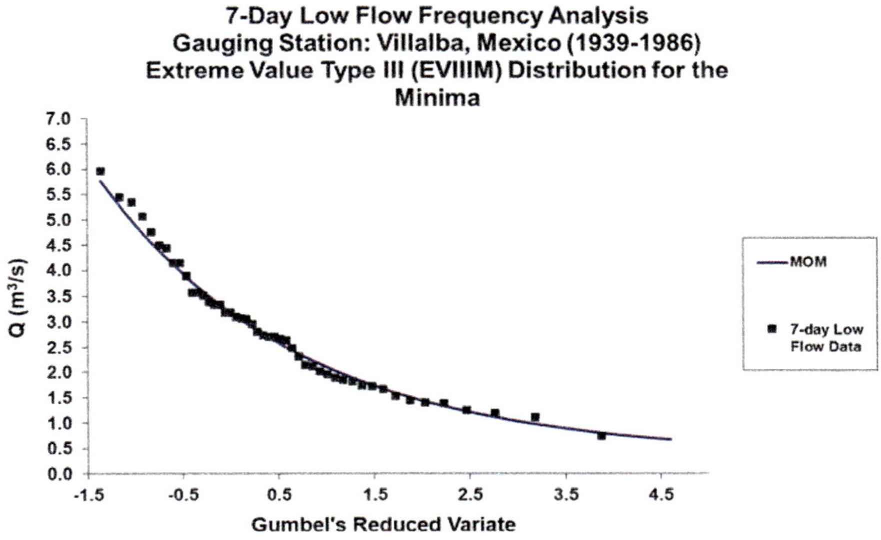

Fig. 13.14 Empirical and MOM-ML theoretical curves of eviiim distribution of 7-day low flow sample of Villalba, Mexico

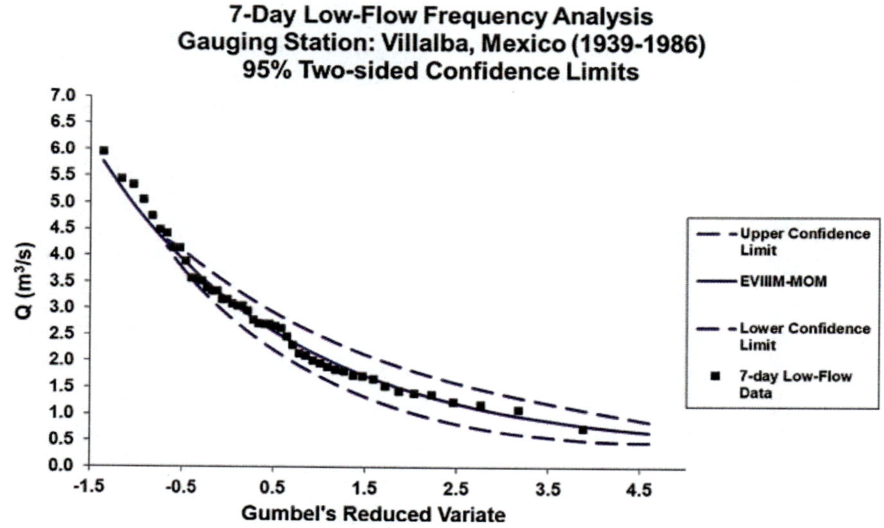

Fig. 13.15 Empirical and MOM-EVIIIM theoretical frequency curves and confidence limits of 7-day low flow sample of Villalba, Mexico

	A	B	C	D
1	EARTHQUAKE EPICENTER DISTANCE FREQUENCY ANALYSIS			
2				
3	*Descriptive Statistics*			
4	Mean	227.2069		
5	Standard Error	11.6856		
6	Median	248.1883		
7	Mode	93.6438		
8	Standard Deviation	90.5161		
9	Sample Variance	8193.1719		
10	Kurtosis	-1.1039		
11	Skewness	-0.0732		
12	Range	347.5440		
13	Minimum	36.8461		
14	Maximum	384.3901		
15	Sum	13632.4134		
16	Count	60.0000		
17	Largest(1)	384.3901		
18	Smallest(1)	36.8461		
19	Confidence Level(95.0%)	23.3828		

Fig. 13.16 Descriptive statistics for the earthquake epicenter distance sample data

13.9.3 Earthquake Epicenter Distance Frequency Analysis

By using the earthquake epicenter distance data, Castillo (1988), the following descriptive statistics were obtained and are shown in Fig. 13.16.

	A	B	C	D	E	F	G	H	I
1	EARTHQUAKE EPICENTER DISTANCE FREQUENCY ANALYSIS								
2	Extreme Value Type III (EVIIIM) Distribution for the Minima								
3	Method: MOM								
4									
5	Estimation of Parameters and Goodness of Fit Tests								
6	1+1/k =	1.2543	1+2/k =	1.5086	1+3/k =	1.7628	λ_1 =	1.10E+04	
7	Γ(1+1/k) =	0.9055	Γ(1+2/k) =	0.8708	Γ(1+3/k) =	0.9215	λ2 =	1.95E+06	
8	ψ(1+1/k) =	-0.2223	ψ(1+2/k) =	0.0445	ψ(1+3/k) =	0.2572			
9	μ_z =	0.9055	σ_z =	0.2254	A_k =	0.4191	B_k =	4.4362	
10	Location Parameter =	-136.4070	km						
11	Scale Parameter =	3.9327							
12	Shape Parameter =	265.1404							
13	SEF =	18.18		MARD =	9.04				
14									
15	Design Values and Confidence Limits								
16	Return Period	Exceedance		Frequency		Squared Root	Two-sided	Design	Two-sided
17		Probability		Factor		Stand. Error	95%	Values	95%
18	Tr	Π(x)	y	K_T	$\partial K_T/\partial \gamma$	S_T	Lower Limit	D_T	Upper Limit
19	(years)					(Km)	(Km)	(Km)	(Km)
20	2	0.5	0.6931	0.0243	1.1358E-06	0.0203	229.3690	229.4087	229.4485
21	5	0.2	1.6094	0.9897	1.6848E-06	0.0238	137.7650	137.8116	137.8582
22	10	0.1	2.3026	1.4671	1.4969E-06	0.0242	90.1281	90.1755	90.2230
23	20	0.05	2.9957	1.8466	1.1552E-06	0.0239	52.2301	52.2770	52.3239
24	25	0.04	3.2189	1.9547	1.0285E-06	0.0237	41.5872	41.6338	41.6803
25	50	0.02	3.9120	2.2583	6.0601E-07	0.0230	12.4298	12.4749	12.5199
26	100	0.01	4.6052	2.5241	1.5835E-07	0.0220	-11.7870	-11.7440	-11.7009
27	Two-sided Confidence Limits:		z_α =	1.96					

Fig. 13.17 MOM estimators for the parameters, standard errors, quantiles, and confidence limits of the EVIIIM distribution for the earthquake epicenter distance sample data

13.9.3.1 MOM Method

By using an Excel© spreadsheet, the results contained in Fig. 13.17, were obtained by the MOM method applied to the EVIIIM distribution for the earthquake epicenter distance data.

13.9.3.2 ML Method

By using an Excel© spreadsheet, the results contained in Figs. 13.18 and 13.19, were obtained by the ML method applied to the EVIIIM distribution for the earthquake epicenter distance data.

In Fig. 13.20 a comparison is made between the histogram and MOM-EVIIIM theoretical density for the earthquake epicenter distance sample data.

A graphical comparison between the empirical and theoretical frequency curves from the results provided by the two methods shown before is shown in Fig. 13.21.

A graphical depiction of the empirical and theoretical frequency curves and the confidence limits of the best fit provided, in this case that of MOM, is shown in Fig. 13.22.

	A	B	C	D	E	F	G
1	EPICENTER DISTANCE FREQUENCY ANALYSIS						
2	Extreme Value Type III (EVIIIM) Distribution for the Minima						
3	Method: ML						
4							
5	Estimation of Parameters and Goodness of Fit Tests						
6	Initial Moment Estimators:			1-1/k =	0.7455	ϕ_{11}=	-48.1368
7	Location Parameter =	-136.4070	m³/s	Γ(1−1/k) =	1.2314	ϕ_{22}=	-4.6797
8	Scale Parameter =	3.9327	m³/s	ψ(1−1/k) =	-1.0972	ϕ_{33}=	-3.4499
9	Shape Parameter =	265.1404		1-2/k =	0.4911	ϕ_{12}=	13.9709
10	Iterative Procedure:			Γ(1−2/k) =	1.8041	ϕ_{13}=	3.2438
11	Location Parameter =	-81.0000		ψ(1−2/k) =	-2.0081	ϕ_{23}=	-14.8049
12	Scale Parameter =	3.9300		2-1/k =	1.7455	2-2/k =	1.4911
13	Shape Parameter =	265.0000		(2-1/k)-1 =	0.7455	(2-2/k)-1 =	0.4911
14	$F_1(\varepsilon,k)$ =	-8.2598E-02		Γ(2−1/k) =	0.9181	Γ(2−2/k) =	0.8860
15	$F_2(\varepsilon,k)$ =	7.9071E-02		ψ(2−1/k) =	0.2441	ψ(2−1/k) =	0.0281
16	Final MV Estimators:			A =	1.2441	J =	0.0893
17	Location Parameter =	-81.0000		C =	1.0028	M =	-0.0342
18	Scale Parameter =	3.9300		Log-likelihood function =		-1343.8676	
19	Shape Parameter =	265.0000					
20	SEF =	16.88	MARD =	8.17	AIC =	2693.7351	
21							
22	Variance-Covariance Matrix						
23	var(ε) =	-96045.7301		cov(ε,k) =	80.5655		
24	var(k) =	-1.2046		cov(ε,v)	6472.2191		
25	var(v) =	-6883.3965		cov(k,v) =	-85.3747		

Fig. 13.18 ML Estimators for the parameters of the EVIIIM distribution for the earthquake epicenter distance sample data

	A	B	C	D	E	F	G	H	I
27	Design Values and Confidence Limits								
28–31	Return Period Tr (years)	Exceedance Probability Π(x)	∂x_T/∂ε	∂x_T/∂k	∂x_T/∂v	Squared Root Stand. Error S_T (km)	Two-sided 95% Lower Limit (km)	Design Values D_T (km)	Two-sided 95% Upper Limit (km)
32	2	0.5	0.0890	7.4796	0.9110	#¡NUM!	#¡NUM!	234.1909	#¡NUM!
33	5	0.2	-0.1287	-12.0332	1.1287	#¡NUM!	#¡NUM!	155.2226	#¡NUM!
34	10	0.1	-0.2364	-23.1015	1.2364	#¡NUM!	#¡NUM!	114.1613	#¡NUM!
35	20	0.05	-0.3220	-32.4953	1.3220	#¡NUM!	#¡NUM!	81.4977	#¡NUM!
36	25	0.04	-0.3464	-35.2618	1.3464	#¡NUM!	#¡NUM!	72.3254	#¡NUM!
37	50	0.02	-0.4149	-43.2375	1.4149	#¡NUM!	#¡NUM!	47.1984	#¡NUM!
38	100	0.01	-0.4749	-50.4598	1.4749	#¡NUM!	#¡NUM!	26.3310	#¡NUM!
39	Two-sided Confidence Limits:	z_α =	1.96						

Fig. 13.19 ML estimators of the standard errors, quantiles, and confidence limits of the EVIIIM distribution for the earthquake epicenter distance sample data

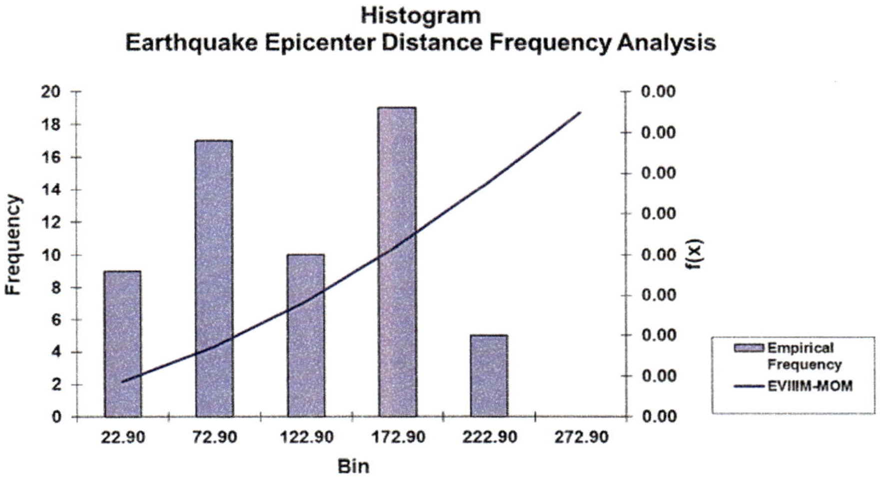

Fig. 13.20 Histogram and MOM-EVIIIM theoretical density for the earthquake epicenter distance sample data

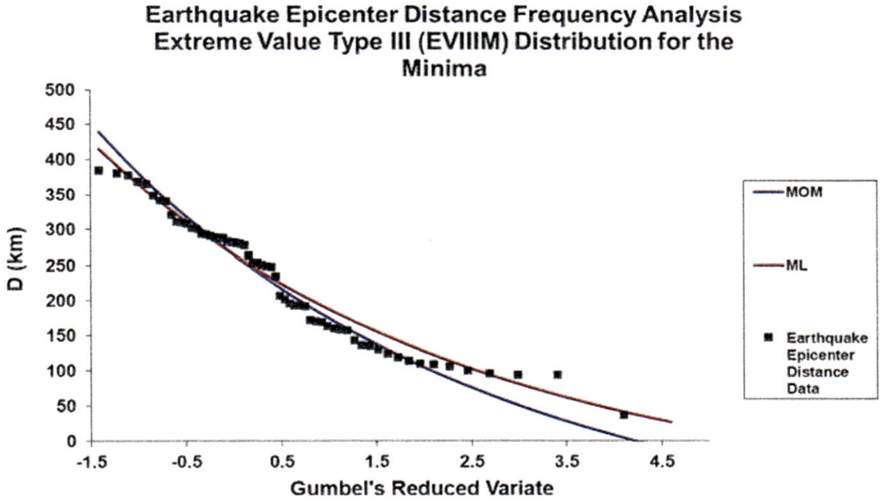

Fig. 13.21 Empirical and MOM-ML theoretical curves of EVIIIM distribution of earthquake epicenter distance sample data

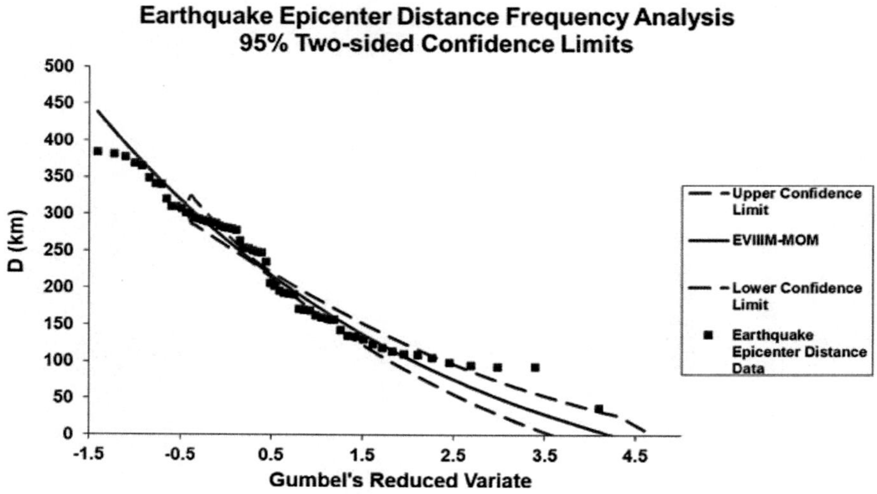

Fig. 13.22 Empirical and MOM-EVIIIM theoretical frequency curves and confidence limits of earthquake epicenter distance sample data

14 General Extreme Value Distribution for the Minima

Success isn't about the end result; is about what you learn along the way. V. Wang.

14.1 Introduction

The General Extreme Value (GEV) distribution is the general solution, found by Jenkinson (1955), to the Stability Postulate that all the extremes must comply with. The GEV distribution has been under study since 1955 and it has experienced a growing acceptance by the practicing engineers as the computing devices have improved every single year since the 1980's. Raynal (1994, 1995 and 1996) has used the general extreme value distribution with much better results that those produced by the extreme value type III distribution for the minima in several Mexican rivers. Caruso (2000) analyzed 21 rivers in the Otago Region, South Island, New Zealand by performing low flow frequency analysis using Three-parameters Log-Normal, extreme value type I, extreme value type III, and general extreme value distributions, and he reported that log-Normal distribution gave overestimated values while extreme value I gave underestimated values when compared with the results produced by the extreme value type III distribution. In some cases, the general extreme value distribution was the best option. Zaidman et al. (2003) used the general extreme value, the generalized Pareto, generalized Logistic and Pearson type III to characterize low flows in British rivers and they found that the general extreme value distribution was the model with more applicability to different conditions. More recently, Hewa et al. (2007) explored the application of LH-moments to low flow frequency analysis using the general extreme value distribution.

J. A. Raynal Villaseñor, *Frequency Analyses of Natural Extreme Events*,
Earth and Environmental Sciences Library,
https://doi.org/10.1007/978-3-030-86390-6_14

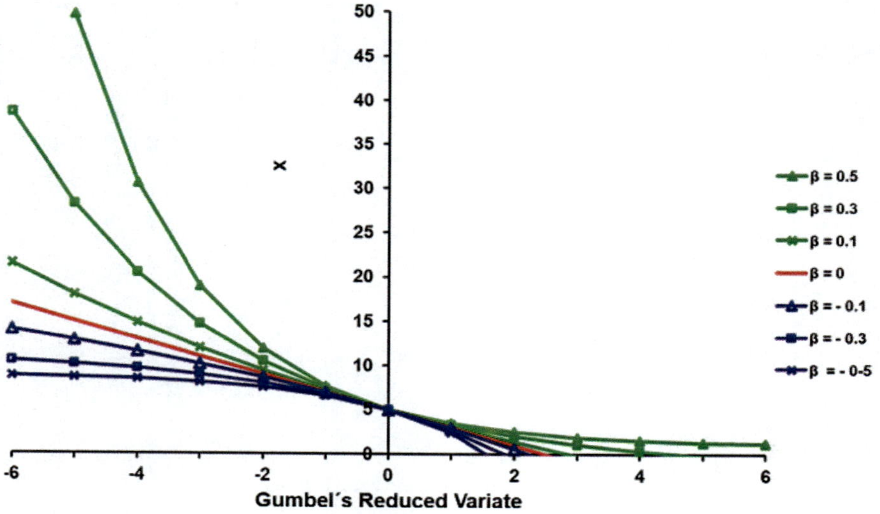

Fig. 14.1 The general extreme value for the minima family of distributions

The GEV distribution is, as a matter of fact, a family of distributions, see Fig. 14.1, can directly represent the Extreme Value types II (EVII) and III (EVIII) distributions and when taking the limit when $\beta \to 0$, the GEV distribution can also represent the Extreme Value type I (EVI) distribution.

14.2 Chapter Objectives

After reading this chapter, you will know how to:

1. Recognize the distribution and density functions of the GEVM distribution
2. Estimate the parameters of the GEVM distribution
3. Compute the quantiles and confidence limits of the GEVM distribution
4. Make a graphic display of your data and the GEVM distribution
5. Develop an application of all the above using Excel® spreadsheets.

14.3 Probability Distribution and Density Functions

The probability distribution function of the general extreme value (GEVM) distribution for the minima is, Raynal-Villasenor (1994):

$$\Pi(x) = \exp\left\{-[1 - \beta(\omega - x)/\alpha]^{\frac{1}{\beta}}\right\} \qquad (14.1)$$

where ω, α and β are the location, scale, and shape parameters, respectively. $\Pi(x)$ is the probability distribution function of the random variable x and for the case of low flow frequency analysis is equal to the exceedance probability, $Pr(X > x)$. The scale parameter must meet the condition that $\alpha > 0$. The domain of variable x in GEVM distribution is as follows:

For $\beta < 0$:

$$-\infty < x \leq \omega - \alpha/\beta \tag{14.2}$$

For $\beta > 0$:

$$\omega - \alpha/\beta \leq x < \infty \tag{14.3}$$

The probability density function for the GEVM distribution is, Raynal-Villasenor (1994):

$$\pi(x) = \frac{1}{\alpha} exp\left\{-\left[1 - \frac{\beta(\omega - x)}{\alpha}\right]^{\frac{1}{\beta}}\right\}\left[1 - \frac{\beta(\omega - x)}{\alpha}\right]^{\frac{1}{\beta}-1} \tag{14.4}$$

where $\pi(x)$ is the probability density distribution of random variable x.

14.4 Estimation of the Parameters

14.4.1 MOM Method

The population mean, variance and skewness of the GEVM distribution may be expressed in terms of its reduced variates for $\beta < 0$ and $\beta > 0$:

1) For $\beta < 0$:

The reduced variate is:

$$-y_2 = 1 - \frac{(\omega - x)\beta}{\alpha} \tag{14.5}$$

and its probability distribution and density functions are:

$$G(y_2) = \exp\left[-(-y_2)^{\frac{1}{\beta}}\right] \tag{14.6}$$

and:

$$g(y_2) = -\frac{(-y_2)^{\frac{1}{\beta}-1}}{\beta} \exp\left[-(-y_2)^{\frac{1}{\beta}}\right] \tag{14.7}$$

the domain is now $-\infty < y_2 \leq 0$. Now, the mean, variance and skewness may be expressed in terms of the reduced variate y_2 as follows:

$$E(y_2) = -\Gamma(1 + \beta) \tag{14.8}$$

where Γ() is the complete Gamma function with argument (), and:

$$\mathrm{var}(y_2) = \Gamma(1 + 2\beta) - \Gamma^2(1 + \beta) \tag{14.9}$$

so, the skewness coefficient is:

$$\gamma = -\frac{\left[\Gamma(1 + 3\beta) - 3\Gamma(1 + 2\beta)\Gamma(1 + \beta) + 2\Gamma^3(1 + \beta)\right]}{\left[\Gamma(1 + 2\beta) - \Gamma^2(1 + \beta)\right]^{\frac{3}{2}}} \tag{14.10}$$

So, the mean and variance of the actual variable x, can be expressed as:

$$\mu = \omega + \frac{\alpha}{\beta}[\Gamma(1 + \beta) - 1] \tag{14.11}$$

and:

$$\sigma^2 = \left(\frac{\alpha}{\beta}\right)^2 \left[\Gamma(1 + 2\beta) - \Gamma^2(1 + \beta)\right] \tag{14.12}$$

the skewness coefficient remains as shown in Eq. (14.10). The skewness coefficient when $\beta < 0$ is always lesser than -1.1396.

2) For $\beta > 0$:

The reduced variate is:

$$y_3 = 1 - \frac{(\omega - x)\beta}{\alpha} \tag{14.13}$$

and its probability distribution and density functions are:

$$G(y_3) = \exp\left[-(y_3)^{\frac{1}{\beta}}\right] \tag{14.14}$$

and:

$$g(y_3) = \frac{(y_3)^{\frac{1}{\beta}-1}}{\beta} \exp\left[-(y_3)^{\frac{1}{\beta}}\right] \tag{14.15}$$

the domain is now $0 \leq y_3 < \infty$. Now, the mean, variance and skewness may be expressed in terms of the reduced variate y_3 as follows:

$$E(y_3) = \Gamma(1 + \beta) \tag{14.16}$$

and:

$$\mathrm{var}(y_3) = \Gamma(1 + 2\beta) - \Gamma^2(1 + \beta) \tag{14.17}$$

and:

$$\gamma = \left\{ \frac{\Gamma(1+3\beta) - 3\Gamma(1+2\beta)\Gamma(1+\beta) + \Gamma^3(1+\beta)}{\left[\Gamma(1+2\beta) - \Gamma^2(1+\beta)\right]^{\frac{3}{2}}} \right\} \tag{14.18}$$

So, the mean and variance of the actual variable x, can be expressed as in the same way when $\beta > 0$:

$$\mu = \omega + \frac{\alpha}{\beta}[\Gamma(1+\beta) - 1] \tag{14.19}$$

and:

$$\sigma^2 = \left(\frac{\alpha}{\beta}\right)^2 \left[\Gamma(1+2\beta) - \Gamma^2(1+\beta)\right] \tag{14.20}$$

the skewness coefficient remains as shown in Eq. (14.18). The skewness coefficient when $\beta > 0$ is always bigger than -1.1396.

For the GEVM distribution, using the method of moments, the estimators of the parameters are obtained by first equating the population moments with the sample moments, and then simultaneously solving the resulting system of equations:

$$\mu = \overline{x} \tag{14.21}$$

$$\sigma^2 = \hat{\sigma}^2 = s^2 \tag{14.22}$$

and:

$$\gamma = \hat{\gamma} = g \tag{14.23}$$

then using Eqs. (14.11) and (14.12) and Eqs. (14.10) or (14.18) in the left- hand side of Eqs. (14.21) to (14.23), respectively, and the expressions for the sample mean, sample variance and sample skewness coefficient, given in Chap. 2, in the right-hand side of Eqs. (14.21) to (14.23), respectively, we obtain the following expressions:

$$\omega + \frac{\alpha}{\beta}[\Gamma(1+\beta) - 1] = \frac{1}{N}\sum_{i=1}^{N} x_i \tag{14.24}$$

$$\left(\frac{\alpha}{\beta}\right)^2 \left[\Gamma(1+2\beta) - \Gamma^2(1+\beta)\right] = \frac{1}{N}\sum_{i=1}^{N} (x - \overline{x})^2 \tag{14.25}$$

and for $\beta < 0$:

$$-\left\{\frac{\Gamma(1+3\beta)-3\Gamma(1+2\beta)\Gamma(1+\beta)+\Gamma^3(1+\beta)}{\left[\Gamma(1+2\beta)-\Gamma^2(1+\beta)\right]^{\frac{3}{2}}}\right\}$$
$$=\frac{N}{(N-1)(N-2)\hat{\sigma}^3}\sum_{i=1}^{N}\left(x_i-\hat{\mu}\right)^3 \tag{14.26}$$

and for $\beta > 0$:

$$\left\{\frac{\Gamma(1+3\beta)-3\Gamma(1+2\beta)\Gamma(1+\beta)+\Gamma^3(1+\beta)}{\left[\Gamma(1+2\beta)-\Gamma^2(1+\beta)\right]^{\frac{3}{2}}}\right\}$$
$$=\frac{N}{(N-1)(N-2)\hat{\sigma}^3}\sum_{i=1}^{N}\left(x_i-\hat{\mu}\right)^3 \tag{14.27}$$

The solution to the system of equations formed by Eqs. (14.24) and (14.25) and (14.26) or (14.27), provide the estimators of the method of moments for the GEVM distribution:

$$\hat{\omega}=\hat{A}+\frac{\hat{\alpha}}{\hat{\beta}} \tag{14.28}$$

$$\hat{\alpha}=\hat{B}\left|\hat{\beta}\right| \tag{14.29}$$

where:

$$\hat{A}=\hat{\mu}-\frac{\hat{\alpha}}{\hat{\beta}}\Gamma\left(1+\hat{\beta}\right) \tag{14.30}$$

$$\hat{B}=\left(\frac{\hat{\sigma}^2}{\sigma_z^2}\right)^{\frac{1}{2}}=\frac{\hat{\sigma}}{\sigma_z} \tag{14.31}$$

The values of the mean and standard deviation of the z's are provided in Eqs. (14.8) or (14.16), depending on the sign of the shape parameter, and the squared root of Eq. (14.9), respectively. Equations (14.26) and (14.27) have been inverted to produce the following polynomial relationships:

1) For $\beta < 0$ and $-19.0 < \hat{\gamma} \leq -1.1396$:

$$\hat{\beta}=0.24662+0.286678\hat{\gamma}+0.072454\hat{\gamma}^2+0.010176\hat{\gamma}^3$$
$$+0.000816\hat{\gamma}^4+0.000037\hat{\gamma}^5 \tag{14.32}$$

2) For $\beta > 0$ and $-1.1396 \le \hat{\gamma} < 11.35$:

$$\hat{\beta} = 0.279434 + 0.333535\hat{\gamma} + 0.048305\hat{\gamma}^2 + 0.024414\hat{\gamma}^3 + 0.003765\hat{\gamma}^4 - 0.000263\hat{\gamma}^5 \quad (14.33)$$

where $\hat{\mu}$ or $\overline{x}$ is the sample mean, $\hat{\sigma}$ or s is the sample standard deviation, $\hat{\gamma}$ or g is the sample skewness coefficient and N is the sample size. As it may be seen, the branch of the GEVM distribution with shape parameter less than zero (β < 0) it is not useful for low flow frequency analysis; so, it will not be pursued any longer in this chapter.

14.4.1.1 Example of Application of Estimation of the Parameters of the GEVM Distribution Using the MOM Method

With the one-day low flow sample data of gauging station Villalba, Mexico (1939–1991), contained in Appendix A, the following statistics have already been obtained:

$$\hat{\mu} = \overline{x} = \frac{1}{N}\sum_{i=1}^{N} x_i = 0.3306\text{m}^3/\text{s}$$

$$\hat{\sigma} = s = \left[\frac{1}{N}\sum_{i=1}^{N}(x_i - \hat{\mu})^2\right]^{1/2} = 0.1465\ \text{m}^3/\text{s}$$

$$\hat{\gamma} = \frac{N}{(N-1)(N-2)\hat{\sigma}^3}\sum_{i=1}^{N}(x_i - \hat{\mu})^3 = 0.5853$$

So, the moment estimators of the parameters of the GEVM distribution are as follows:

$$\hat{\beta} = 0.279434 + 0.333535\hat{\gamma} + 0.048305\hat{\gamma}^2 + 0.024414\hat{\gamma}^3 + 0.003765\hat{\gamma}^4 - 0.000263\hat{\gamma}^5$$

$$\hat{\beta} = 0.279434 + 0.333535(0.5853) + 0.048305(0.5853)^2 + 0.024414(0.5853)^3 + 0.003765(0.5853)^4 - 0.000263(0.5853)^5 = 0.4965$$

$$\hat{\omega} = A + \frac{\hat{\alpha}}{\hat{\beta}} = 0.0486 + \frac{0.1580}{(0.4965)} = 0.3368\ \text{m}^3/\text{s}$$

$$\hat{\alpha} = B\left|\hat{\beta}\right| = 0.3182(|0.4965|) = 0.1580\ \text{m}^3/\text{s}$$

where:

$$A = \hat{\mu} - B\hat{\mu}_z = 0.3306 - 0.3182(0.8861) = 0.0486$$

$$B = \left(\frac{\sigma^2}{\sigma_z^2}\right)^{\frac{1}{2}} = \frac{\sigma}{\sigma_z} = \frac{0.1465}{\left[\Gamma(1+2(0.4965)) - \Gamma^2(1+(0.4965))\right]^{\frac{1}{2}}} = \frac{0.1465}{\left[0.9971 - (0.8861)^2\right]^{\frac{1}{2}}} = 0.3182$$

$$\mu_z = \Gamma(1+\beta) = \Gamma(1.4965) = 0.8861$$

$$\sigma_z = \left[\Gamma(1+2\beta) - \Gamma^2(1+\beta)\right]^{\frac{1}{2}} = \left[\Gamma(1.993) - \Gamma^2(1.4965)\right]^{\frac{1}{2}} = \left[0.9971 - (0.8861)^2\right]^{\frac{1}{2}} = 0.4603$$

So, the moment estimators of the parameters of the GEVM distribution are:
location parameter:

$$\hat{x}_0 = 0.3668 \text{ m}^3/\text{s}$$

scale parameter:

$$\hat{\alpha} = 0.1580 \text{ m}^3/\text{s}$$

and:
shape parameter:

$$\hat{\beta} = 0.4965$$

14.4.2 ML Method

The likelihood function for the GEVM distribution is as follows:

$$L(x, \omega, \alpha, \beta) = \prod_{i=1}^{N} \left\{ \frac{1}{\alpha} \exp\left\{ -\left[1 - \frac{\beta(\omega - x_i)}{\alpha} \right]^{\frac{1}{\beta}} \right\} \left[1 - \frac{\beta(\omega - x_i)}{\alpha} \right]^{\frac{1}{\beta} - 1} \right\} \tag{14.34}$$

By taking the natural logarithm of the previous equation, the log-likelihood function for the GEVM distribution is obtained as:

$$LL(x, \omega, \alpha, \beta) = -NLn(\alpha) - \sum_{i=1}^{N} \left[1 - \frac{\beta(\omega - x_i)}{\alpha} \right]^{\frac{1}{\beta}} + \left(\frac{1}{\beta} - 1 \right) \sum_{i=1}^{N} Ln\left[1 - \frac{\beta(\omega - x_i)}{\alpha} \right] \tag{14.35}$$

Now, the classical approach to the method of maximum likelihood requires the computation of the first-order partial derivatives of the log-likelihood function with respect to each of its parameters, equating them equal to zero and then solving the resulting system of equations. So, the first-order partial derivatives are obtained as follows:

$$\frac{\partial LnL}{\partial \omega} = \frac{1}{\alpha} \left\{ \sum_{i=1}^{N} \left[1 - \frac{\beta(\omega - x_i)}{\alpha} \right]^{\frac{1}{\beta} - 1} + (\beta - 1) \sum_{i=1}^{N} \left[1 - \frac{\beta(\omega - x_i)}{\alpha} \right]^{-1} \right\} = 0 \tag{14.36}$$

$$\frac{\partial LL}{\partial \alpha} = \frac{1}{\alpha} \left\{ -N - \sum_{i=1}^{N} \left[1 - \frac{\beta(\omega - x_i)}{\alpha} \right]^{\frac{1}{\beta} - 1} \frac{(\omega - x_i)}{\alpha} \right.$$

$$\left\{-(\beta-1)\sum_{i=1}^{N}\left[1-\frac{\beta(\omega-x_i)}{\alpha}\right]^{-1}\frac{(\omega-x_i)}{\alpha}=0\right. \quad (14.37)$$

$$\frac{\partial LL}{\partial\beta}=\frac{1}{\beta}\left\{\sum_{i=1}^{N}\left[1-\frac{\beta(\omega-x_i)}{\alpha}\right]^{\frac{1}{\beta}}\left[\frac{1}{\beta}Ln\left(1-\frac{\beta(\omega-x_i)}{\alpha}\right)\right.\right.$$

$$\left.+\left[1-\frac{\beta(\omega-x_i)}{\alpha}\right]^{-1}\frac{(\omega-x_i)}{\alpha}\right]$$

$$\left.+(\beta-1)\sum_{i=1}^{N}\left[1-\frac{\beta(\omega-x_i)}{\alpha}\right]^{-1}\frac{(\omega-x_i)}{\alpha}-\frac{1}{\beta}\sum_{i=1}^{N}Ln\left[1-\frac{\beta(\omega-x_i)}{\alpha}\right]\right\}=0 \quad (14.38)$$

The exact solution to the system formed by equations from (14.36) to (14.38) is not known, so an iterative procedure is needed to evaluate the maximum likelihood estimators of the parameters of the GEVM distribution. The iterative procedure is as follows:

1) Define a reduced variate as:

$$y_i=\frac{1}{\beta}Ln\left[1-\frac{(\omega-x_i)\beta}{\alpha}\right] \quad (14.39)$$

2) Define parameters P, Q and R as follows:

$$P=N-\sum_{i=1}^{N}\exp(y_i) \quad (14.40)$$

$$Q=\sum_{i=1}^{N}\exp[(1-\beta)y_i]+(\beta-1)\sum_{i=1}^{N}\exp(-\beta y_i) \quad (14.41)$$

$$R=-N-\sum_{i=1}^{N}y_i+\sum_{i=1}^{N}y_i\exp(y_i) \quad (14.42)$$

3) Define the iterative procedure by:

$$(\omega)_{i+1}=(\omega)_i+\left(\delta_{x0}\right)_i \quad (14.43)$$

$$(\alpha)_{i+1}=(\alpha)_i+(\delta_\alpha)_i \quad (14.44)$$

$$(\beta)_{i+1}=(\beta)_i+\left(\delta_\beta\right)_i \quad (14.45)$$

where the sub-index i refers to the iteration stage and δ are the differences between the estimator at iteration i and the true value for the ML estimator for such parameter.

The relationship between the differences between the estimator at iteration i and the true value for the ML estimator for such parameter (δ's) and the first partial derivatives of the log-likelihood function with respect to the parameters of the GEVM distribution has the following form, Raynal-Villasenor (1994):

$$\begin{bmatrix} -\delta_\omega \\ -\delta\alpha \\ -\delta\beta \end{bmatrix}_i = \begin{bmatrix} E\left(-\frac{\partial^2 LL}{\partial\omega^2}\right) & E\left(-\frac{\partial^2 LL}{\partial\omega\partial\alpha}\right) & E\left(-\frac{\partial^2 LL}{\partial\omega\partial\beta}\right) \\ E\left(-\frac{\partial^2 LL}{\partial\alpha\partial\omega}\right) & E\left(-\frac{\partial^2 LL}{\partial\alpha^2}\right) & E\left(-\frac{\partial^2 LL}{\partial\alpha\partial\beta}\right) \\ E\left(-\frac{\partial^2 LL}{\partial\beta\partial\omega}\right) & E\left(-\frac{\partial^2 LL}{\partial\beta\partial\alpha}\right) & E\left(-\frac{\partial^2 LL}{\partial\beta^2}\right) \end{bmatrix}_i^{-1} \begin{bmatrix} -\frac{\partial LL}{\partial\omega} \\ -\frac{\partial LL}{\partial\alpha} \\ -\frac{\partial LL}{\partial\beta} \end{bmatrix}_i \quad (14.46)$$

The first matrix in the right-hand side of the previous equation is the Fisher's information matrix, it can be stated as:

$$[I] = \begin{bmatrix} E\left(-\frac{\partial^2 LL}{\partial\omega^2}\right) & E\left(-\frac{\partial^2 LL}{\partial\omega\partial\alpha}\right) & E\left(-\frac{\partial^2 LL}{\partial\omega\partial\beta}\right) \\ E\left(-\frac{\partial^2 LL}{\partial\alpha\partial\omega}\right) & E\left(-\frac{\partial^2 LL}{\partial\alpha^2}\right) & E\left(-\frac{\partial^2 LL}{\partial\alpha\partial\beta}\right) \\ E\left(-\frac{\partial^2 LL}{\partial\beta\partial\omega}\right) & E\left(-\frac{\partial^2 LL}{\partial\beta\partial\alpha}\right) & E\left(-\frac{\partial^2 LL}{\partial\beta^2}\right) \end{bmatrix} \quad (14.47)$$

The expected values inside the Fisher's information matrix have been obtained by Raynal-Villasenor (1996) for the interval $-0.50 < \beta < 0.50$, as:

$$E\left(-\frac{\partial^2 LL}{\partial\omega^2}\right) = \frac{N}{\alpha^2} p \quad (14.48)$$

$$E\left(-\frac{\partial^2 LL}{\partial\alpha^2}\right) = \frac{N}{\alpha^2\beta^2}[1 - 2(1-\beta)\,\Gamma(1-\beta) + p] \quad (14.49)$$

$$E\left(-\frac{\partial^2 LL}{\partial\beta^2}\right) = \frac{N}{\beta^2}\left[\frac{\pi^2}{6} + \left(1 - \varepsilon - \frac{1}{\beta}\right)^2 + \frac{2q}{\beta} + \frac{p}{\beta^2}\right] \quad (14.50)$$

$$E\left(-\frac{\partial^2 LL}{\partial\omega\partial\alpha}\right) = \frac{N}{\alpha^2\beta}[(1-\beta)\Gamma(1-\beta) - p] \quad (14.51)$$

$$E\left(-\frac{\partial^2 LL}{\partial\omega\partial\beta}\right) = \frac{N}{\alpha\beta}\left[\frac{p}{\beta} + q\right] \quad (14.52)$$

$$E\left(-\frac{\partial^2 LL}{\partial\alpha\partial\beta}\right) = \frac{N}{\alpha\beta^2}\left[1 - \varepsilon - \frac{[1-(1-\beta)\Gamma(1-\beta)]}{\beta} - \frac{p}{\beta} - q\right] \quad (14.53)$$

where:

$$p = (1-\beta)^2\Gamma(1-2\beta) \quad (14.54)$$

$$q = (1-\beta)\Gamma(1-\beta)\left[\psi(1-\beta) - \frac{(1-\beta)}{\beta}\right] \quad (14.55)$$

and $\Gamma(.)$ is the complete Gamma function, $\psi(.)$ is the Digamma function and ε is the Euler's constant (equal to 0.5772157).

The variance–covariance matrix of the parameters of the GEVM distribution has the following form, Raynal-Villasenor (1996):

$$[V] = \begin{bmatrix} Var(\omega) & Cov(\omega, \alpha) & Cov(\omega, \beta) \\ Cov(\alpha, \omega) & Var(\alpha) & Cov(\alpha, \beta) \\ Cov(\beta, \omega) & Cov(\beta, \alpha) & Var(\beta) \end{bmatrix}$$

$$= \begin{bmatrix} E\left(-\frac{\partial^2 LL}{\partial \omega^2}\right) & E\left(-\frac{\partial^2 LL}{\partial \omega \partial \alpha}\right) & E\left(-\frac{\partial^2 LL}{\partial \omega \partial \beta}\right) \\ E\left(-\frac{\partial^2 LL}{\partial \alpha \partial \omega}\right) & E\left(-\frac{\partial^2 LL}{\partial \alpha^2}\right) & E\left(-\frac{\partial^2 LL}{\partial \alpha \partial \beta}\right) \\ E\left(-\frac{\partial^2 LL}{\partial \beta \partial \omega}\right) & E\left(-\frac{\partial^2 LL}{\partial \beta \partial \alpha}\right) & E\left(-\frac{\partial^2 LL}{\partial \beta^2}\right) \end{bmatrix}^{-1} \tag{14.56}$$

An alternative way to express Eq. (14.57) is, Raynal-Villasenor (1996):

$$[V] = \begin{bmatrix} Var(\omega) & Cov(\omega, \alpha) & Cov(\omega, \beta) \\ Cov(\alpha, \omega) & Var(\alpha) & Cov(\alpha, \beta) \\ Cov(\beta, \omega) & Cov(\beta, \alpha) & Var(\beta) \end{bmatrix} = \frac{1}{N} \begin{bmatrix} \alpha^2 b & \alpha^2 h & \alpha f \\ \alpha^2 h & \alpha^2 a & \alpha g \\ \alpha f & \alpha g & c \end{bmatrix} \tag{14.57}$$

where a, b, c, f, g, and h are the variance–covariance matrix coefficients for the GEVM distribution. These coefficients have been evaluated as a function of the shape parameter, and their values are shown in Table 14.1.

Now, Eq. (14.47) can be modified as:

$$\begin{bmatrix} \delta_\omega \\ \delta\alpha \\ \delta\beta \end{bmatrix}_i = -\frac{1}{N} \begin{bmatrix} \alpha^2 b & \alpha^2 h & \alpha f \\ \alpha^2 h & \alpha^2 a & \alpha g \\ \alpha f & \alpha g & c \end{bmatrix}_i \begin{bmatrix} \frac{Q}{\alpha} \\ -\frac{(P+Q)}{\alpha\beta} \\ \frac{\left[R + \frac{(P+Q)}{\beta}\right]}{\beta} \end{bmatrix}_i \tag{14.58}$$

Then, the values of the differences between the estimator at iteration i and the true values for the ML estimator for such parameter (δ's) are:

$$\delta_i(x_0) = -\frac{\alpha}{N}\left[b(Q)_i - \frac{h(P+Q)_i}{\beta} + \frac{f}{\beta}\left(R + \frac{(P+Q)}{\beta}\right)_i\right] \tag{14.59}$$

Table 14.1 Exact coefficients of the variance–covariance matrix of the parameters of the GEVM distribution

β	a	b	c	f	g	h
0.0	0.7723	1.0790	0.5463	−0.2077	0.2849	-0.3300
0.1	0.6082	1.2271	0.4004	−0.2419	0.1848	-0.2155
0.2	0.5839	1.2017	0.3303	−0.2201	0.2139	-0.0919
0.3	0.5795	1.1727	0.2653	−0.1933	0.2333	-0.0347
0.4	0.5945	1.1413	0.2058	−0.1623	0.2422	0.1644
0.45	0.6091	1.1250	0.1781	−0.1456	0.2424	0.2205

$$\delta_i(\alpha) = -\frac{\alpha}{N}\left[h(Q)_i - \frac{a(P+Q)_i}{\beta} + \frac{g}{\beta}\left(R + \frac{(P+Q)}{\beta}\right)_i\right] \tag{14.60}$$

$$\delta_i(\beta) = -\frac{1}{N}\left[f(Q)_i - \frac{g(P+Q)_i}{\beta} + \frac{c}{\beta}\left(R + \frac{(P+Q)}{\beta}\right)_i\right] \tag{14.61}$$

4) Define a set of criteria of convergence in the following form:

$$\left|-\frac{\partial LL(x,\omega,\alpha,\beta)}{\partial\omega}\right| = \left|\frac{Q}{\alpha}\right| \approx 10^{-6} \tag{14.62}$$

$$\left|-\frac{\partial LL(x,\omega,\alpha,\beta)}{\partial\alpha}\right| = \left|-\frac{(P+Q)}{\alpha\beta}\right| \approx 10^{-6} \tag{14.63}$$

$$\left|-\frac{\partial LL(x,\omega,\alpha,\beta)}{\partial\beta}\right| = \left|\frac{1}{\beta}\left[R + \frac{(P+Q)}{\beta}\right]\right| \approx 10^{-6} \tag{14.64}$$

When conditions established by Eqs. (14.62)–(14.64) are met simultaneously, then the values of such parameters will correspond to the ML estimators of the parameters of the GEVM distribution.

14.4.2.1 Example of Application of Estimation of the Parameters of the GEVM Distribution Using the ML Method

With the one-day low flow sample data of gauging station Villalba, Mexico (1939–1991), contained in Appendix A, the following statistics have already been obtained:

$$\hat{\mu} = \overline{x} = \frac{1}{N}\sum_{i=1}^{N} x_i = 0.3356\ \text{m}^3/\text{s}$$

$$\hat{\sigma} = s = \left[\frac{1}{N}\sum_{i=1}^{N}(x_i - \hat{\mu})^2\right]^{1/2} = 0.1535\ \text{m}^3/\text{s}$$

Using the moment estimators as initial values of the iterative scheme for the maximum likelihood estimators for the parameters of the GEVM distribution, so the initial values are:

$$\hat{\omega} = 0.3668\ \text{m}^3/\text{s}$$

$$\hat{\alpha} = 0.1580\ \text{m}^3/\text{s}$$

$$\hat{\beta} = 0.4965$$

Iteration 1

Defining an initial reduced variate as:

$$y_i = \frac{1}{(0.4965)} Ln\left[1 - \frac{(0.3668 - x_i)(0.4965)}{0.1580}\right]$$

The parameters P, Q and R for the initial values are:

$$P = N - \sum_{i=1}^{N} \exp(y_i) = 53 - 53.001 = -0.001$$

$$Q = \sum_{i=1}^{N} \exp[(1-\beta)y_i] + (\beta - 1)\sum_{i=1}^{N} \exp(-\beta y_i)$$
$$= 46.9804 + (0.4965 - 1)(88.9249) = 2.2026$$

$$R = -N - \sum_{i=1}^{N} y_i + \sum_{i=1}^{N} y_i \exp(y_i) = -53 - (-30.1858) + 21.6115 = -1.2027$$

Then, the initial deviations between the estimator at iteration 1 and the true value for the maximum likelihood estimator for such parameters are:

$$\delta_i(\omega) = \frac{\alpha}{N}\left[b(Q)_i - \frac{h(P+Q)_i}{\beta} + \frac{f}{\beta}\left(R + \frac{(P+Q)}{\beta}\right)_i\right]$$

$$\delta_i(\omega) = \frac{0.1580}{53}\left[(1.1297)(2.2026) + \frac{(0.2443)(-0.001 + 2.2026)}{0.4965} + \frac{(-0.2415)}{0.4965}\right.$$
$$\left.\left(-1.2027 + \frac{(-0.001 + 2.2026)}{0.4965}\right)\right] = -0.0014$$

$$\delta_i(\alpha) = \frac{\alpha}{N}\left[h(Q)_i - \frac{a(P+Q)_i}{\beta} + \frac{g}{\beta}\left(R + \frac{(P+Q)}{\beta}\right)_i\right]$$

$$\delta_i(\alpha) = \frac{0.1580}{53}\left[(0.2443)(2.2026) - \frac{(0.6684)(-0.001 + 2.2026)}{0.4965} + \frac{0.3279}{0.4965}\right.$$
$$\left(-1.2027 + \frac{(-0.001 + 2.2026)}{0.4965}\right) = -0.0009$$

$$\delta_i(\beta) = \frac{1}{N}\left[f(Q)_i - \frac{g(P+Q)_i}{\beta} + \frac{c}{\beta}\left(R + \frac{(P+Q)}{\beta}\right)_i\right]$$

$$\delta_i(\beta) = -\frac{1}{53}\left[(-0.2415)(2.2026) - \frac{(0.3279)(-0.001 + 2.2026)}{0.4965} + \frac{0.41}{0.4965}\cdot\right.$$
$$\left(-1.2027 + \frac{(-0.001 + 2.2026)}{0.4965}\right) = 0.017$$

The new parameters are:

$$(\omega)_{i+1} = (\omega)_i + (\delta_\omega)_i = 0.3668 - 0.0014 = 0.3654$$

$$(\alpha)_{i+1} = (\alpha)_i + (\delta_\alpha)_i = 0.1580 - 0.0009 = 0.1571$$

$$(\beta)_{i+1} = (\beta)_i + \left(\delta_\beta\right)_i = 0.4965 + 0.017 = 0.5135$$

The criteria of convergence are not met at this iteration:

$$\left| -\frac{\partial LL(x,\omega,\alpha,\beta)}{\partial\omega} \right| = \left| \frac{Q}{\alpha} \right| = \left| \frac{2.2026}{0.1571} \right| = 14.020 > 10^{-6}$$

$$\left| -\frac{\partial LL(x,\omega,\alpha,\beta)}{\partial\alpha} \right| = \left| \frac{(P+Q)}{\alpha\beta} \right| = \left| \frac{(-0.001 + 2.2026)}{(0.1571)(0.5135)} \right| = 27.2911 > 10^{-6}$$

$$\left| -\frac{\partial LL(x,\omega,\alpha,\beta)}{\partial\beta} \right| = \left| \frac{1}{\beta}\left[R + \frac{(P+Q)}{\beta} \right] \right|$$

$$= \left| \frac{1}{(0.5135)}\left[-1.2027 + \frac{(-0.001 + 2.2026)}{0.5135} \right] \right|$$

$$= 6.0148 > 10^{-6}$$

After five iterations, the final values of the procedure are as follows:
Iteration 5.
Defining a final reduced variate as:

$$y_i = -\frac{1}{(0.5135)} Ln\left[1 - \frac{(0.3654 - x_i)(0.5135)}{0.1571} \right]$$

The parameters P, Q and R for the final values are:

$$P = N - \sum_{i=1}^{N} \exp(y_i) = 53 - 53.0030 = -0.003$$

$$Q = \sum_{i=1}^{N} \exp[(1-\beta)y_i] + (\beta - 1)\sum_{i=1}^{N} \exp(-\beta y_i)$$
$$= 46.9892 - (0.5135 - 1)(96.9020)$$
$$Q = 0.0596$$

$$R = -N - \sum_{i=1}^{N} y_i + \sum_{i=1}^{N} y_i \exp(y_i) = 53 - 31.0702 + 21.8874 = -0.0424$$

Then the final deviations between the estimator at iteration twenty-five and the true value for the ML estimator for such parameters are:

$$\begin{aligned}
\delta_i(\omega) =& \frac{0.1576}{53}\left[(1.1251)(0.0596) - \frac{(0.2488)(-0.003+0.0596)}{0.5158}\right. \\
&\left. + \frac{(-0.2560)}{(0.5301)}\left(-0.0424 + \frac{(-0.003+0.0596)}{0.5158}\right)\right] = 1.8549\times 10^{-5} \\
\delta_i(\alpha) =& \frac{0.1576}{53}\left[(0.2488)(0.5096) - \frac{(0.7021)(-0.003+0.0596)}{0.5158}\right. \\
&+ \frac{(0.3611)}{(0.5301)}\left(-0.0424 + \frac{(-0.003+0.596)}{0.5158}\right) = -4.4718\times 10^{-5} \\
\delta_i(\beta) =& \frac{1}{53}\left[(-0.2560)(0.0596) + \frac{(0.3611)(-0.003+0.0596)}{0.5158}\right. \\
&+ \frac{0.4484}{(0.5301)}\left(-0.0424 + \frac{(-0.003+0.0596)}{0.5158}\right) = 7.0203\times 10^{-5}
\end{aligned}$$

The final parameters are:

$$\begin{aligned}
(\omega)_{i+1} =& (\omega)_i + (\delta_\omega)_i = 0.3654 + 1.8549\times 10^{-5} = 0.3654 \\
(\alpha)_{i+1} =& (\alpha)_i + (\delta_\alpha)_i = 0.1571 - 4.4718\times 10^{-5} = 0.1571 \\
(\beta)_{i+1} =& (\beta)_i + (\delta_\beta)_i = 0.5135 + 7.0203\times 10^{-5} = 0.5135
\end{aligned}$$

The criteria of convergence are met for all parameters:

$$\begin{aligned}
\left|-\frac{\partial LL(x,\omega,\alpha,\beta)}{\partial\omega}\right| =& \left|\frac{0.0506}{0.1571}\right| = 0.378 > 10^{-6} \\
\left|-\frac{\partial LL(x,\omega,\alpha,\beta)}{\partial\alpha}\right| =& \left|\frac{-0.003+.0506}{(0.1571)(0.5135)}\right| = 0.6964 > 10^{-6} \\
\left|-\frac{\partial LL(x,\omega,\alpha,\beta)}{\partial\beta}\right| =& \left|\frac{1}{(0.5135)}\left[-0.0424 - \frac{(-0.003+0.0506)}{0.5135}\right]\right| \\
=& 0.1307 > 10^{-6}
\end{aligned}$$

No further improvement is possible. So, the maximum likelihood estimators of the parameters of the GEVM distribution are:
location parameter:

$$\hat{\omega} = 0.3654 \ \ \mathrm{m^3/s}$$

scale parameter:

$$\hat{\alpha} = 0.1571\ \mathrm{m^3/s}$$

shape parameter:

$$\hat{\beta} = 0.5135$$

14.5 Estimation of Quantiles for the GEVM Distribution

The quantiles for the GEVM distribution are obtained by using the inverse form of the GEVM distribution function:

$$x = \omega + \frac{\alpha}{\beta}\{[-Ln(\Pi(x))]^{\beta} - 1\} \tag{14.65}$$

where x_T is the quantile value for a certain value of the distribution function $\Pi(x)$.

The term Q_T is more frequently used in engineering instead of x_T and this also applies to T_r instead of $\Pi(x)$; so, the following expression is widely used when it is needed to relate an event Q_T to a specific return period T_r:

$$Q_T = \omega + \frac{\alpha}{\beta}\left\{\left[-Ln\left(1 - \frac{1}{T_r}\right)\right]^{\beta} - 1\right\} \tag{14.66}$$

where Q_T is a design value corresponding to a specific return period T_r.

14.5.1 Examples of Estimation of MOM, and ML Quantiles for the GEVM Distribution

The estimation of MOM, and ML quantiles for the GEVM distribution is made by using the parameters estimated in the preceding sections and then inserted one pair of parameters at a time in the Eq. (14.66). Table 14.2 summarizes these results.

Table 14.2 Estimation of MOM and ML quantiles for the GEVM distribution

Tr (years)	MOM	ML
	Q_T (m^3/s)	Q_T (m^3/s)
2	0.3139	0.3123
5	0.1997	0.2004
10	0.1527	0.1551
20	0.1214	0.1255
25	0.1136	0.1181
50	0.0945	0.1003
100	0.0810	0.0879

14.6 Goodness of Fit Test

The standard error of fit (SEF) for the GEVM distribution has the following form:

$$SEF = \left[\frac{\sum_{i=1}^{N} (xi - yi)^2}{(N - 3)} \right]^{1/2} \tag{14.67}$$

while the mean absolute relative deviation (MARD) remains the same as it was defined in Chap. 2:

$$MARD = \frac{100}{N} \sum_{i=1}^{N} \left| \frac{(x_i - y_i)}{x_i} \right| \tag{14.68}$$

where x_i are the sample historical values, y_i are the distribution function values, corresponding to the same return periods of the historical values, N is the sample size.

14.6.1 Examples of Application of the SEF and MARD to the MOM and ML Estimators of the Parameters of the GEVM Distribution

Find the values of the SEF and MARD of the MOM and ML estimators of the parameters of the GEVM distribution for the one-day low flow sample data of gauging station Villalba, Mexico (1939–1991), Appendix A.

The values of SEF and MARD for the MOM and ML estimators of the parameters of the GEVM distribution for the one-day low flow sample data of gauging station Villalba, Mexico, have been obtained through the application of Eqs. (14.67) and (14.68) using the parameters obtained in previous examples. Table 14.3 contains a summary of both measures of goodness of fit tests.

For the one-day low flow sample data at gauging station Villalba, Mexico, the best choice according to SEF measure is either the MOM method or the ML method. When using the MARD measure, the choice is the MOM method. Finally, the method with the best overall performance, for this sample of one-day low flow data, is the MOM method.

Table 14.3 SEF and MARD Measures for MOM and ML estimators of the parameters of the GEVM distribution for the one-day low flow sample data of gauging station Villalba, Mexico

Goodness of fit test	Method	
	MOM	ML
SEF	**0.02**	**0.02**
MARD	**4.36**	4.68

14.7 Estimation of Confidence Limits for the GEVM Distribution

By assuming that the quantiles are normally distributed, the following form can be used to set the confidence limits on such quantiles:

$$x_l = x_T \pm z_\alpha S_T \tag{14.69}$$

where x_l is the upper or lower confidence limit, depending on the sign in the preceding formula (+) for the upper confidence limit and (−) for the lower confidence limit, z_α is the standard normal variate corresponding to a confidence level α, and S_T is squared root of the standard error of the estimate. The evaluation procedures of the standard errors of the estimates will be described in the following subsections.

14.8 Estimation of Standard Errors for the GEVM Distribution

14.8.1 MOM Method

The general form of the MOM estimator of the standard error of the estimate of a three-parameter distribution is:

$$\begin{aligned} S_T^2 =& \left(\frac{\partial x_T}{\partial m_1'}\right)^2 \mathrm{var}(m_1') + \left(\frac{\partial x_T}{\partial m_2}\right)^2 \mathrm{var}(m_2) \\ &+ \left(\frac{\partial x_T}{\partial m_3}\right)^2 \mathrm{var}(m_3) + 2\left(\frac{\partial x_T}{\partial m_1'}\right)\left(\frac{\partial x_T}{\partial m_2}\right)\mathrm{cov}(m_1', m_2) \\ &+ 2\left(\frac{\partial x_T}{\partial m_1'}\right)\left(\frac{\partial x_T}{\partial m_3}\right)\mathrm{cov}(m_1', m_3) + 2\left(\frac{\partial x_T}{\partial m_2}\right)\left(\frac{\partial x_T}{\partial m_3}\right)\mathrm{cov}(m_2, m_3) \end{aligned} \tag{14.70}$$

and Eq. (14.70) can be simplified in terms of the frequency factor, K_T, as, Kite (1988):

$$\begin{aligned} S_T^2 =& \frac{\mu_2}{N}\left\{1 + K_T\hat{\gamma} + \frac{K_T^2}{4}(\hat{\kappa} - 1) + \frac{\partial K_T}{\partial \hat{\gamma}}\left[2\hat{\kappa} - 3\hat{\gamma}^2 - 6 + K_T\left(\hat{\lambda}_1 - \frac{6\hat{\gamma}\hat{\kappa}}{4} - \frac{10\hat{\gamma}}{4}\right)\right]\right. \\ &\left. + \left(\frac{\partial K_T}{\partial \gamma}\right)^2\left[\hat{\lambda}_2 - 3\hat{\gamma}\hat{\lambda}_1 - 6\hat{\gamma} + \frac{9\hat{\gamma}^2\hat{\kappa}}{4} + \frac{35\hat{\gamma}^2}{4} + 9\right]\right\} \end{aligned} \tag{14.71}$$

where K_T is the frequency factor given by:

$$K_T = B_K\left[\left(-Ln\left(\frac{1}{T_r}\right)\right)^\beta - A_K\right] \tag{14.72}$$

and:

$$A_K = \Gamma(1+\beta) \tag{14.73}$$

$$B_K = \frac{1}{\left[\Gamma(1+2\beta) - \Gamma^2(1+\beta)\right]^{\frac{1}{2}}} \tag{14.74}$$

and:

$$\hat{\lambda}_1 = \frac{\frac{\left[\sum_{i=1}^{N}(x_i-\hat{\mu})^5\right]}{N}}{\left(\sigma^2\right)^{2.5}} \tag{14.75}$$

$$\hat{\lambda}_2 = \frac{\frac{\left[\sum_{i=1}^{N}(x_i-\hat{\mu})^6\right]}{N}}{\left(\sigma^2\right)^{3}} \tag{14.76}$$

The first-order partial derivative of the frequency factor, K_T, with respect to the skewness coefficient, γ, can be evaluated by means of the chain rule for differentiation:

$$\frac{\partial K_T}{\partial \gamma} = \left(\frac{\partial K_T}{\partial \beta}\right)\left(\frac{\partial \beta}{\partial \gamma}\right) \tag{14.77}$$

and:

$$\frac{\partial K_T}{\partial \beta} = \left\{\frac{\left[y^{\beta} Ln(y) - G_1 P_1\right] - \frac{1}{2}\left[y^{\beta} - G_1\right]\left[G_2 - G_1^2\right]^{-1}\left[G_2 P_2 - 2G_1^2 P_1\right]}{\left[G_2 - G_1^2\right]^{\frac{1}{2}}}\right\} \tag{14.78}$$

$$\frac{\partial \gamma}{\partial \beta} = \left\{\frac{\left[G_3 P_3 + 3G_1\left(P_1\left(G_1^2 - G_2\right) - G_2 P_2\right)\right] - \frac{3}{2}\left[G_2 - G_1^2\right]^{-1}\left[G_3 - 3G_1 G_2 + 2G_1^3\right]\left[G_2 P_2 - 2G_1^2 P_1\right]}{\left[G_2 - G_1^2\right]^{\frac{3}{2}}}\right\} \tag{14.79}$$

where:

$$y = -Ln\left(\frac{1}{T_r}\right) \tag{14.80}$$

$$G_r = \Gamma(1 + r\beta) \tag{14.81}$$

$$P_r = \psi(1 + r\beta) \tag{14.82}$$

and $\Gamma(.)$ and $\psi(.)$ are the complete Gamma and Digamma functions, previously defined.

Then, the first-order partial derivative of the frequency factor, K_T, with respect to the skewness coefficient, γ, is:

$$\frac{\partial K_T}{\partial \gamma} = \frac{\left[G_2 - G_1^2\right]\left\{\left[y^{\beta} Ln(y) - G_1 P_1\right]\right\} - \frac{1}{2}\left[y^{\beta} - G_1\right]\left[G_2 P_2 - 2G_1^2 P_1\right]}{\left\{\left[G_3 P_3 + 3G_1\left(P_1\left(G_1^2 - G_2\right) - G_2 P_2\right)\right] - \frac{3}{2}\left[G_2 - G_1^2\right]^{-1}\left[G_3 - 3G_1 G_2 + 2G_1^3\right]\left[G_2 P_2 - 2G_1^2 P_1\right]\right\}} \tag{14.83}$$

The substitution of Eqs. (14.72) to (14.83) into Eq. (14.71) will provide the moment estimator of the standard error of the variate for the GEVM distribution.

14.8.1.1 Example of Estimation of MOM Standard Errors and the Two-Sided 95% Confidence Limits for the GEVM Distribution

Find the MOM estimators of the standard errors and the two-sided 95% confidence limits for the GEVM distribution for the one-day low flow sample data of gauging station Villalba, Mexico (1939–1991), contained in Appendix A.

The estimation of MOM estimators of the standard errors and the two-sided 95% confidence limits for the GEVM distribution is made by using the MOM estimators of the parameters and then inserted in the Eqs. (14.72) to (14.81) and then insert these results into Eq. (14.71), using selected values of the return intervals. Table 14.4 summarizes these results.

14.8.2 ML Method

The general form of the ML estimator of the standard error of the estimate of a three-parameter distribution is, Kite (1988):

$$S_T^2 = \left(\frac{\partial x_T}{\partial \alpha}\right)^2 \text{var}(\alpha) + \left(\frac{\partial x_T}{\partial \beta}\right)^2 \text{var}(\beta) + \left(\frac{\partial x_T}{\partial \gamma}\right)^2 \text{var}(\gamma)$$

Table 14.4 Estimation of MOM standard errors and the two-sided 95% confidence Limits for the GEVM distribution for gauging Station Villalba, Mexico

Tr (years)	S_T (m^3/s)	95% Lower limit (m^3/s)	Q_T (m^3/s)	95% Upper limit (m^3/s)
2	0.0239	0.2671	0.3139	0.3607
5	0.0177	0.1650	0.1997	0.2345
10	0.0052	0.1426	0.1527	0.1629
20	0	0.1214	0.1214	0.1214
25	0	0.1136	0.1136	0.1136
50	0	0.0945	0.0945	0.0945
100	0	0.0810	0.0810	0.0810

Two-sided Limits: $z_{\alpha} = 1.96$

$$+2\left(\frac{\partial x_T}{\partial \alpha}\right)\left(\frac{\partial x_T}{\partial \beta}\right)\mathrm{cov}(\alpha,\beta)+2\left(\frac{\partial x_T}{\partial \alpha}\right)\left(\frac{\partial x_T}{\partial \gamma}\right)\mathrm{cov}(\alpha,\gamma)$$
$$+2\left(\frac{\partial x_T}{\partial \beta}\right)\left(\frac{\partial x_T}{\partial \gamma}\right)\mathrm{cov}(\beta,\gamma) \tag{14.84}$$

and the variance–covariance matrix of the parameters for the GEVM distribution is known to be, Raynal-Villasenor (1995):

$$[V]=\begin{bmatrix} Var(\omega) & Cov(\omega,\alpha) & Cov(\omega,\beta) \\ Cov(\alpha,\omega) & Var(\alpha) & Cov(\alpha,\beta) \\ Cov(\beta,\omega) & Cov(\beta,\alpha) & Var(\beta) \end{bmatrix}=\frac{1}{N}\begin{bmatrix} \alpha^2 b & \alpha^2 h & \alpha f \\ \alpha^2 h & \alpha^2 a & \alpha g \\ \alpha f & \alpha g & c \end{bmatrix} \tag{14.85}$$

and from Eq. (14.82):

$$\frac{\partial x_T}{\partial \omega}=1 \tag{14.86}$$

$$\frac{\partial x_T}{\partial \alpha}=\frac{1}{\beta}\left\{\left[-Ln\left(\frac{1}{Tr}\right)\right]^{\beta}-1\right\} \tag{14.87}$$

$$\frac{\partial x_T}{\partial \beta}=\frac{\alpha}{\beta}\left\{\left[-Ln\left(\frac{1}{T_r}\right)\right]^{\beta}\left[Ln\left(-Ln\left(\frac{1}{T_r}\right)\right)\right]-\frac{1}{\beta}\left[\left(-Ln\left(\frac{1}{T_r}\right)\right)^{\beta}-1\right]\right\} \tag{14.88}$$

So, Eq. (14.82) becomes to:

$$S_T^2=\frac{1}{N}\left[\alpha^2 b+\alpha^2 a\left(\frac{\partial x_T}{\partial \alpha}\right)^2+c\left(\frac{\partial x_T}{\partial \beta}\right)^2+2\alpha^2 h\left(\frac{\partial x_T}{\partial \alpha}\right)\right.$$
$$\left.+2\alpha f\left(\frac{\partial x_T}{\partial \beta}\right)+2\alpha g\left(\frac{\partial x_T}{\partial \alpha}\right)\left(\frac{\partial x_T}{\partial \beta}\right)\right] \tag{14.89}$$

14.8.2.1 Example of Estimation of Maximum Likelihood Standard Errors and the Two-Sided 95% Confidence Limits for the GEVM Distribution

Find the maximum likelihood estimators of the standard errors and the two-sided 95% confidence limits for the GEVM distribution for the one-day low flow sample data of gauging station Villalba, Mexico (1939–1991), Appendix A.

The estimation of maximum likelihood estimators of the standard errors and the two-sided 95% confidence limits for the GEVM distribution is made by using the maximum likelihood estimators of the parameters and then inserted in the Eqs. (14.85) to (14.88) and into Eq. (14.89), using selected values of the return intervals. Table 14.5 summarizes these results.

Table 14.5 Estimation of ML Standard Errors and the Two-sided 95% confidence limits for the GEVM distribution for gauging station Villalba, mexico

Tr (years)	S_T (m^3/s)	95% Lower limit (m^3/s)	Q_T (m^3/s)	95% Upper limit (m^3/s)
2	0.0010	0.3104	0.3123	0.3142
5	0.0018	0.1969	0.2004	0.2038
10	0.0036	0.1482	0.1551	0.1621
20	0.0052	0.1153	0.1255	0.1357
25	0.0057	0.1069	0.1181	0.1293
50	0.0072	0.0862	0.1003	0.1144
100	0.0086	0.0711	0.0879	0.1048

Two-sided Limits: $z_\alpha = 1.96$.

14.9 Examples of Application for the GEVM Distribution Using Excel© Spreadsheets

14.9.1 One-Day Low Flow Frequency Analysis

By using the one-day low flow data from gauging station Villalba, Mexico (1939–1991) the following descriptive statistics were obtained and are shown in Fig. 14.2.

ONE-DAY LOW FLOW FREQUENCY ANALYSIS
Gauging Station: Villalba, Mexico (1939-1991)

Descriptive Statistics	
Mean	0.3306
Standard Error	0.0201
Median	0.3280
Mode	0.4850
Standard Deviation	0.1465
Sample Variance	0.0214
Kurtosis	0.1906
Skewness	0.5853
Range	0.6250
Minimum	0.0790
Maximum	0.7040
Sum	17.5214
Count	53.0000
Largest(1)	0.7040
Smallest(1)	0.0790
Confidence Level(95.0%)	0.0404

Fig. 14.2 Descriptive statistics for the one-day low flow sample data at Villalba, Mexico

	A	B	C	D	E	F	G	H	I
1	ONE-DAY LOW-FLOW FREQUENCY ANALYSIS								
2	Gauging Station: Villalba, Mexico (1939-1991)								
3	General Extreme Value (GEVM) for the Minima								
4	Method: MOM								
5	Estimation of Parameters and Goodness of Fit Tests								
6	$1+\beta =$	1.4965	$1+2\beta =$	1.9930	$1+3\beta =$	2.4895			
7	$\Gamma(1+\beta) =$	0.8861	$\Gamma(1+2\beta) =$	0.9971	$\Gamma(1+3\beta) =$	1.3196			
8	$\psi(1+\beta) =$	0.0332	$\psi(1+2\beta) =$	0.4183	$\psi(1+3\beta) =$	0.6980			
9	$\mu_z =$	0.8861	$\sigma_z =$	0.4603	$\lambda_1 =$	0.0354			
10	A =	0.0486	B =	0.3182	$\lambda_2 =$	0.0425			
11	Location Parameter =		0.3668	m^3/s	Scale Parameter =		0.1580	m^3/s	
12	Shape Parameter =		0.4965	SEF =	0.02	MARD =	4.36		
13	Design Values and Confidence Limits								
14	Return Period	Exceedance		Frequency		Squared Root	Two-sided	Design	Two-sided
15		Probability		Factor		Stand. Error	95%	Values	95%
16	Tr	$\Pi(x)$	y	K_T	$\partial K_T/\partial\gamma$	S_T	Lower Limit	Q_T	Upper Limit
17	(years)					(m^3/s)	(m^3/s)	(m^3/s)	(m^3/s)
18	2	0.5	0.6931	-0.1141	-0.0643	0.0239	0.2671	0.3139	0.3607
19	5	0.2	1.6094	0.8265	0.0731	0.0177	0.1650	0.1997	0.2345
20	10	0.1	2.3026	1.3620	0.1814	0.0052	0.1426	0.1527	0.1629
21	20	0.05	2.9957	1.8209	0.2880	#¡NUM!	#¡NUM!	0.1214	#¡NUM!
22	25	0.04	3.2189	1.9569	0.3217	#¡NUM!	#¡NUM!	0.1136	#¡NUM!
23	50	0.02	3.9120	2.3516	0.4250	#¡NUM!	#¡NUM!	0.0945	#¡NUM!
24	100	0.01	4.6052	2.7124	0.5256	#¡NUM!	#¡NUM!	0.0810	#¡NUM!
25	Two-sided Confidence Limits:		$z_\alpha =$	1.96					

Fig. 14.3 Moments estimators for the parameters and standard errors, Quantiles and confidence limits of the GEVM distribution for the one-day low flow sample data at Villalba, Mexico

14.9.1.1 MOM Method

By using an Excel© spreadsheet, the results contained in Fig. 14.3, were obtained by the MOM method applied to the GEVM distribution one-day low flow data from gauging station Villalba, Mexico.

14.9.1.2 ML Method

By using an Excel© spreadsheet, the results contained in Figs. 14.4 and 14.5, were obtained by the ML method applied to the GEVM distribution one-day low flow data from gauging station Villalba, Mexico.

In Fig. 14.6 a comparison is made between the histogram and MOM-GEVM theoretical density for the one-day low flow sample of gauging station Villalba, Mexico.

A graphical depiction of the empirical and theoretical frequency curves, in this case that of MOM and ML, is shown in Fig. 14.7.

A graphical depiction of the empirical and theoretical frequency curves and the confidence limits of the best fit provided, in this case that of MOM, is shown in Fig. 14.8.

14.9.2 7-day Low Flow Frequency Analysis

By using the 7-day low flow data from gauging station Villalba, Mexico (1939–1986) the following descriptive statistics were obtained and are shown in Fig. 14.9.

A	B	C	D	E	F	G
ONE-DAY LOW-FLOW FREQUENCY ANALYSIS						
Gauging Station: Villalba, Mexico (1939-1991)						
General Extreme Value (GEVM) for the Minima						
Method: ML						
Estimation of Parameters and Goodness of Fit Tests						
Initial Moment Estimators:			$(1-\beta)-1$ =	-0.5158	$(1-2\beta)-1$ =	-1.0316
Location Parameter =	0.3668	m^3/s	$1-\beta$ =	0.4842	$1-2\beta$ =	-0.0316
Scale Parameter =	0.1580	m^3/s	$\Gamma(1-\beta)$ =	1.7875	$\Gamma(1-2\beta)$ =	6.4236
Shape Parameter =	0.4965		$\psi(1-\beta)$ =	-2.0436		
Iterative Procedure:						
i-th iteration				(i+1)-th iteration		
Location Parameter =	0.3649		a =	0.7023	f =	-0.2561
Scale Parameter =	0.1576		b =	1.1251	g =	0.3613
Shape Parameter =	0.5158		c =	0.4486	h =	0.2488
P =	-9.7852E-04		$\delta(x_0)$ =	2.0095E-05		
Q =	3.6219E-02		$\delta(\alpha)$ =	-4.4608E-05		
R =	-3.4104E-02		$\delta(\beta)$ =	-7.9318E-05		
ABS($\partial LL/\partial\omega$) =	2.2981E-01		Location Parameter=		0.3649	
ABS($\partial LL/\partial\alpha$) =	4.3351E-01		Scale Parameter =		0.1576	
ABS($\partial LL/\partial\beta$) =	6.6338E-02		ShapeParameter =		0.5157	
Final MV Estimators:						
Location Parameter =	0.3649		Log-likelihood function =		29.8776	
Scale Parameter =	0.1576					
Shape Parameter =	0.5157					
SEF =	0.02	MARD =	4.68	AIC =	-54	
Variance-Covariance Matrix						
var(ω) =	0.0005		cov(ω,α) =	0.0001		
var(α) =	0.0003		cov(ω,β)	-0.0008		
var(β) =	0.0085		cov(α,β) =	0.0011		

Fig. 14.4 Maximum likelihood estimators for the parameters of the GEVM distribution for the one-day low flow sample data at Villalba, Mexico

A	B	C	D	E	F	G	H
Design Values and Confidence Limits							
Return Period Tr (years)	Exceedance Probability $\Pi(x)$	$\partial x_T/\partial\alpha$	$\partial x_T/\partial\beta$	Squared Root Stand. Error S_T (m^3/s)	Two-sided 95% Lower Limit (m^3/s)	Design Values Q_T (m^3/s)	Two-sided 95% Upper Limit (m^3/s)
2	0.5	-0.3340	-0.0093	0.0010	0.3104	0.3123	0.3142
5	0.2	0.5394	-0.0210	0.0018	0.1969	0.2004	0.2038
10	0.1	1.0421	-0.0734	0.0036	0.1481	0.1551	0.1621
20	0.05	1.4755	-0.1395	0.0052	0.1153	0.1254	0.1356
25	0.04	1.6044	-0.1625	0.0057	0.1069	0.1181	0.1293
50	0.02	1.9793	-0.2374	0.0072	0.0862	0.1003	0.1144
100	0.01	2.3232	-0.3158	0.0086	0.0710	0.0879	0.1048
Two-sided Confidence Limits:	z_α =	1.96					

Fig. 14.5 Maximum likelihood estimators of the Standard Errors, Quantiles and confidence limits for the one-day low flow sample data at Villalba, Mexico

14.9.2.1 MOM Method

By using an Excel© spreadsheet, the results contained in Fig. 14.10, were obtained by the MOM method applied to the GEVM distribution 7-day low flow data from gauging station Villalba, Mexico.

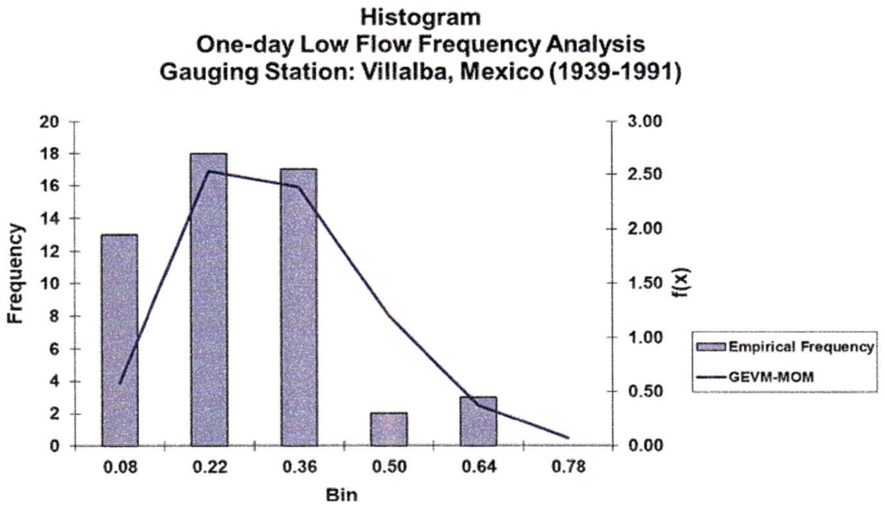

Fig. 14.6 Histogram and MOM-GEVM theoretical density for the one-day low flow sample of Villalba, Mexico

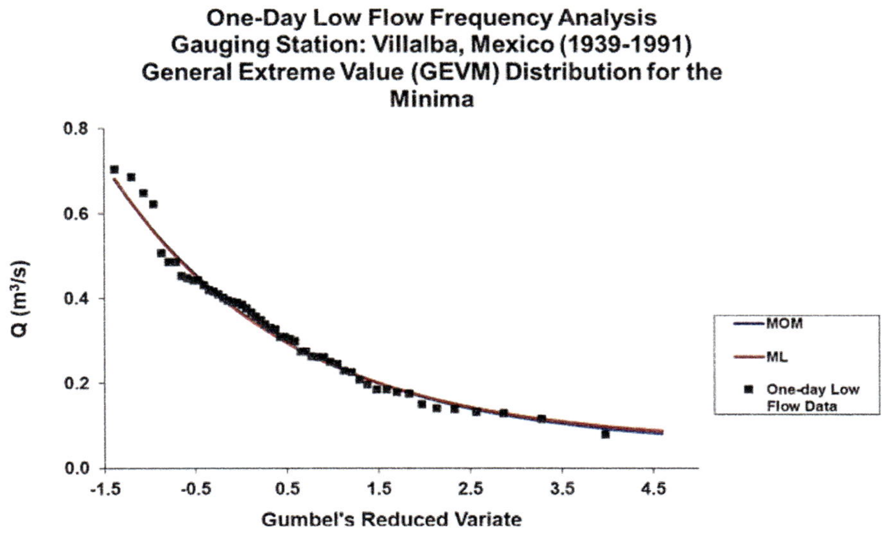

Fig. 14.7 Empirical and MOM-ML theoretical curves of GVEM distribution of one-day low flow sample of Villalba, Mexico

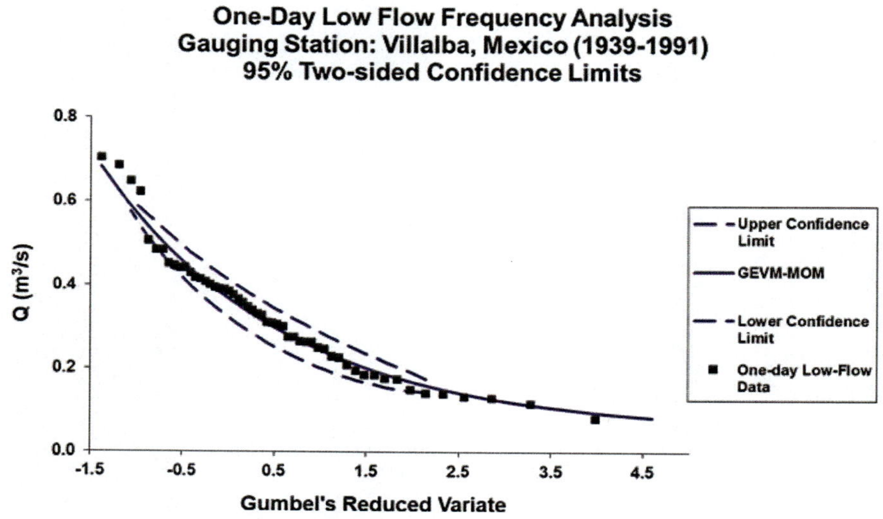

Fig. 14.8 Empirical and MOM-GVEM theoretical frequency curves and confidence limits of one-day low flow sample of Villalba, Mexico

	A	B	C
1	7-DAY LOW FLOW FREQUENCY ANALYSIS		
2	Gauging Station: Villalba, Mexico (1939-1986)		
3			
4	*Descriptive Statistics*		
5	Mean	2.8483	
6	Standard Error	0.1813	
7	Median	2.7125	
8	Mode	#N/D	
9	Standard Deviation	1.2561	
10	Sample Variance	1.5779	
11	Kurtosis	-0.2429	
12	Skewness	0.5782	
13	Range	5.2120	
14	Minimum	0.7450	
15	Maximum	5.9570	
16	Sum	136.7200	
17	Count	48.0000	
18	Largest(1)	5.9570	
19	Smallest(1)	0.7450	
20	Confidence Level(95.0%)	0.3647	

Fig. 14.9 Descriptive statistics for the 7-day low flow sample data at Villalba, Mexico

7-DAY LOW FLOW FREQUENCY ANALYSIS
Gauging Station: Villalba, Mexico (1939-1986)
General Extreme Value (GEVM) for the Minima
Method: MOM

Estimation of Parameters and Goodness of Fit Tests							
$1+\beta =$	1.4936	$1+2\beta =$	1.9871	$1+3\beta =$	2.4807		
$\Gamma(1+\beta) =$	0.8860	$\Gamma(1+2\beta) =$	0.9946	$\Gamma(1+3\beta) =$	1.3115		
$\psi(1+\beta) =$	0.0305	$\psi(1+2\beta) =$	0.4145	$\psi(1+3\beta) =$	0.6937		
$\mu_z =$	0.8860	$\sigma_z =$	0.4578	$\lambda_1 =$	5.8804		
A =	0.4170	B =	2.7441	$\lambda_2 =$	19.4983		
Location Parameter =	3.1610	m^3/s	Scale Parameter =		1.3544	m^3/s	
Shape Parameter =	0.4936	SEF =	0.12	MARD =	3.54		

Return Period Tr (years)	Exceedance Probability $\Pi(x)$	y	Frequency Factor K_T	$\partial K_T/\partial\gamma$	Squared Root Stand. Error S_T (m^3/s)	Two-sided 95% Lower Limit (m^3/s)	Design Values Q_T (m^3/s)	Two-sided 95% Upper Limit (m^3/s)
2	0.5	0.6931	-0.1125	-0.0633	0.2181	2.2795	2.7070	3.1344
5	0.2	1.6094	0.8273	0.0731	0.2090	1.3161	1.7258	2.1354
10	0.1	2.3026	1.3615	0.1802	0.2735	0.7846	1.3207	1.8568
20	0.05	2.9957	1.8188	0.2854	0.3650	0.3350	1.0504	1.7659
25	0.04	3.2189	1.9544	0.3187	0.3967	0.2054	0.9829	1.7604
50	0.02	3.9120	2.3474	0.4205	0.4973	-0.1578	0.8169	1.7916
100	0.01	4.6052	2.7065	0.5195	0.5985	-0.4727	0.7003	1.8734
Two-sided Confidence Limits:		$z_\alpha =$	1.96					

Fig. 14.10 Moments estimators for the parameters and standard errors, Quantiles and confidence limits of the GEVM distribution for the 7-day low flow sample data at Villalba, Mexico

14.9.2.2 ML Method

By using an Excel© spreadsheet, the results contained in Figs. 14.11 and 14.12, were obtained by the ML method applied to the GEVM distribution 7-day low flow data from gauging station Villalba, Mexico.

In Fig. 14.13 a comparison is made between the histogram and MOM-GEVM theoretical density for the 7-day low flow sample from gauging station Villalba, Mexico.

A graphical depiction of the empirical and theoretical frequency curves, in this case that of MOM and ML, is shown in Fig. 14.14.

A graphical depiction of the empirical and theoretical frequency curves and the confidence limits of the best fit provided, in this case that of MOM, is shown in Fig. 14.15.

14.9.3 Earthquake Epicenter Distance Frequency Analysis

By using the earthquake epicenter distance data, Castillo (1988), the following descriptive statistics were obtained and are shown in Fig. 14.16.

14.9.3.1 MOM Method

By using an Excel© spreadsheet, the results contained in Fig. 14.17, were obtained by the MOM method applied to the GEVM distribution for the earthquake epicenter distance data.

	A	B	C	D	E	F	G
1	7-DAY LOW FLOW FREQUENCY ANALYSIS						
2	Gauging Station: Villalba, Mexico (1939-1986)						
3	General Extreme Value (GEVM) for the Minima						
4	Method: ML						
5							
6	Estimation of Parameters and Goodness of Fit Tests						
7	Initial Moment Estimators:			$(1-\beta)-1$ =	-0.1078	$(1-2\beta)-1$ =	-0.2156
8	Location Parameter =	3.1610	m³/s	$1-\beta$ =	0.8922	$1-2\beta$ =	0.7844
9	Scale Parameter =	1.3544	m³/s	$\Gamma(1-\beta)$ =	1.0750	$\Gamma(1-2\beta)$ =	1.1818
10	Shape Parameter =	0.4936		$\psi(1-\beta)$ =	-0.7700		
11	Iterative Procedure:						
12	i-th iteration				(i+1)-th iteration		
13	Location Parameter =	3.3791		a =	0.6312	f=	-0.2662
14	Scale Parameter =	1.1059		b =	1.2298	g =	0.2144
15	Shape Parameter =	0.1078		c =	0.4755	h =	-0.2071
16	P =	-1.6326E+00		$\delta(x_0)$ =	-2.3006E+00		
17	Q =	4.6698E+00		$\delta(\alpha)$ =	1.6365E+00		
18	R =	1.6958E+01		$\delta(\beta)$ =	3.9960E+00		
19	ABS(∂LL/ $\partial\omega$) =	4.2226E+00		Location Parameter=		1.0785	
20	ABS(∂LL/ $\partial\alpha$) =	2.5476E+01		Scale Parameter =		2.7424	
21	ABS(∂LL/ $\partial\beta$) =	4.1867E+02		ShapeParameter =		4.1038	
22	Final MV Estimators:			No convergence was attained			
23	Location Parameter =	1.0785		Log-likelihood function =		-154.0050	
24	Scale Parameter =	2.7424					
25	Shape Parameter =	4.1038					
26	SEF =	2.78	MARD =	70.98	AIC =	314	
27							
28	Variance-Covariance Matrix						
29	var(ω) =	0.1745		cov(ω,α) =	-0.0294		
30	var(α) =	0.0896		cov(ω,β)	-0.0138		
31	var(β) =	0.0090		cov(α,β) =	0.0111		

Fig. 14.11 Maximum likelihood estimators for the parameters of the GEVM distribution for the 7-day low flow sample data at Villalba, Mexico

	A	B	C	D	E	F	G	H
33	Design Values and Confidence Limits							
34	Return Period	Exceedance			Squared Root	Two-sided	Design	Two-sided
35		Probability			Stand. Error	95%	Values	95%
36	Tr	$\Pi(x)$	$\partial x_T/\partial\alpha$	$\partial x_T/\partial\beta$	S_T	Lower Limit	Q_T	Upper Limit
37	(years)				(m³/s)	(m³/s)	(m³/s)	(m³/s)
38	2	0.5	-0.1895	-0.0722	0.0141	0.5310	0.5588	0.5865
39	5	0.2	1.4741	-1.2567	0.1104	0.1954	0.4117	0.6280
40	10	0.1	7.2256	-12.2555	0.6966	-0.9551	0.4103	1.7758
41	20	0.05	21.7496	-51.6418	2.5894	-4.6650	0.4103	5.4855
42	25	0.04	29.2911	-75.1130	3.6859	-6.8140	0.4103	7.6345
43	50	0.02	65.5091	-202.1898	9.5036	-18.2168	0.4103	19.0374
44	100	0.01	128.1800	-452.1986	20.7428	-40.2456	0.4103	41.0662
45	Two-sided Confidence Limits:	z_α =	1.96					

Fig. 14.12 Maximum likelihood estimators of the standard errors, Quantiles and confidence limits for the 7-day low flow sample data at Villalba, Mexico

14.9.3.2 ML Method

By using an Excel© spreadsheet, the results contained in Figs. 14.18 and 14.19, were obtained by the ML method applied to the GEVM distribution for the earthquake epicenter distance data.

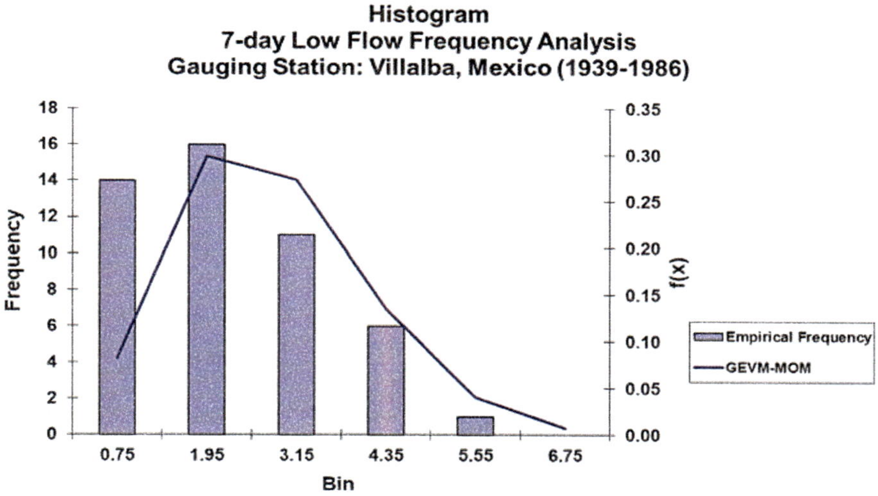

Fig. 14.13 Histogram and MOM-GEVM theoretical density for the 7-Day low flow sample of Villalba, Mexico

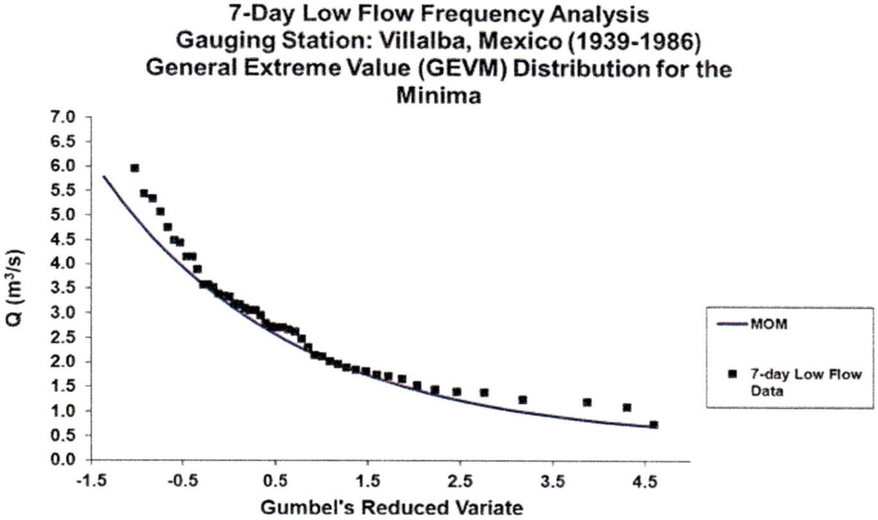

Fig. 14.14 Empirical and MOM-ML theoretical curves of GVEM distribution of 7-day low flow sample of Villalba, Mexico

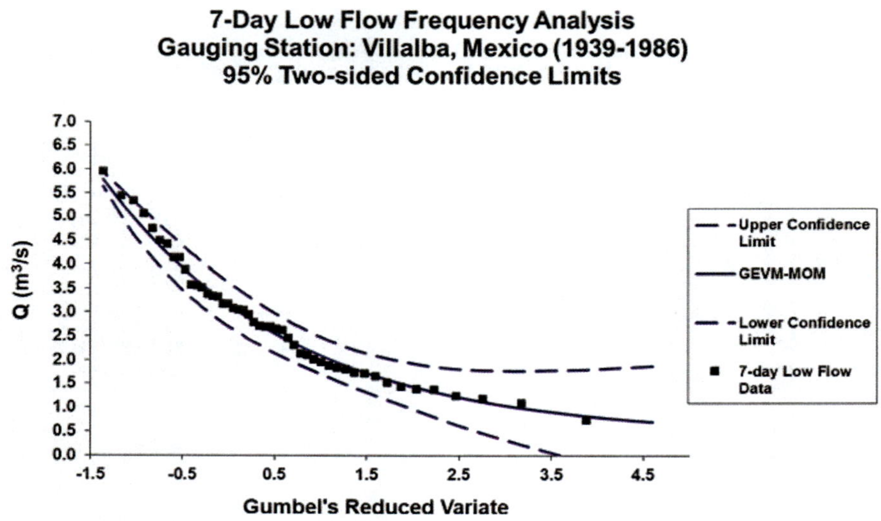

Fig. 14.15 Empirical and MOM-GVEM theoretical frequency curves and confidence limits of 7-day low flow sample of Villalba, Mexico

	A	B	C	D
1	EARTHQUAKE EPICENTER DISTANCE FREQUENCY ANALYSIS			
2				
3	*Descriptive Statistics*			
4	Mean	227.2069		
5	Standard Error	11.6856		
6	Median	248.1883		
7	Mode	93.6438		
8	Standard Deviation	90.5161		
9	Sample Variance	8193.1719		
10	Kurtosis	-1.1039		
11	Skewness	-0.0732		
12	Range	347.5440		
13	Minimum	36.8461		
14	Maximum	384.3901		
15	Sum	13632.4134		
16	Count	60.0000		
17	Largest(1)	384.3901		
18	Smallest(1)	36.8461		
19	Confidence Level(95.0%)	23.3828		

Fig. 14.16 Descriptive statistics for the earthquake epicenter distance sample data

EARTHQUAKE EPICENTER DISTANCE FREQUENCY ANALYSIS
General Extreme Value (GEVM) for the Minima
Method: MOM

Estimation of Parameters and Goodness of Fit Tests					
$1+\beta$ =	1.3041	$1+2\beta$ =	1.6082	$1+3\beta$ =	1.9123
$\Gamma(1+\beta)$ =	0.8969	$\Gamma(1+2\beta)$ =	0.8945	$\Gamma(1+3\beta)$ =	0.9660
$\psi(1+\beta)$ =	-0.1645	$\psi(1+2\beta)$ =	0.1331	$\psi(1+3\beta)$ =	0.3646
μ_z =	0.8969	σ_z =	0.3002	λ_1 =	-0.2639
A =	43.2236	B =	301.5313	λ_2 =	4.5379
Location Parameter =		258.3077	km		
Scale Parameter =		91.6969	km		
Shape Parameter =		0.3041			
SEF =	15.77		MARD =	7.16	

Design Values and Confidence Limits

Return Period T_r (years)	Exceedance Probability $\Pi(x)$	y	Frequency Factor K_T	$\partial K_T/\partial\gamma$	Squared Root Stand. Error S_T (km)	Two-sided 95% Lower Limit (km)	Design Values D_T (km)	Two-sided 95% Upper Limit k(m)
2	0.5	0.6931	-0.0078	-0.0161	12.4618	202.08	227	250.93
5	0.2	1.6094	0.8623	0.0522	4.5017	139.04	148	156.69
10	0.1	2.3026	1.3053	0.0940	#¡NUM!	#¡NUM!	109	#¡NUM!
20	0.05	2.9957	1.6630	0.1308	#¡NUM!	#¡NUM!	79	#¡NUM!
25	0.04	3.2189	1.7657	0.1418	#¡NUM!	#¡NUM!	71	#¡NUM!
50	0.02	3.9120	2.0562	0.1740	#¡NUM!	#¡NUM!	49	#¡NUM!
100	0.01	4.6052	2.3127	0.2037	#¡NUM!	#¡NUM!	31	#¡NUM!
Two-sided Confidence Limits:		z_α =	1.96					

Fig. 14.17 Moments estimators for the parameters, standard errors, quantiles, and confidence limits of the GEVM distribution for the earthquake epicenter distance sample data

In Fig. 14.20 a comparison is made between the histogram and MOM-GEVM theoretical density for the earthquake epicenter distance sample data.

A graphical comparison between the empirical and theoretical frequency curves from the results provided by the MOM method, given that the ML method did not produced results is shown in Fig. 14.21.

A graphical depiction of the empirical and theoretical frequency curves and the confidence limits of the best fit provided, in this case that of MOM, is shown in Fig. 14.22.

	A	B	C	D	E	F	G
1	EARTHQUAKE EPICENTER DISTANCE FREQUENCY ANALYSIS						
2	General Extreme Value (GEVM) Distribution for the Minima						
3	Method: ML						
4							
5	Estimation of Parameters and Goodness of Fit Tests						
6	Initial Moment Estimators:			$(1-\beta)-1$ =	-0.7603	$(1-2\beta)-1$ =	-1.5206
7	Location Parameter =	260.3151	km	$1-\beta$ =	0.2397	$1-2\beta$ =	-0.5206
8	Scale Parameter =	89.2748	km	$\Gamma(1-\beta)$ =	3.0290	$\Gamma(1-2\beta)$ =	29.0010
9	Shape Parameter =	0.2553		$\psi(1-\beta)$ =	-4.4117		
10	Iterative Procedure:						
11	i-th iteration				(i+1)-th iteration		
12	Location Parameter =	208.0876		a =	2.6894	f =	-1.2576
13	Scale Parameter =	89.5738		b =	0.9993	g =	2.6703
14	Shape Parameter =	0.7603		c =	3.0234	h =	-0.5758
15	P =	#¡NUM!		$\delta(x_0)$ =	#¡NUM!		
16	Q =	#¡NUM!		$\delta(\alpha)$ =	#¡NUM!		
17	R =	#¡NUM!		$\delta(\beta)$ =	#¡NUM!		
18	ABS(∂LL/ ∂ω) =	#¡NUM!		Location Parameter=		#¡NUM!	
19	ABS(∂LL/ ∂α) =	#¡NUM!		Scale Parameter =		#¡NUM!	
20	ABS(∂LL/ ∂β) =	#¡NUM!		ShapeParameter =		#¡NUM!	
21	Final MV Estimators:				No convergence was attained		
22	Location Parameter =	#¡NUM!	km	Log-likelihood function =		#¡NUM!	
23	Scale Parameter =	#¡NUM!	km				
24	Shape Parameter =	#¡NUM!					
25	SEF =	#¡NUM!	MARD =	#¡NUM!	AIC =	#¡NUM!	
26							
27	Variance-Covariance Matrix						
28	var(ω) =	#¡NUM!		cov(ω,α) =	#¡NUM!		
29	var(α) =	#¡NUM!		cov(ω,β)	#¡NUM!		
30	var(β) =	0.0504		cov(α,β) =	#¡NUM!		

Fig. 14.18 Maximum likelihood estimators for the parameters of the GEVM distribution for the earthquake epicenter distance sample data

	A	B	C	D	E	F	G	H
32	Design Values and Confidence Limits							
33–34	Return Period	Exceedance Probability			Squared Root Stand. Error	Two-sided 95%	Design Values	Two-sided 95%
35	Tr	Π(x)	$\partial x_T/\partial\alpha$	$\partial x_T/\partial\beta$	S_T	Lower Limit	D_T	Upper Limit
36	(years)				(km)	(km)	(km)	(km)
37	2	0.5	#¡NUM!	#¡NUM!	#¡NUM!	#¡NUM!	#¡NUM!	#¡NUM!
38	5	0.2	#¡NUM!	#¡NUM!	#¡NUM!	#¡NUM!	#¡NUM!	#¡NUM!
39	10	0.1	#¡NUM!	#¡NUM!	#¡NUM!	#¡NUM!	#¡NUM!	#¡NUM!
40	20	0.05	#¡NUM!	#¡NUM!	#¡NUM!	#¡NUM!	#¡NUM!	#¡NUM!
41	25	0.04	#¡NUM!	#¡NUM!	#¡NUM!	#¡NUM!	#¡NUM!	#¡NUM!
42	50	0.02	#¡NUM!	#¡NUM!	#¡NUM!	#¡NUM!	#¡NUM!	#¡NUM!
43	100	0.01	#¡NUM!	#¡NUM!	#¡NUM!	#¡NUM!	#¡NUM!	#¡NUM!
44	Two-sided Confidence Limits:	z_α =	1.96					

Fig. 14.19 Maximum likelihood estimators of the standard errors, quantiles and confidence limits for the earthquake epicenter distance sample data

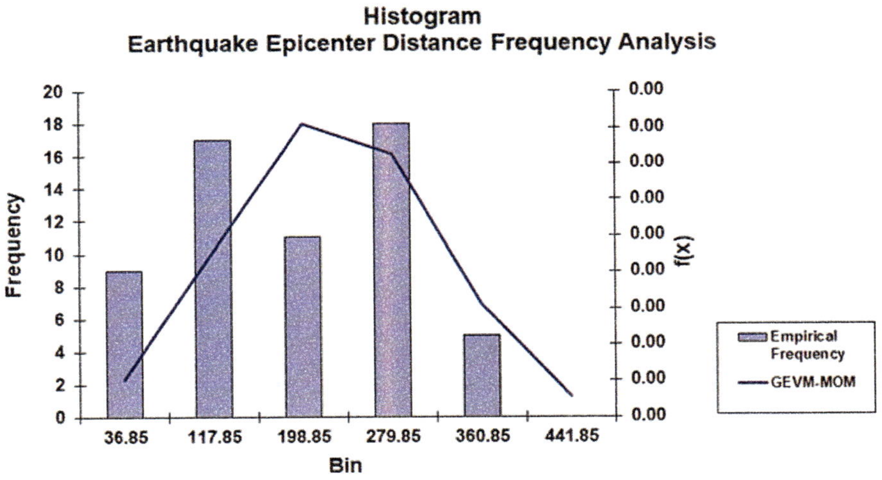

Fig. 14.20 Histogram and MOM-GEVM theoretical density for the earthquake epicenter distance sample data

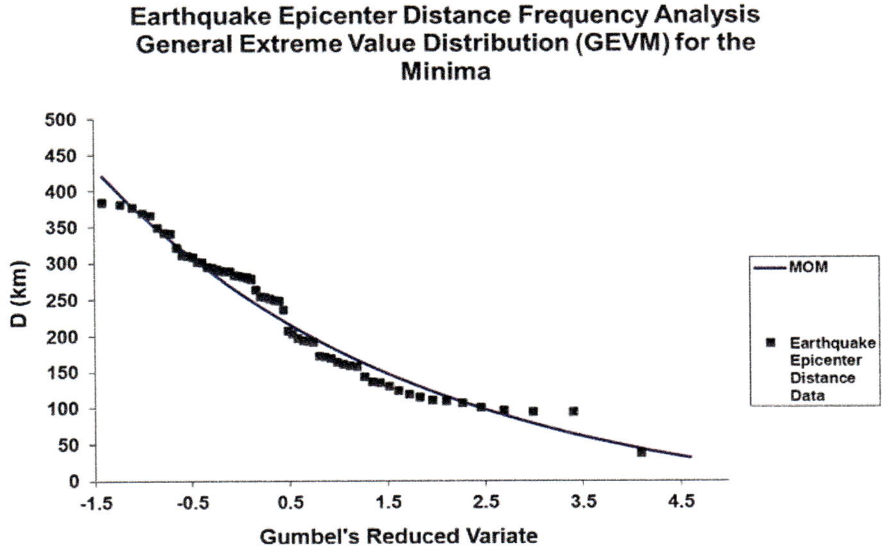

Fig. 14.21 Empirical and MOM-ML theoretical curves of GVEM distribution of earthquake epicenter distance sample data

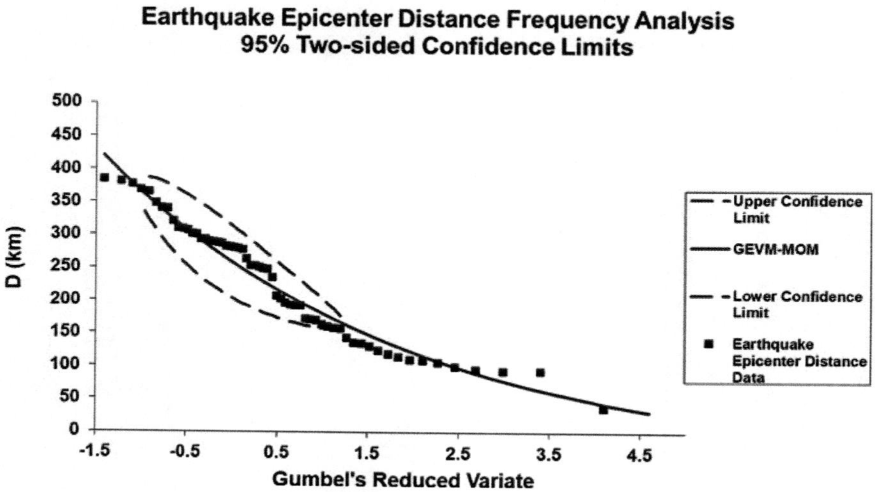

Fig. 14.22 Empirical and MOM-GVEM theoretical frequency curves and confidence limits of earthquake epicenter distance sample data

Samples of Natural Extreme Value Data

A

Samples of Maximum Values Data

In this appendix, there are several groups of maximum natural extreme values data. There are three sets of flood data, one from gauging station Huites (1942–1992) and another from gauging station Villalba, Mexico (1939–1991). There is a third one from gauging station St. Mary's River at Stillwater, Canada (1915–1986), Kite (1988).

Two of maximum annual rainfall coming from meteorological stations Chihuahua, Mexico (1960–2005) and Boquilla, Mexico (1915–1986).

Finally, one of annual maximum wind speed velocity and another from maximum significant wave height data, both coming from Castillo (1988).

Samples of Flood Data

Three samples of flood data are contained in this section:

Flood data from gauging station Huites, Mexico (1942–1992) shown in Table A.1

Flood data from gauging station Villalba, Mexico (1939–1991) shown in Table A.2

Flood data from gauging station St. Mary's River at Stillwater, Canada (1915–1986), Kite (1988), shown in Table A.3

Samples of 24 h Annual Maximum Rainfall

Two samples of 24 h annual maximum rainfall data are contained in this section:

J. A. Raynal Villaseñor, *Frequency Analyses of Natural Extreme Events*, Earth and Environmental Sciences Library,
https://doi.org/10.1007/978-3-030-86390-6

Table A.1 Flood data at gauging station Huites, Mexico (1942–1992)

Year	Q (m^3/s)	Year	Q (m^3/s)	Year	Q (m^3/s)
1942	1894.7	1959	1627.5	1976	1627.7
1943	8600.4	1960	8991.8	1977	1071.3
1944	2115.5	1961	1314.9	1978	3977.1
1945	1363.6	1962	1176.1	1979	4346.9
1946	842.5	1963	2509.4	1980	1351.9
1947	1001.3	1964	1133.9	1981	3110.3
1948	1820.2	1965	1414.7	1982	1775.7
1949	6758.4	1966	2050.5	1983	6478.6
1950	2722.4	1967	2193.3	1984	4082.8
1951	475.3	1968	1411.8	1985	2660.5
1952	915.7	1969	1084.4	1986	1298.3
1953	728.6	1970	1312.7	1987	948.5
1954	853.6	1971	2018.8	1988	1702.5
1955	4007.3	1972	1910.9	1989	1342.3
1956	629.7	1973	5801.1	1990	10,129.3
1957	421.8	1974	2908.6	1991	2014.0
1958	2613.1	1975	991.1	1992	1913.8

24 h annual maximum rainfall from meteorological station Chihuahua, Mexico (1960–2005) is shown in Table A.4

24 h annual maximum rainfall form meteorological station Boquilla, Mexico (1915–1986) is shown in Table A.5

Samples of Maximum Wind Speed Data

One sample of annual maximum wind speed data is contained in this section and is shown in Table A.6

Samples of Maximum Wave Height Data

One sample of maximum wave height data is contained in this section and is shown in Table A.7

Table A.2 Flood data at gauging station Villalba, Mexico (1939–1991)

Year	Q (m^3/s)	Year	Q (m^3/s)	Year	Q (m^3/s)
1939	207.8	1957	125.9	1975	401.0
1940	133.5	1958	539.9	1976	404.9
1941	356.2	1959	446.7	1977	184.9
1942	427.0	1960	313.9	1978	2234.3
1943	217.3	1961	105.7	1979	276.1
1944	278.6	1962	152.0	1980	414.4
1945	298.6	1963	190.9	1981	469.4
1946	279.0	1964	78.3	1982	19.4
1947	173.8	1965	252.2	1983	381.1
1948	45.4	1966	644.0	1984	303.5
1949	394.3	1967	230.4	1985	208.7
1950	125.5	1968	508.6	1986	410.9
1951	13.01	1969	76.1	1987	148.2
1952	581.0	1970	134.8	1988	144.2
1953	156.6	1971	418.5	1989	256.0
1954	227.3	1972	267.2	1990	257.0
1955	334.8	1973	287.5	1991	258.0
1956	177.4	1974	1697.9		

Samples of Minimum Values Data

There are four groups of minimum values data in this section, three for drought data, two coming from gauging station Villalba, Mexico (1939–1991) and (1939–1986), and another one from gauging station St. Mary's River at Stillwater, Canada (1915–1986), Kite (1988).

The fourth one is related with earthquake epicenter distances from power plants, Castillo (1988).

Samples of Drought Data

Three samples of low flow data are contained in this section:

(1) One-day low flow from Gauging station Villalba, Mexico (1939–1991) is shown in Table A.8
(2) 7-day low flow from Gauging station Villalba, Mexico (1939–1986) is shown in Table A.9
(3) One-day low flow from Gauging station St. Mary's River at Stillwater, Canada (1915–1986), Kite (1988), is shown in Table A.10

Table A.3 Flood data gauging station St. Mary's River at Stillwater, Canada (1915–1986), Kite (1988)

Year	Q (m^3/s)	Year	Q (m^3/s)	Year	Q (m^3/s)	Year	Q (m^3/s)
1915	565	1933	385	1951	328	1969	725
1916	294	1934	351	1952	564	1970	232
1917	303	1935	518	1953	527	1971	974
1918	569	1936	365	1954	510	1972	456
1919	232	1937	515	1955	371	1973	289
1920	405	1938	280	1956	824	1974	348
1921	228	1939	289	1957	292	1975	564
1922	232	1940	255	1958	345	1976	479
1923	394	1941	334	1959	442	1977	303
1924	238	1942	456	1960	360	1978	603
1925	524	1943	479	1961	371	1979	514
1926	368	1944	334	1962	544	1980	377
1927	464	1945	394	1963	552	1981	318
1928	411	1946	348	1964	651	1982	342
1929	368	1947	428	1965	190	1983	593
1930	487	1948	337	1966	202	1984	378
1931	394	1949	311	1967	405	1985	255
1932	337	1950	453	1968	583	1986	292

Table A.4 24 h annual maximum rainfall data gauging station Chihuahua, Mexico (1960–2005)

Year	P_{24} (mm)	Year	P_{24} (mm)	Year	P_{24} (mm)	Year	P_{24} (mm)
1960	398.7	1972	499.0	1984	375.4	1996	496.1
1961	304.5	1973	413.2	1985	548.6	1997	456.0
1962	282.8	1974	406.1	1986	714.2	1998	317.5
1963	552.0	1975	267.1	1987	637.6	1999	426.9
1964	245.3	1976	506.9	1988	294.7	2000	554.5
1965	269.7	1977	385.5	1989	293.8	2001	272.6
1966	500.9	1978	616.4	1990	647.0	2002	427.1
1967	476.2	1979	384.8	1991	563.9	2003	352.8
1968	457.6	1980	504.8	1992	512.5	2004	668.1
1969	361.8	1981	643.0	1993	220.9	2005	374.8
1970	304.9	1982	273.2	1994	171.2		
1971	323.5	1983	302.7	1995	334.1		

Table A.5 24 h annual maximum rainfall data gauging station Boquilla, Mexico (1957–2005)

Year	P_{24} (mm)	Year	P_{24} (mm)	Year	P_{24} (mm)	Year	P_{24} (mm)
1957	202.4	1970	297.1	1983	193.7	1996	264.8
1958	473.9	1971	320.8	1984	414.8	1997	233.4
1959	259.9	1972	455.7	1985	259.7	1998	144.1
1960	314.2	1973	499.8	1986	343.4	1999	306.9
1961	284.4	1974	365.7	1987	326.5	2000	184.7
1962	294.8	1975	228.8	1988	308.9	2001	145.7
1963	180.7	1976	454.8	1989	187.9	2002	305.9
1964	254.0	1977	266.1	1990	383.2	2003	334.0
1965	403.6	1978	362.4	1991	374.6	2004	350.4
1966	469.9	1979	207.9	1992	300.0	2005	144.5
1967	374.9	1980	419.9	1993	231.6		
1968	467.2	1981	462.3	1994	112.4		
1969	183.9	1982	131.4	1995	171.6		

Table A.6 Annual maximum wind speed data, Castillo (1988)

	V (m/s)		V (m/s)		V (m/s)		V (m/s)		V (m/s)
1	2.91	11	3.74	21	4.09	31	5.88	41	6.42
2	6.93	12	7.21	22	7.92	32	8.26	42	8.79
3	9.17	13	9.50	23	9.62	33	10.00	43	10.14
4	10.28	14	10.45	24	10.77	34	11.65	44	11.65
5	11.82	15	12.27	25	12.68	35	13.28	45	13.46
6	13.88	16	13.98	26	14.32	36	14.38	46	14.46
7	14.86	17	15.03	27	15.30	37	16.07	47	16.23
8	17.36	18	18.68	28	18.72	38	19.44	48	20.09
9	21.06	19	21.13	29	21.53	39	21.80	49	23.15
10	24.75	20	25.45	30	28.13	40	29.95	50	37.19

Sample of Earthquake Epicenter Distance to Power Plants

One sample of earthquake epicenter distance from power plants, Castillo (1988) is contained in Table A.11.

Table A.7 Maximum significant wave height data, Castillo (1988)

	H (m)		H (m)		H (m)		H (m)		H (m)
1	0.89	11	1.14	21	1.25	31	1.79	41	1.96
2	2.11	12	2.20	22	2.41	32	2.52	42	2.68
3	2.80	13	2.90	23	2.93	33	3.05	43	3.09
4	3.13	14	3.19	24	3.28	34	3.55	44	3.55
5	3.60	15	3.74	25	3.86	35	4.05	45	4.10
6	4.23	16	4.26	26	4.36	36	4.38	46	4.41
7	4.53	17	4.58	27	4.66	37	4.90	47	4.95
8	5.29	18	5.69	28	5.71	38	5.93	48	6.12
9	6.42	19	6.44	29	6.56	39	6.64	49	7.06
10	7.54	20	7.76	30	8.57	40	9.13	50	11.33

Table A.8 One-day low flow data gauging station Villalba, Mexico (1939–1991)

Year	Q (m^3/s)	Year	Q (m^3/s)	Year	Q (m^3/s)	Year	Q (m^3/s)
1939	0.622	1953	0.132	1967	0.339	1981	0.704
1940	0.685	1954	0.140	1968	0.328	1982	0.443
1941	0.452	1955	0.245	1969	0.348	1983	0.367
1942	0.305	1956	0.150	1970	0.208	1984	0.177
1943	0.648	1957	0.079	1971	0.309	1985	0.310
1944	0.485	1958	0.115	1972	0.377	1986	0.391
1945	0.447	1959	0.401	1973	0.395	1987	0.442
1946	0.485	1960	0.262	1974	0.274	1988	0.384
1947	0.431	1961	0.262	1975	0.409	1989	0.274
1948	0.392	1962	0.185	1976	0.507	1990	0.175
1949	0.226	1963	0.229	1977	0.416	1991	0.129
1950	0.264	1964	0.357	1978	0.250		
1951	0.195	1965	0.139	1979	0.419		
1952	0.184	1966	0.330	1980	0.300		

Table A.9 7-day low flow data gauging station Villalba, Mexico (1939–1986)

Year	Q (m^3/s)	Year	Q (m^3/s)	Year	Q (m^3/s)	Year	Q (m^3/s)
1939	5.446	1951	1.739	1963	1.720	1975	4.490
1940	3.89	1952	1.663	1964	3.090	1976	4.138
1941	3.570	1953	1.100	1965	1.378	1977	3.563
1942	4.427	1954	1.245	1966	2.662	1978	2.720
1943	4.754	1955	1.888	1967	2.705	1979	5.957
1944	4.143	1956	1.528	1968	2.626	1980	2.469
1945	3.332	1957	0.745	1969	3.177	1981	5.337
1946	3.511	1958	1.189	1970	1.844	1982	3.169
1947	3.328	1959	3.387	1971	2.309	1983	2.703
1948	3.057	1960	2.018	1972	3.046	1984	2.791
1949	1.811	1961	2.141	1973	2.952	1985	5.061
1950	1.955	1962	1.435	1974	2.109	1986	1.402

Table A.10 One-day low flow data gauging station St. Mary's River at Stillwater, Canada (1915–1986), Kite (1988)

Year	Q (m^3/s)	Year	Q (m^3/s)	Year	Q (m^3/s)	Year	Q (m^3/s)
1915	4.05	1933	1.59	1951	2.10	1969	0.84
1916	1.05	1934	0.45	1952	0.73	1970	4.96
1917	1.05	1935	1.04	1953	0.73	1971	2.34
1918	2.52	1936	3.45	1954	1.15	1972	6.43
1919	3.09	1937	0.56	1955	1.50	1973	1.92
1920	3.09	1938	3.96	1956	1.17	1974	1.21
1921	0.65	1939	0.67	1957	1.01	1975	0.24
1922	2.78	1940	1.04	1958	2.35	1976	0.72
1923	2.01	1941	2.24	1959	7.36	1977	6.88
1924	1.59	1942	0.15	1960	0.22	1978	0.71
1925	1.22	1943	3.03	1961	0.54	1979	3.01
1926	1.05	1944	0.59	1962	5.44	1980	0.84
1927	6.71	1945	1.28	1963	3.37	1981	2.84
1928	1.78	1946	0.59	1964	7.65	1982	2.30
1929	2.24	1947	0.45	1965	1.96	1983	5.21
1930	1.15	1948	3.34	1966	0.96	1984	1.90
1931	2.75	1949	1.78	1967	3.74	1985	1.90
1932	1.70	1950	0.41	1968	0.51	1986	2.72

Table A.11 Epicenter of earthquakes distance to power plants, Castillo (1988)

	D		D		D		D		D
	(km)		(km)		(km)		(km)		(km)
1	58.2	13	58.2	25	59.5	37	61.8	49	65.8
2	67.8	14	68.5	26	70.9	38	73.7	50	77.0
3	80.8	15	83.7	27	84.3	39	89.0	51	97.6
4	98.3	16	99.6	28	101.4	40	105.1	52	105.8
5	106.7	17	119.1	29	119.5	41	119.9	53	121.9
6	125.7	18	128.4	30	146.1	42	153.9	54	154.6
7	155.8	19	157.4	31	157.7	43	163.7	55	172.7
8	173.9	20	174.2	32	175.1	44	176.0	56	178.7
9	179.1	21	179.5	33	180.7	45	182.1	57	182.7
10	186.7	22	187.5	34	191.0	46	192.6	58	193.0
11	199.4	23	211.6	35	212.1	47	216.8	59	22.9
12	227.3	24	229.4	36	234.5	48	236.8	60	238.9

Tutorial For the Construction of a Frequency Analysis Excel® Spreadsheet

B

Tutorial for Building a Frequency Analysis Spreadsheet

To build a spreadsheet like the ones contained in this book, it is just a matter of organizing the data for using it for frequency analysis purposes. Figure B.1 illustrates how to construct a frequency analysis Excel® spreadsheet for the Normal (NOR) distribution, it is just needed to plug the appropriate formulas contained in the text of the book in the adequate cells of the spreadsheet to have the frequency analysis spreadsheet ready to be used.

It is needed to have a set of descriptive statistics that Excel® provides as a utility function within the library of Data Analysis of any Excel® spreadsheet. The descriptive statistics that Excel® provides is shown in Fig. B.2.

Furthermore, it is needed a working Excel® spreadsheet to produce several of the data needed to construct the Excel® spreadsheet shown in Fig. B.1 and the

	A	B	C	D	E	F	G
1	RAINFALL FREQUENCY ANALYSIS						
2	Meteorological Station: Chihuahua, Mexico (1960-2005)						
3	Normal Distribution						
4	Methods: MOM-ML						
5							
6	Estimation of Parameters and Goodness of Fit Tests						
7	Location Parameter =		42.0454	Equation 3.6			
8	Scale Parameter =		13.4184	Equation 3.7			
9	Log-Likelihood function =		-292.4109				
10	SEF =	1.98	Equation 3.20	MARD =	5.12	Equation 3.19	589
11							
12	Design Values and Confidence Limits						
13				Squared Root	Two-sided	Design	Two-sided
14				Standard Error	95%	Values	95%
15	Tr (years)	F(x)	z	S_T	Lower Limit	P_T	Upper Limit
16			Equation 3.14	Equation 3.24	Equation 3.21	Equation 3.17	Equation 3.21
17	2	0.5	0.0000	1.9784	38	42	46
18	5	0.8	0.8416	2.3023	49	53	58
19	10	0.9	1.2816	2.6699	54	59	64
20	20	0.95	1.6449	3.0347	58	64	70
21	50	0.98	2.0537	3.4884	63	70	76
22	100	0.99	2.3263	3.8087	66	73	81
23	Two-sided Confidence Limits:		z_α =	1.96			

Fig. B.1 Tutorial for building a frequency analysis spreadsheet

J. A. Raynal Villaseñor, *Frequency Analyses of Natural Extreme Events*, Earth and Environmental Sciences Library,
https://doi.org/10.1007/978-3-030-86390-6

	A	B
1	*Column1*	
2		
3	Mean	42.045435
4	Standard Error	1.9784355
5	Median	40.24
6	Mode	#N/A
7	Standard Deviation	13.418402
8	Sample Variance	180.05351
9	Kurtosis	-0.7252436
10	Skewness	0.3591164
11	Range	54.3
12	Minimum	17.12
13	Maximum	71.42
14	Sum	1934.09
15	Count	46
16	Largest(1)	71.42
17	Smallest(1)	17.12
18	Confidence Level(95.0%)	3.9847735
19		

Fig. B.2 Descriptive statistics

graphs shown in the book. Part of such a working Excel® spreadsheet is shown in Fig. B.3.

The other graphs showed in Chaps. 3–14 of the several distributions contained in the book, are being built using the graph construction tools provided by the Excel® spreadsheet platform.

24 Hr. Annual Maximum Rainfall Frequency Analysis
Meteorological Station: Chihuahua, Mexico (1960-2005)
Normal (NOR) Distribution

	A	B	C	D	E	F	G	H	I	J	K	L	M	N
5											MOM-ML			
6–8	Year	P (cm)	Pord (cm)	m	Tr	Gumbel's Red. Var.	F(x)	z_T	S_T	Lower Confidence Limit	y_i	Upper Confidence Limit	SEF $(x_i-y_i)^2/(N-N_p)$	MARD $Abs((x_i-y_i)/x_i)/N$
9					100	4.600149	0.9900	2.3263	3.8087	66	73.2613	81		
10	1960	39.87	71.42	1	47.0000	3.839414	0.9787	2.0281	3.4589	62	69.2589	76	1.5757	0.0658
11	1961	30.45	66.81	2	23.5000	3.135337	0.9574	1.7218	3.1171	59	65.1492	71	0.1427	0.0540
12	1962	28.28	64.7	3	15.6667	2.718738	0.9362	1.5234	2.9079	57	62.4870	68	0.0983	0.0744
13	1963	55.2	64.3	4	11.7500	2.419709	0.9149	1.3715	2.7560	55	60.4490	66	0.0967	0.1302
14	1964	24.53	63.76	5	9.4000	2.184998	0.8936	1.2460	2.6368	54	58.7647	64	0.1259	0.1703
15	1965	26.97	61.64	6	7.8333	1.990877	0.8723	1.1375	2.5390	52	57.3092	62	0.1468	0.1527
16	1966	50.09	56.39	7	6.7143	1.824687	0.8511	1.0410	2.4566	51	56.0141	61	0.1280	0.0145
17	1967	47.62	55.45	8	5.8750	1.678863	0.8298	0.9533	2.3860	50	54.8375	60	0.0353	0.0240
18	1968	45.76	55.2	9	5.2222	1.548524	0.8085	0.8724	2.3246	49	53.7519	58	0.0386	0.0570
19	1969	36.18	54.86	10	4.7000	1.430331	0.7872	0.7969	2.2709	48	52.7380	57	0.0560	0.0841
20	1970	30.49	51.25	11	4.2727	1.321898	0.7660	0.7256	2.2236	47	51.7818	56	0.0700	0.0226
21	1971	32.36	50.69	12	3.9167	1.22146	0.7447	0.6578	2.1820	47	50.8727	55	0.0086	0.0078
22	1972	49.9	50.48	13	3.6154	1.127669	0.7234	0.5930	2.1453	46	50.0023	54	0.0156	0.0206
23	1973	41.32	50.09	14	3.3671	1.039476	0.7021	0.5306	2.1131	45	49.1643	53	0.0299	0.0402
24	1974	40.61	49.9	15	3.1333	0.956041	0.6809	0.4701	2.0849	44	48.3532	52	0.0395	0.0674
25	1975	26.71	49.61	16	2.9375	0.876685	0.6596	0.4113	2.0604	44	47.5645	52	0.0531	0.0896
26	1976	50.69	47.62	17	2.7647	0.800843	0.6383	0.3539	2.0394	43	46.7944	51	0.0641	0.0377
27	1977	38.55	45.76	18	2.6111	0.728046	0.6170	0.2977	2.0218	42	46.0396	50	0.0360	0.0133
28	1978	61.64	45.6	19	2.4737	0.65789	0.5957	0.2423	2.0073	41	45.2974	49	0.0105	0.0144
29	1979	38.48	42.71	20	2.3500	0.59003	0.5745	0.1878	1.9958	41	44.5649	48	0.0235	0.0944

Fig. B.3 Working excel® spreadsheet

References

Abramowitz M, Stegun IA (1965) Handbook of mathematical functions. Dover Publications, New York, N. Y.

Akaike H (1974) A new look at the statistical model identification. IEEE Trans on Automatic Control 19:716–723

Anderson TW (1957) Maximum likelihood estimates for a multivariable normal distribution when some observations are missing. J Am Stat Assoc 52:220–230

Bobée B (1973) Sample error of T-year events computed by fitting a pearson type 3 distribution. Wat Res Res 9(5):1264–1270

Bobée B, DesGroseiliers L (1985) Adjustment des distributions pearson type 3, gamma, gamma généraliseée et log pearson. INRS-EAU, Raport Scientifique No. 105

Caruso BS (2000) Evaluation of low-flow frequency analysis methods. J Hydrol (NZ) 39(1):19–47

Castillo E (1988) Extreme value theory in engineering. Academic Press Inc., San Diego, California

Chow VT (1954) The log-probability law and its engineering applications. Proc ASCE 80:1–25

Chow VT (1959) Determination of the frequency factor. ASCE, J. Hyd. Div., pp 93–98

Chow VT (1964) Handbook of applied hydrology. McGraw-Hill Book Co., New York, N. Y.

Cohen AC, Witten BJ (1988) Parameter estimation in reliability and life span models, statistics, textbooks and monographs, vol 96. Marcel Dekker, New York, N. Y.

Condie R, Nix G (1975) Modeling of low flow frequency distributions and parameters estimation. In: International water resources association proceedings symposium on water for arid lands, Iran

Condie R (1977) The log-pearson type 3 distribution: the T-year event and its asymptotic standard error by maximum likelihood theory. Wat. Res. Res. 13(6):987–991

Condie R (1979) Reply. Wat Res Res 15(1):191–192

Deininger RA, Westfield JD (1969) Estimation of Gumbel's third asymptotic distribution by different methods. Wat Res Res 5(6):1238–1243

Durrans SR, Tomic S (2007) Comparison of parametric tail estimators for low-flow frequency analysis. J Amer Water Resour Assoc 37(5):1203–1214

EM-DAT CRED / UCLouvain, Brussels, Belgium—www.emdat.be (D. Guha-Sapir): The OFDA/CRED international disaster Database (2019), Disasters 2018: a year in review. CRED Crunch. Issue 54, April 2019

EM-DAT CRED / UCLouvain, Brussels, Belgium—www.emdat.be (D. Guha-Sapir): The OFDA/CRED International disaster database (2019), Disaster year in review 2019, CRED Crunch, Issue 58, April 2020

Fisher RA, Tippett LHC (1928) Limiting forms of the frequency distribution of the largest or smallest member of a sample. Proc Camb Philos Soc 24:180–190

Fréchet M (1927) Sur la Loi de Probabilite de l'ecart maximum. Annales De La Société Polonaise De Mathematique, Cracovie 6:93–116

J. A. Raynal Villaseñor, *Frequency Analyses of Natural Extreme Events*, Earth and Environmental Sciences Library,
https://doi.org/10.1007/978-3-030-86390-6

Fuller WE (1914) Flood flows. Trans Am Soc Civ Eng 77:564

Greenwood JA, Durand D (1960) Aids for fitting the gamma distribution. Technometrics 2(1):55–66

Greenwood JA, Lanwher JM, Matalas NC, Wallis JR (1979) Probability weighted moments: definition, and relation to parameters of several distributions expressable in inverse form. Wat Res Res 15:1049–1054

Grubbs F, Beck G (1972) Extension of sample sizes and percentage points for significance tests of outlying observations. Technometrics 14(4):847–854

Gumbel EJ (1958) Statistics of extremes. Columbia University Press, New York, N. Y., p 8

Gumbel EJ (1962) Statistical theory of extremes (Main Results), Chapter 6. In: Sarhan AS, Greenberg BG (eds) Contributions to order statistics. Wiley, New York, N. Y., pp 59–93

Haan CT (1977) Statistical methods in hydrology. The Iowa State University Press, Ames, Iowa, p 63

Hazen A (1914) Storage to be provided in impounding reservoirs for municipal water supply. Trans ASCE vol 77, Paper 1308, ASCE, New York, N. Y

Hewa GA, Wang QJ, McMahon TA, Nathan RJ, Peel MC (2007) Generalized extreme value distribution fitted by LH moments for low-flow frequency analysis. Water Resour Res 43(6):W06301

Hoshi K, Burges SJ (1981) Sampling properties of parameter estimates for the log pearson type 3 distribution, using moments in real space. J Hydrol 53:305–316

Hoshi K, Stedinger JR, Burges SJ (1989) Estimation of log-normal quantiles: monte carlo results and first order approximation. J Hydrol 71:1–30

Hosking JRM, Wallis JR, Wood EF (1985) Estimation of the generalized extreme-value distribution by the method of probability weighted moments estimation of parameters:method of probability weighted moments. Technometrics 27:251–261

Hosking JRM, Wallis JR (1987) Parameter and quantile estimation for the generalized pareto distribution. Technometrics 29:339–349

Hosking JRM (1990) L-moments: analysis and estimation of distribution using linear combination of order statistics. J R Statist Soc B 52(1):105–124

Jain D, Singh VP (1987) Estimating parameters of EV1 distribution for flood frequency analysis. Wat Res Bull 23(1):59–71

Jenkinson AF (1955) The frequency distribution of the annual, maximum (or minimum) values of meteorological elements. Quart J R Meteor Soc 87:158–171

Jenkinson AF (1969) Estimation of maximum floods, chapter 5. Technical Note 98, 183–227. WMO, Geneva, Switzerland

Johnson NL, Kotz S, Balakrishnan N (1994) Continuous univariate distributions. vol 1, 2nd ed. Wiley, Inc., New York, N.Y

Kilmarin RF, Peterson JR (1972) Rainfall-runoff regression with logarithmic transforms and zeros in the data. Wat Res Res 8(4):1096–1099

Kimball BF (1949) An approximation to the sampling variances of an estimated maximum value of a given frequency based on the fit of the doubly exponential distribution of maximum values. Ann Math Stat 20:110–113

Kite GW (1975) Confidence limits for design events. Wat Res Res 11(1):48–53

Kite GW (1988) Frequency y risk analyses in hydrology. Water Resources Publications, Littleton, Colorado

Kotz S, Nadarajah S (2000) Extreme value distributions: theory and applications. Imperial College Press, London, U. K.

Kroll CN, Vogel RM (2002) Probability distribution of low streamflow series in the United States. J Hydrol Eng 7(2):137–146

Landwher JM, Matalas N, Wallis JR (1979) Probability weighted moments compared with some traditional techniques in estimating gumbel parameters and quantiles . Wat Res Res 15:1055–1064

Landwher JM, Matalas N, Wallis JR (1979b) Estimation of parameters and quantiles of wakeby distributions. Wat Res Res 15:1361–1379, correction1672

Mann HB, Whitney DR (1947) On the test whether one of two random variables is stochastically larger than the other. Ann Math Statist 18:50–60
Markovic RD (1965) Probability functions of best fit to distributions of annual precipitation and runoff. Colorado State University Paper No. 8, Fort Collins, CO
Matalas NC (1963) Probability distribution of low flows, geological survey paper No. 434-A, Department of the Interior, pp 27
Matalas NC, Wallis JR (1973) Eureka! it fits a pearson type 3 distribution. Wat Res Res 9(2):281–289
Mood AM, Graybill F, Boes DC (1974) In: Introduction to the theory of statistics, 3rd edn., McGraw-Hill Inc., New York, N. Y, pp 283
NERC, Natural Environment Research Council (1975) Flood Studies Report, I, Hydrologic Studies. Whitefriars Press Ltd., London, U. K, pp 51
Phien HN, Jivajirajah T (1984) The transformed gamma distribution for annual streamflow frequency analysis. In: Proceedings fourth congress IAHR-APD on water resources management and development. vol 2. Chiang Mai, Thailand, pp 1151–1166
Phien HN (1987) A review of methods of parameter estimation for the extreme value type-1 distribution. J Hydrol 90:251–260
Pilon PJ, Adamowski K (1993) Asymptotic variance measures of dispersion:variance of flood quantile in log-pearson Type III distribution with historical information. J Hydrol 143(3–4):481–503
Prakash A (1981) Statistical determination of design low flows. J Hydrol 51:109–118
Prescott P, Walden AT (1980) Maximum likelihood estimation of the parameters of the generalized extreme value distribution. Biometrika 67(3):723–724
Rao AR, Hamed KH (2000) Flood frequency analysis. CRC Press, Boca Raton, Florida, pp 350
Raynal JA, Salas JD (1986) Estimation procedures for the type-1 extreme value distribution. J Hydrol 87:315–336
Raynal-Villasenor JA (1987) Computation of probability weighted moments for the general extreme value distribution (maxima and minima). Hydrol Sci Technol J 3(1–2):47–52
Raynal-Villasenor JA (1995) Maximum likelihood parameter estimators for the general extreme value distribution for the minima. Hydrol Sci Technol J 11(1–4):140–149
Raynal-Villasenor JA (1996) On the use of exact variance xe measures of dispersion:variance - covariance matrix element coefficients for the general extreme value distribution for the minima. Hydrol Sci Technol J 12(1–4):61–170
Raynal-Villasenor JA (2013) Moment estimators of the GEV distribution for the minima. Appl Wat Sci 3:13–18. https://doi.org/10.1007/s13201-012-0052-3
Reich BM (1972) Log-pearson type 3 and gumbel analysis of floods. In: Second international symposium in hydrology. Fort Collins, Colorado, USA, pp 290–303
Rossi F, Florentino M, Versace P (1984) Two component extreme value distribution for flood frequency analysis. Wat Resour Res 20(7):847–856
Salas JD, Smith RA (1980) Computer programs of probability distribution functions in hydrology. Colorado State University, Fort Collins, Colorado, Hydrology and Water Resources Program
Salas JD, Cardenas A, Smith RA (1980) Computer programs of extreme value distribution in hydrology. Colorado State University, Fort Collins, Colorado, Hydrology and Water Resources Program
Salas JD, Smith RA, Tabios GQ (1990) Statistical computer techniques in hydrology and water resources. Colorado State University, Fort Collins, Colorado, Department of Civil Engineering
Sangal BP, Biswas AK (1970) The 3-parameter log-normal distribution and its applications in hydrology. Wat Resour Res 6(2):505–515
Singh KP, Sinclair RA (1972) Two-distribution method for flood frequency analysis. Proc ASCE 98(HY1):29–45
Slack JR, Wallis JR, Matalas NC (1975) On the value of information to flood frequency analysis. Wat Resour Res 11(5):629–647
Stuart A, Ord JK (1994) In: Kendall's advanced theory of statistics. Distribution Theory, vol 1. Arnold, London, U.K.
Todorovic P, Rousselle J (1971) Some problems of flood analysis. Wat Resour Res 7(5):1144–1150

U. S. Environmental Protection Agency (EPA) (2018) Office of water, low flow statistics tools. A How-To Handbook for NPDES Permit Writers. https://www.epa.gov/sites/production/files/201811/documents/low_flow_stats_tools_handbook.pdf Accessed 13 March 2021
U. S. Geological Survey (2019) Hydrologic analysis and interpretation, techniques and methods 4-B5. Guidelines for Determining Flood Flow Frequency, Bulletin 17C, Chapter 5 of Section B, Surface Water, Book 4, Version 1.1, pp 41
U. S. Geological Survey (2020) Statistical methods in water resources, chapter 3, section a, statistical analysis, book 4,. Hydrological Analysis and Interpretation, Techniques and Methods 4-A2. https://pubs.usgs.gov/tm/04/a03/tm4a3.pdf . Accessed 13 March 2021
Wald A, Wolfowitz J (1943) An exact test for randomness in the non-parametric case based in serial correlation. Ann Math Statist 14:378–388
Woodroofe M (1975) In: Probability with applications. McGaw-Hill Book Co.
Yevjevich Y (1972) Probability and statistics in hydrology. Water Resources Publications, Littleton, Colorado
Yevjevich V, Obeysekera JTB (1984) Estimation of skewness of hydrological variables. Wat Res Res 20(7):935–943
Zaidman MD, Keller V, Young AR, Cadman D (2003) Flow-duration-frequency behaviour of british rivers based on annual minima data. J Hydrol 277(3–4):195–213